HIGH-POWER LASERS AND LASER PLASMAS

MOSHCHNYE LAZERY I LAZERNAYA PLAZMA

МОЩНЫЕ ЛАЗЕРЫ И ЛАЗЕРНАЯ ПЛАЗМА

The Lebedev Physics Institute Series

Editors: Academicians D.V. Skobel'tsyn and N.G. Basov

P. N. Lebedev Physics Institute, Academy of Sciences of the USSR

Recent Volumes in this Series

Proceedings (Trudy) of the P. N. Lebedev Physics Institute

Volume 85

High-Power Lasers and Laser Plasmas

Edited by
N. G. Basov

P. N. Lebedev Physics Institute
Academy of Sciences of the USSR
Moscow, USSR

Translated from Russian by
J. George Adashko
New York University

PHYSICS

CONSULTANTS BUREAU
NEW YORK AND LONDON

Library of Congress Cataloging in Publication Data

Main entry under title:

High-power lasers and laser plasmas.

(Proceedings (Trudy) of the P. N. Lebedev Physics Institute; v. 85)
Translation of Moshchnye lazery i lazernaia plazma.
Includes bibliographical references and index.
1. Laser plasmas. 2. Laser beams. 3. Lasers. I. Basov, Nikolaĭ Gennadievich, 1922-   II.
Series: Akademiia nauk SSSR. Fizicheskiĭ institut. Proceedings; v. 85.
QC1.A4114 vol. 85   [QC718.5.L3] 530'.08s [530.4'4]   78-794
ISBN 0-306-10943-3

The original Russian text was published by Nauka Press in Moscow in 1976 for the Academy of Sciences of the USSR as Volume 85 of the Proceedings of the P. N. Lebedev Physics Institute. This translation is published under an agreement with the Copyright Agency of the USSR (VAAP).

© 1978 Consultants Bureau, New York
A Division of Plenum Publishing Corporation
227 West 17th Street, New York, N. Y. 10011

Printed in the United States of America

CONTENTS

Stimulated Mandel'shtam – Brillouin Scattering
Lasers
V. V. Ragul'skii

Compressed-Gas Lasers
V. A. Danilychev, O. M. Kerimov, and I. B. Kovsh

Experimental Investigation of the Reflection and Absorption
of High-Power Radiation in a Laser Plasma
O. N. Krokhin, G. V. Sklizkov, and A. S. Shikanov

Chapter I

Experimental Study of Cumulative Phenomena in a Plasma
   Focus and in a Laser Plasma

V. A. Gribkov, O. N. Krokhin, G. V. Sklizkov, N. V. Filippov,
and T. I. Filippova

STIMULATED MANDEL'SHTAM–BRILLOUIN SCATTERING LASERS

V. V. Ragul'skii

The generation and amplification of light in stimulated Mandel'shtam–Brillouin scattering (SMBS) are investigated. The conditions necessary for effective operation of SMBS lasers, amplifiers, and modulators are ascertained. The possibility of conversion, with efficiency close to 100%, of the pump radiation in lasers and amplifiers at diffraction divergence of the generated light is demonstrated experimentally and theoretically. To choose the optimal medium for the construction of the laser devices, the gains and line widths of a number of compressed gases are determined. The influence of the width of the spectrum of the exciting radiation on the stimulated scattering is investigated. It is shown that the wavefront of the light scattered by SMBS can be inverted relative to the front of the exciting radiation. This effect is used to compensate for the phase distortions in the amplifying medium.

INTRODUCTION

The high intensity and coherence of laser light makes it possible to observe numerous nonlinear effects, particularity different processes of stimulated scattering of light. Soon after the invention of the laser, stimulated Raman scattering (SRS) was observed [1], followed by stimulated Mandel'shtam–Brillouin scattering (SMBS) [2]. A number of other types of stimulated scattering were discovered later [3-8]. As is well known, in these processes the intensity of the scattered light increases nonlinearly with increasing intensity of the exciting radiation and can become comparable with the latter [9]. As a result, stimulated scattering can be used for an effective conversion of the laser frequency, as well as to change other characteristics of its radiation, such as the spatial distribution, the divergence, the width of the spectrum, etc.

Since SMBS is the dominant scattering process in many media [10], it is useful to consider the possiblities of constructing laser devices on its basis. The present article is devoted to a study of this question.

Mandel'shtam–Brillouin scattering is due to fluctuations of the dielectric constant as a result of pressure fluctuations (sound waves) [11]. If the electric field intensity of the exciting radiation is large, then this field (together with the field of the scattered light) causes, on account of the electrostriction effect, an increase in the intensity of the sound wave. This increase leads to an increase of the intensity of the scattered light, which in turns causes an increase of the pressure fluctuations. As a result, the intensity of the scattered light increases nonlinearly as it propagates in the scattering medium. This is precisely the effect known as stimulated Mandel'shtam–Brillouin scattering.

Stimulated Mandel'shtam–Brillouin scattering has been the subject of a tremendous number of works. The reader can find detailed information in the review of Starunov and Fabelinskii [12] (see also [13] and [14]). We present here only the principal relations describing this phenomenon.   According to Tang's theory [15], in the case of a stationary regime the interaction of plane monochromatic waves of the exciting and scattered light traveling opposite to each other in an active medium is described by the equations

$$\frac{dI}{dx} = gI(x)\mathcal{I}(x),$$
$$\frac{d\mathcal{I}}{dx} = gI(x)\mathcal{I}(x),$$
(1)

where I and $\mathcal{I}$ are the intensities of the exciting and scattered radiations; the scattered radiation propagates in the direction +x ($0 \le x \le l$). The gain g is determined by the parameters of the scattering medium:

$$g(\nu') = \frac{\pi\nu^2\rho}{2c^3 n v}\left[\frac{\partial\varepsilon}{\partial\rho}\right]^2 \frac{\delta\nu_0}{(\nu_B - \nu')^2 + (\delta\nu_0/2)^2},$$
(2)

where ν is the frequency of the exciting light, ρ is the density of the scattering medium, ε is its dielectric constant, n is the refractive index, v is the speed of sound in this medium, and c is the speed of light in vacuum. The position of the center of the gain line is given by the expression

$$\nu_B \equiv \nu - \frac{2vn}{c}\,\nu\sin\frac{\theta}{2},$$
(3)

where θ is the scattering angle.

The line width $\delta\nu_0$ is determined by the damping of the hypersound in the active medium: $\delta\nu_0 = 1/(2\pi\tau_s)$, where τ_s is the hypersound damping time.

If I varies little over the length of the scattering region (i.e., there is no saturation), we obtain from (1)

$$\mathcal{I}(l) = \mathcal{I}(0)e^{gIl}.$$
(4)

Equations (1) are valid also for the saturation regime; the corresponding solution is given in Chapter IV. Equation (2) pertains to scattering through 180°. In scattering through an arbitrary angle θ, the gain at the maximum of the line is equal to [16]:

$$g(\nu_B) = 2\pi\frac{\nu^2\rho}{c^3 n v}\left[\frac{\partial\varepsilon}{\partial\rho}\right]^2\frac{\sin(\theta/2)}{\delta\nu_0}.$$
(5)

Prior to the performance of the investigations described in this article, there were only few known publications on SMBS lasers. The experimental studies [17-20] were devoted to the lasing spectrum at different angles between the axis of the laser cavity and the beam of the exciting radiation. The lasing threshold and the output power were determined [21]. Studies were made also of the competition in such lasers between the SMBS and the SRS processes [19, 22]. The initial lasing period was analyzed theoretically by Yariv [23]. He has shown that the intensity of the generated radiation increases exponentially in time if the lasing is only on the first Stokes component of the SMBS and there is no saturation. Stationary lasing on several components was theoretically considered in [24]. Its authors took into account saturation, but the analysis was carried out only for the case when the exciting radiation coincides in frequency with one of the resonator modes. This case has not yet been realized in experiment.

The papers cited above constitute the entire literature on SMBS lasers. We note that the experiments yielded a low (< 1%) efficiency of conversion of the pump radiation into laser radiation. The cause is apparently the nonstationary regime of the operation of the described lasers.

The conditions under which stationary lasing, high conversion efficiency, and high directivity of the generated light are attained have not yet been investigated.

The scattering media in all the described lasers were liquids. At the same time, the use of gases is more promising for the conversion of intense light beams because of the better optical quality and higher self-focusing threshold [25].

To convert the radiation it is advantageous to use a stimulated-scattering laser-amplifier scheme.* In such a scheme the laser can operate with low output power, making the control of its parameters easier. On the other hand, the amplifier, working in the saturation regime [27], ensures conversion of the major fraction of the pump energy. There have been practically no experimental investigations of the operation of SMBS amplifiers in the saturation region.

So far we have referred to SMBS laser systems that require pumping by other light generators. Interesting possibilities for the control of the characteristics of laser radiation are provided by the scattering in a medium placed in the laser resonator together with the active medium. SMBS was observed [28-33] in liquids placed inside the resonator of a ruby or neodymium laser. Several SMBS components were observed in [30] to be excited in succession in the emission spectrum.

In [28-31], SMBS was excited in a laser resonator by light of a giant pulse produced by Q-switching with a saturable filter. As shown by Pohl [32], SMBS can develop in a resonator also when excited with ordinary "free-running" lasing light. In his experiments, in contrast to the studies mentioned above, the radiation was focused in the interior of the scattering liquid. It was established that the strong reflection of the light as a result of the SMBS is equivalent to an increase of the resonator Q. The rapid growth of the intensity in the Stokes components has led to generation of pulses of ~100 MW power. Pohl's laser (with small design modifications) was subsequently duplicated in [33]. In similar systems it is possible to produce Q-switching also by stimulated temperature scattering [34].

The cited Q-switching studies [32-34] suffer from a common shortcoming — the use of focused radiation. It appears that the focusing should impose a definite limit on the generated-radiation power. By way of example we indicate that in the laser described in [34] generation of pulses with power more than 100 MW was impossible because of damage to the cell with the scattering medium. It should also be noted that no effort was made in these studies to determine which of the characteristics of the scattering and amplifying media are necessary to attain the Q-switching regime.

The present article is based on SMBS research performed in 1968-1973 at the Laboratory for Quantum Radiophysics of the Physics Institute of the Academy of Sciences. They have provided answers to the questions raised above. The conditions necessary for effective operation of SMBS lasers, amplifiers, and modulators were determined and realized. The feasibility of conversion of pump radiation in lasers and amplifiers with near-100% efficiency was demonstrated experimentally and theoretically for diffractive divergence of the generated light. To choose the optimal medium for laser devices, the gains and line widths of a number of compressed gases were determined. The influence of the width of the exciting-radiation spectrum on the stimulated scattering was investigated. It was also shown that the wavefront of the light scattered as a result of the SMBS can be inverted relative to the front of the exciting radiation. This effect was used to cancel out phase distortions in the amplifying medium, making it possible to obtain the diffractive divergence of the amplified radiation in an amplifier of low optical quality.

---

*Amplification of light from a stimulated-scattering laser was first observed in [26].

An exposition of the results of this study follows. The author is deeply grateful to all participants.

CHAPTER I

CONDITIONS FOR OBTAINING STATIONARY LASING WITH STIMULATED SCATTERING OF LIGHT

1. Influence of Intensity, Energy Density, and Exciting-Radiation Pulse Duration on the Laser Operation

Of greatest interest for the transformation of radiation with the aid of stimulated-light-scattering lasers is the stationary (more accurately speaking, quasi-stationary) generation regime of these lasers. Until recently, however, it was not clear whether this regime can be realized by pulsed pumping. To ascertain the conditions necessary to reach stationary lasing, we analyzed the operation of the laser under transition conditions [35]. It turned out that the transient can terminate before the end of the pump pulse only if the intensity, energy density, and duration of the exciting pulses exceed definite threshold values.

It follows from a theoretical analysis [23] that at the start of the SMBS laser operation the light intensity in its resonator ($\mathcal{I}$), at single-mode lasing, single-mode pump radiation, and low efficiency, varies with time like $\mathcal{I} \sim e^{\gamma t}$. The expression for γ is

$$\gamma = \frac{1}{2} \sqrt{4 \left(glI \frac{c}{\tau_s \varkappa L} - \frac{1}{\tau \tau_s} \right) + \left(\frac{1}{\tau} + \frac{1}{\tau_s} \right)^2} - \frac{1}{2} \left(\frac{1}{\tau} + \frac{1}{\tau_s} \right) . \tag{1.1}$$

Here g is the gain due to the stimulated scattering in the active medium, l is the length of the active medium, I is the pump radiation intensity, c is the speed of light, and τ_s is the hypersound damping time

$$\varkappa = \begin{cases} 2 & \text{for} \quad \theta \neq \pi/2 \\ 1 & \text{for} \quad \theta = \pi/2 \end{cases},$$

where θ is the angle between the light-scattering direction in a given resonator mode and the propagation of the exciting radiation, L is the optical length of the resonator, and τ is the damping time of the light in the resonator.

It follows from (1.1) that lasing is possible only if

$$I > I_{\text{thr}} \equiv \frac{\varkappa L}{c\tau} \frac{1}{gl} . \tag{1.2}$$

In other words, lasing is possible only when the gain in the active medium (which increases with increasing pump intensity) exceeds the losses in the resonator.

Let the pump pulse be of the form

$$I(t) = \begin{cases} 0 & \text{for} \quad t < 0, \quad t > T, \\ I > I_{\text{thr}} & \text{for} \quad 0 \leqslant t \leqslant T. \end{cases} \tag{1.3}$$

The initial light intensity in the resonator is determined by the thermal Mandel'shtam−Brillouin scattering, so that we can assume

$$\mathcal{I}(t) = \beta I e^{\gamma t}. \tag{1.4}$$

Here β is the probability of the thermal scattering in a given resonator mode. Expression (1.4), which describes the lasing transient, is valid so long as the intensity of the generated radiation is low. The transient terminates when this intensity becomes comparable with the pump intensity. Let us find what is required for this purpose.

Let $\mathscr{I}(t_0) \equiv \mathscr{I}_{\text{thr}}$. We then obtain from (1.4)

$$\gamma t_0 = \ln(1/\beta) - \ln(I/I_{\text{thr}}). \tag{1.5}$$

The numerical value of $\ln(1/\beta)$ depends little on the choice of the active medium [11] and on the resonator geometry. When liquids and compressed gases are used as active media we have $\ln(1/\beta) = 20\text{-}30$ [19]. We therefore always have in practice $\ln(1/\beta) > \ln(I/I_{\text{thr}})$ and we can put

$$\gamma t_0 \approx \ln(1/\beta). \tag{1.6}$$

It is easy to show that at $I > I_{\text{thr}}$ the expression

$$\gamma \leqslant \frac{cgl}{\varkappa L}(I - I_{\text{thr}}) \tag{1.7}$$

is valid, in which the equality is satisfied with great accuracy at $I/I \ll \tau/\tau_s$. From (1.6) and (1.7) we have

$$\frac{cgl}{\varkappa L}(I - I_n)t_0 \gg \ln\frac{1}{\beta}$$

or

$$It_0 \gg \frac{\varkappa L \ln(1/\beta)}{cgl(1 - I_{\text{thr}}/I)}. \tag{1.8}$$

It is clear from (1.8) that to attain stationary lasing it is necessary to satisfy, besides the usual requirement $I > I_{\text{thr}}$, also the condition

$$E \gg E_{\text{thr}} \equiv \frac{\varkappa L \ln(1/\beta)}{l}\frac{1}{cg}, \tag{1.9}$$

where E is the pump-pulse energy density.

Production of a laser beam with small divergence calls for a definite time, since the lasing evolves from thermal-scattering light which has no sharp directivity. The formation of the directivity pattern is clear from an examination of Fig. 1.

Following Fox and Li [37] we represent a laser resonator made up of flat mirrors in the form of a sequence of diaphragms of equal size (d). After traversing a path $\mathscr{L}$, the diver-

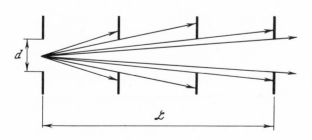

Fig. 1.   Formation of the directivity pattern of laser radiation.

gence of the scattered light is

$$\delta\theta_g = d/\mathscr{L}.$$

$$(1.10)$$

The geometric approximation remains valid until the geometric angle at a certain $\mathscr{L}_{thr}$ becomes comparable with the diffraction angle:

$$d/\mathscr{L}_{thr} \sim \lambda/d,$$

where λ is the lasing wavelength. The time required to negotiate the path $\mathscr{L}_{thr}$ is

$$\mathscr{L}_{thr}/c \sim d^2/c\lambda.$$

$$(1.11)$$

Consequently, for the diffractive directivity pattern to be formed the pump pulse duration must satisfy the inequality

$$T \gtrsim T_{thr} \equiv d^2/c\lambda.$$

$$(1.12)$$

The directivity pattern can also be formed by frequency selection of the angular oscillation modes. Such a selection is realizable only under the condition

$$T > 1/\Delta\nu_1,$$

$$(1.13)$$

where $\Delta\nu_1$ is the difference between the frequencies of the axial mode and the closest angular mode. Assume for the sake of argument that the resonator mirrors are circular. In this case a theoretical calculation [38] yields $\Delta\nu_1 \approx 0.45(c\lambda/d^2)$. Substituting this expression in (1.13) we again arrive at the condition (1.12).*

Thus, stationary lasing can be obtained only if the pump radiation intensity I, its energy density E, and pulse duration T satisfies simultaneously the inequalities $I > I_{thr}$, $E > E_{thr}$, $T > T_{thr}$. We note that this requirement is valid for lasers with any type of stimulated scattering.

If $I > I_{thr}$, then high efficiency and poor directivity can be expected at $E > E_{thr}$ and $T < T_{thr}$, low efficiency and good directivity at $E < E_{thr}$ and $T > T_{thr}$, and low efficiency and poor directivity at $E < E_{thr}$ and $T < T_{thr}$.

2. Experimental Verification of the Conditions
for Stationary Lasing

The conditions derived in Sec. 1 were tested with the aid of two SMBS lasers specially developed for this purpose. The active medium was liquid ether in one laser and carbon disulfide in the other.

The ether laser resonator was made up of two flat mirrors spaced 1.7 cm apart (Fig. 2). The mirrors served simultaneously as the internal walls of a cell filled with the ether.† The pump radiation passes through the cell walls perpendicular to the mirrors, the distance between which is 2 cm. To prevent lasing, the angle between them is made 5°. A diaphragm measuring 1.3 cm in the direction of the resonator axis and 0.1 cm in the perpendicular direction is placed in front of the cell.

---

*Starting from the foregoing considerations, it is easy to estimate T_{thr} also for resonators having different configurations. For example, for a telescopic resonator we obtain $T_{thr} \sim (L/c)[\ln(d^2/\lambda L)/\ln\gamma]$, where γ is the resonator gain.

†Generation in a laser with xenon gas at a pressure 35 atm was obtained by us in a similar scheme.

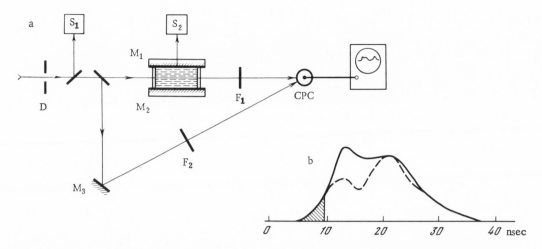

Fig. 2. Setup for the investigation of the ether laser (a) and superimposed oscillograms of the pump (solid curve) and of the radiation passing through the ether (dashed curve) (b). D, diaphgram 1.3×0.1 cm; M_1 and M_2, dielectric mirrors with 4% transmission at $\lambda = 0.694 \mu$m; M_3, delay line mirror; F_1 and F_2, filters; CPC, FÉK photocell; S_1 and S_2, systems for the measurement of the pump-radiation and lasing parameters.

The values of I_{thr}, E_{thr}, T_{thr} for this scheme are listed in Table 1, which gives also the parameters of the active medium (g, $\delta\nu_0$) and of the pump radiation ($\delta\nu$, I, E, T).

The gain was obtained from the data of [12, 39] with account taken of Eq. (5). The line width was calculated on the basis of the results of [39]. It is assumed here and below that $\ln(1/\beta) = 30$. It is also assumed that the damping of the light in the resonator is determined by the transmission of the mirrors, and that the generated laser beam has dimensions $d' = 0.1$ cm and $d'' = 2$ cm. These dimensions correspond to two times, T'_{thr} and T''_{thr}.

We note that in this scheme it suffices to know only the pump characteristics averaged over the beam. Indeed, the inhomogeneities of the pump field in the direction of the resonator axis has no effect on the lasing, while in the perpendicular direction the intensity of the exciting radiation is practically constant because the dimension of the diaphragm is this direction is much smaller than the ruby-laser beam diameter.

We have observed in the Brillouin laser of the described system lasing with emission energy 0.04 J, peak power 8 MW, pulse duration 5 nsec, and divergences $\delta\theta'_g$ rad in the plane perpendicular to the pump propagation direction and $\delta\theta''_g = 5 \cdot 10^{-2}$ rad in the other plane (we indicate for comparison that the divergence of the pump radiation is $\delta\theta = 2 \cdot 10^{-3}$ rad). It should be noted that the probability of axial-mode lasing in this resonator is quite small, since the distance between the modes is $c/2L \gg \delta\nu_0$, and consequently the lasing was as a rule on angular modes. This effect, however, could not be the cause of so large a divergence. The

TABLE 1

Active medium	g, cm/MW	$\delta\nu_0$, MHz	$\delta\nu$, MHz	I_{thr}, MW/cm²	I,* MW/cm²	E_{thr}, J/cm²	E, J/cm²	T_{thr}, nsec	T,† nsec
Ether	0,034	118	<300	1,0	140	0,05	1,6	$5(T_n')$ $2000(T_n'')$	15

*Peak intensity of the exciting radiation.
†Pulse duration at half height.

emission spectrum of lasers of this type is well known [21]. It consists of a series of equidistant lines, the intensity of which decreases with decreasing frequency. In our case there were five such lines. These lines are due to the successive Mandel'shtam − Brillouin scattering through 180°, and consequently the time of lasing on individual lines (T_L) is less than the registered pulse duration (5 nsec), i.e., less than T'_{thr} and T''_{thr}. According to the foregoing analysis, no diffractive directivity pattern can be produced within this time. It can therefore be assumed that the radiation divergence is determined by the geometrical angle

$$\delta\theta_g = d/L_L \equiv d/cT_L, \tag{1.14}$$

where L_L is the path traversed by the light during the lasing time. It follows from (1.14) that

$$T_L = d/c\delta\theta_g. \tag{1.15}$$

If our reasoning is valid, then the time T_L calculated from formula (1.15) should not depend on the plane in which the laser beam and its divergence are measured, and should also be shorter than 5 nsec. Substituting in (1.15) the numerical values of d', $\delta\theta'_g$ and d", $\delta\theta''_g$ we obtain $T'_L \approx 1.1$ nsec and $T''_L \approx 1.3$ nsec. These times are shorter than 5 nsec and agree with each other within the accuracy of the experimental error, thus indicating that our interpretation of the process that governs the directivity pattern is correct.

To study the dynamics of the lasing we registered simultaneously the wave forms of the pump pulse on entering the active medium of the laser and after passing through it, using the scheme shown in Figure 2a. Part of the pump pulse was made to lag, by using an optical delay line, the pulse passing through the cell by 46 nsec. Consequently both pulses were registered in a single oscillogram without superposition. To calibrate the measuring system, the filter F_1 was placed in front of the cell, in which case the pump intensity was below threshold and no lasing was produced. The relative sensitivity of the system to the investigated pulses was determined from the oscillogram obtained under these conditions.

In Fig. 2b, the two pulses are superimposed with allowance for this sensitivity. The shaded part corresponds to an energy density equal to E_{thr}. It is seen that a significant conversion starts only at $E \sim E_{thr}$ (it subsequently reaches 50%).

We have thus verified that at $T < T_{thr}$ the divergence of the generated light greatly exceeds the diffractive divergence, and also that the laser efficiency is low at $E < E_{thr}$.

To investigate the case $T \gtrsim T_{thr}$, we used a carbon disulfide laser [40]. The resonator of this laser was made up of two flat mirrors spaced 1.7 m apart (Fig. 3). A rectangular diaphragm measuring 9×9 mm was placed in the resonator. The carbon disulfide filled a glass cell 24 cm long, 1 cm wide, and 2 cm high. The windows through which the generated radiation passed made an angle of 94° with the sidewalls and were nonreflecting. A system of 17 total-internal-reflection glass prisms ensured multiple passage of the pump light perpendicular to the resonator axis. A thin glycerine layer was placed between the prisms and the cell, and since the refractive indices of carbon disulfide, glass, and glycerine are close to one another, the reflection coefficients at the separation boundaries were small ($\leq 0.1\%$). Therefore the exciting-radiation losses were determined mainly by the absorption in the glass and did not exceed 50%.

Let us determine the pump-source parameters needed to obtain stationary lasing. In the exact calculation it is necessary to take into account the spatial distribution of the exciting radiation, as well as the decrease of its intensity by the absorption in the glass and by the broadening of the beam as a result of the finite divergence. For estimates, however, it can be assumed that the length of the active medium is $l \sim 10$ cm and the pump intensity is constant

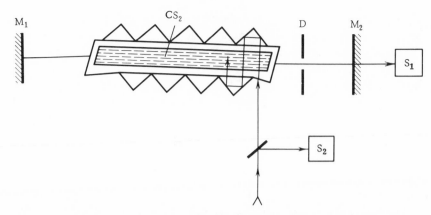

Fig. 3. Installation for the investigation of carbon disulfide laser. M_1, M_2, dielectric mirrors with transmissions 2 and 20% respectively; D, diaphragm; only a few of the prisms are shown in the cell with the carbon disulfide; S_1 and S_2, systems for the measurement of the pump-radiation and lasing parameters.

TABLE 2

Active medium	g, cm/MW	$\delta\nu_0$, MHz	$\delta\nu$, MHz	I_{thr}, MW/cm²	I,* MW/cm²	E_{thr}, J/cm²	E, J/cm²	T_{thr}, μsec	T,† μsec	$\delta\theta$, mrad
Carbon disulfide	0.17	40	40	0.06	0.55	0.11	0.33	0.4	0.8	4.5

*Peak intensity of exciting radiation.
†Pulse duration at half-height.

over this length. It can also be assumed that in this scheme the generated light beam measures ~9 mm.

The values of I_{thr}, E_{thr}, T_{thr} are shown in Table 2. The gain was obtained from the data of [41, 42]. The line width was calculated on the basis of the results of [42]. The pump was a ruby-laser specially developed by us [43]. Its parameters are also listed in the table.

As seen from the table, the main parameters of the pump satisfy the conditions required to obtain stationary lasing.

We observed in the described system generation of light with divergence* $\delta\theta_g = 3 \cdot 10^{-4}$ rad at a beam cross section 0.9×0.5 cm². The spectrum of this radiation consists of three lines; the distance between them corresponds to SMBS through 180°.

It has thus been established that under the condition $T \gtrsim T_{thr}$ the generated radiation has high directivity. The small deviation of the laser light divergence from the diffractive value is apparently due to the inadequate optical quality of the active medium, and also with lasing on several SMBS components.

Owing to the high directivity, the brightness of the emission of this laser greatly exceeds the pump brightness. Indeed, the brightness ratio Q of the converted and excited radiation can be estimated from the formula

$$Q = \frac{E_g}{E} \frac{T}{T_g} \left(\frac{\delta\theta}{\delta\theta_g}\right)^2, \tag{1.16}$$

* The radiation divergence was determined by the procedure described in the Appendix.

where E_g and T_g are respectively the energy density and pulse duration of the generated radiation. Substituting the numerical values and assuming $T_\Gamma \leq T$, we obtain $Q \geq 9$. We note that an increase in the brightness was observed earlier only in the stimulated Raman scattering lasers [44-48].

CHAPTER II

GAINS AND LINE WIDTHS FOR SMBS IN GASES

The media used in all the SMBS lasers reported in the literature are various liquids. In a number of cases, however, it is preferable to use compressed gases [25] because of their better optical quality and higher self-focusing threshold. SMBS was observed many times in compressed gases (see, e.g., [12, 49, 50]), but information on the line widths and the gains are highly limited. Calculations of these parameters are available only for hydrogen and nitrogen [49]. To choose the most suitable medium, we have determined the gains and the line widths of a number of gases [51]. This work was performed with the experimental setup illustrated in Fig. 4.

Light from a ruby-laser passes through an optical insulator and falls into the cell with gas.* The gas temperature is approximately 20°C. The cross section area of the exciting-radiation beam is practically constant over the entire length of the cell and is approximately equal to 0.3 mm². The radiation intensity in the cell (~100 MW/cm²) is sufficient for the development of SMBS † at 180°.

The coaxial photocells CPC$_1$ and CPC$_2$ registered the laser radiation entering and leaving the cell, respectively. The signals from the photocells were fed to a cathode ray tube capable of operating in two modes. In mode I the vertical deflecting system of the CRT receives a

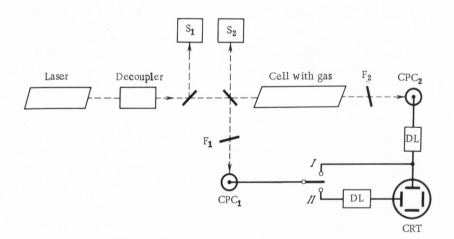

Fig. 4. Diagram of experimental setup. F$_1$ and F$_2$, attenuating filters; CPC$_1$ and CPC$_2$, FÉK-09 photocells; DL, delay lines; CRT, traveling-wave cathode ray tube of I2-7 oscilloscope; S$_1$ and S$_2$, system for the registration of the laser and scattered light radiation.

---

* The diagram of the laser, decoupler, and cell are given in Secs. 2-4 of the Appendix.

† No other types of stimulated scattering were observed in our experiments at a measuring apparatus sensitivity capable of registering these types at a level of 1% of the exciting light.

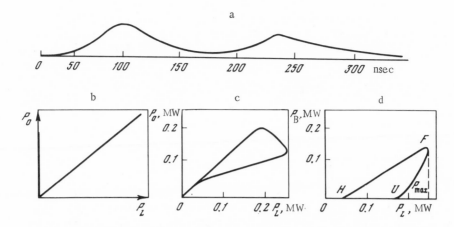

Fig. 5. Experimental oscillograms (a–c) and dependence of the instantaneous power of the scattered light on the instantaneous power of the exciting radiation (d).

signal from CPC_1, as well as the signal from CPC_2 with a delay of 150 nsec. The horizontal system receives at the same time the time sweep. In mode II, the signals from both photocells are applied simultaneously, by using delay lines, to the horizontal and vertical deflecting systems, and consequently the tube screen displays the dependence of the instantaneous radiation power passing through the cell on the instantaneous radiation power entering the cell.*

An oscillogram typical of mode I is shown in Fig. 5a. The first pulse gives the dependence of the power of the exciting radiation P_L on the time, while the second gives the same dependence for the radiation power leaving the cell $P_0 \equiv P_L - P_B$, where P_B is the power of the scattered radiation. As seen from the figure, the second pulse differs substantially from the first, owing to the scattering of a noticeable fraction of the laser light. To calibrate the measuring system, the attenuator F_2, which has a transmission of ~5%, is placed in front of the cell. Owing to the nonlinear character of the SMBS, the power of the scattered radiation becomes much less than the power of the exciting radiation, and the latter passes through the cell with practically no attenuation. This leads to a linear dependence of the power of the light passing through the cell on the power of the exciting radiation. The corresponding oscillogram (obtained in mode II) is shown in Fig. 5b.†

Figure 5c shows an oscillogram obtained in mode II in the presence of strong scattering. The reason for its ambiguity is discussed below.

From Figs. 5b and 5c we readily obtain the dependence of P_B on P_L (Fig. 5d). A similar dependence is obtained when reducing the oscillogram of Fig. 5a by the procedure used in [53]. The curve on Fig. 5d consists of a curvilinear section (UF) and a straight line section (FH). The first section corresponds to the front of the pulse, and the second to its fall-off. The curves obtained at other amplitudes of laser pulse are similar in form and their linear sections

---

*A registration system based on this principle is described in [52].

† The vertical line on the oscillogram is used for an exact determination of the origin and the direction of the coordinates. To attain this line, the matched load (75 Ω) of the traveling-wave CRT was replaced by a cable having the same wave resistance and loaded by a resistor of 1 kΩ. A pulse passing through the traveling-wave reflecting system is reflected from the unmatched resistor and returns to the photocell. Reflected from the latter, it travels again to the deflecting system. Since the cable is long (50 m), the reflected pulse arrives at a time after the pulse in the second deflection system has terminated. The electron beam is therefore reflected only along one coordinate.

lie on the line FH. A similar dependence of the scattered radiation power on the incident power was observed earlier in liquids [53] and was explained by the fact that the scattering on the front of the laser pulse is under transient conditions and on the fall-off under stationary conditions [54].

It follows from the results of [53] that

$$Hgl/S = \text{const},$$

(2.1)

where H is the threshold power (Fig. 5d) and S is the cross-sectional area of the exciting-light beams. The rest of the notation is the same as before. The constant is practically the same, ≈ 25, for all compressed gases (with accuracy not worse than 10%). Equation (2.1) enables us to find the gain g, but to this end it is necessary to determine exactly the absolute value of the threshold power H, and also the spatial distribution of the laser beam. Since it was difficult to do so with high accuracy, we chose another method.

If g_0 and g_i are the gains in the different media and H_0 and H_i are the corresponding threshold powers, then, as follows from (2.1), we have, apart from a factor of the order of unity,

$$g_i = \frac{H_0}{H_i} g_0.$$

(2.2)

By measuring the ratio H_0/H_i we can determine g_i from (2.2) if g_0 is known. Although this method enables us to determine, in fact, only the relative values of the gain, it is attractive because of its simplicity and is advantageous, since the theoretical calculation of the absolute values can frequently not be performed for the lack of the necessary thermodynamic data.

Sufficiently complete and apparently reliable data are available for methane [55]. We were therefore able to calculate the gain for this substance as a function of the pressure.

By determining from the oscillograms the ratio H_0/H_i and using for g_0 the calculated gain in methane at P = 105 atm ($g_0 = 0.072$ cm/MW), we used formula (2.2) to calculate the corresponding values of g_i. The results for methane are given in Fig. 6, and for nitrogen, xenon, and sulfur hexafluoride in Table 3. In the case of methane we see good agreement between the experimental and theoretical data in the character of the dependence of the gain on the pressure.* The value of g for nitrogen agrees well with the calculations in [49]. There are no published data for Xe and SF_6.

Using a Fabry−Perot etalon with a free spectral range $3.33 \cdot 10^{-2}$ cm^{-1}, we compared the spectra of the exciting and scattered light and calculated the hypersound velocity from the frequency shifts. The velocity for SF_6 agrees well with the published data [50]. To our knowledge, no measurements in Xe at high pressures were made before.

By measuring the gains and the speeds of sound we were able to determine also the corresponding line widths. We used Eq. (5), according to which

$$g = \Psi(n, \rho, v)/\delta\nu_0,$$

(2.3)

where Ψ is a function of the refractive index n, the density ρ, and the sound velocity v; $\delta\nu_0$ is the width of the gain line. Knowing g and v, and also n and ρ, we determined $\delta\nu_0$ (see Table 3).

---

*The error in the measurement of H_0/H_i was determined mainly by the thickness of the oscilloscope beam and was < 15%. The total error of g_i therefore does not exceed 25%.

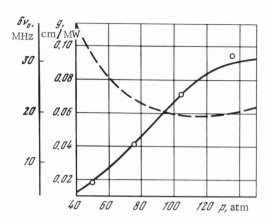

Fig. 6. Pressure dependence of the gain (solid line, calculation; points, experiment) and of the line width (dashed curve) in methane.

TABLE 3

Gas	p, atm	g, cm/MW	$\delta \nu_0$, MHz	ν, m/sec
N_2	135	0.030	15	390 [49]
Xe	39	0.044	11	149 ± 6
SF_6	22	0.035	14	113 ± 6

Let us point out another independent possibility of finding the line width. The effective gain (g^*) on the front of a pulse of exciting radiation was determined in [54] under the assumption that the pulse is of the form $P_L(t) = P_{max} \exp[-(t/T)^2]$. At the instant when the power of the exciting radiation reaches the threshold level (point U on Fig. 5d) we have

$$\frac{g^*}{g} = \Phi(y) \equiv \frac{\sqrt{1+4y}-1-\ln(1+\sqrt{1+4y})+\ln 2}{y}, \tag{2.4}$$

where

$$y \approx 1.2 \frac{P_{max} g l}{2\pi \delta\nu_0 TS}.$$

On the other hand, a relation of the type (2.1) is valid also for g^*, and consequently $g^*/g \equiv H/U$. Therefore

$$\Phi(y) = H/U. \tag{2.5}$$

From (2.1) it follows also that $gl/S \approx 25/H$. From this we get

$$y \approx \frac{4.8}{\delta\nu_0 T} \frac{P_{max}}{H} \quad \text{or} \quad \delta\nu_0 \approx \frac{1}{y} \frac{4.8}{T} \frac{P_{max}}{H}. \tag{2.6}$$

By measuring the ratio H/U on the oscillogram we can, using (2.5), obtain the corresponding value of the parameter y and then use (2.6) to determine $\delta\nu_0$ from the ratio P_{max}/H measured on the same oscillogram and from the pulse duration T. For methane at a pressure 105 atm, the value $\delta\nu_0 = 17 \pm 4$ MHZ obtained in this manner is in good agreement with the calculated value 20 MHZ (see Fig. 6).

Thus, the simple procedures described above make it possible to determine important characteristics of SMBS. We have investigated mainly gases with centrosymmetrical molecules, since one can expect for them the highest self-focusing threshold, but it is clear that the same method can be used to investigate other media. The largest gain was registered by us for methane, and this is why this gas was subsequently used as a rule for laser construction.

CHAPTER III

SINGLE-FREQUENCY SMBS RING LASER

1. Feasibility of Effective Conversion of Pump Radiation

As noted earlier, generation in a SMBS laser is usually on several Stokes components produced by the successive scattering in the resonator. This process increases the time required for the steady state to set in and lowers the laser efficiency. On the other hand, the presence of several components greatly hinders the construction of an effectively operating Brillouin laser-amplifier, inasmuch as only the first component is amplified in the amplifier.

It is therefore expedient to suppress the lasing at the higher components. We use for this purpose a ring resonator with a Faraday decoupler [56]. In such a resonator, the radiation can propagate only in one direction, and consequently lasing due to successive scattering through 180° is impossible.

In the stationary case, the interaction of the pump radiation and the lasing in the active medium of such a laser is approximately described by Eqs. (1)

$$\frac{dI}{dx} = gI\mathcal{J}, \qquad 0 \leqslant x \leqslant l.$$
$$\frac{d\mathcal{J}}{dx} = gI\mathcal{J},$$

In the stationary regime, it is necessary also to satisfy the usual condition

$$\mathcal{J}(l)/\mathcal{J}(0) = K, \tag{3.1}$$

where $\mathcal{J}(l)/\mathcal{J}(0)$ is the gain in the cell and K is the attenuation of the light in the resonator $(1 < K < \infty)$. It follows from (1) that

$$I - \mathcal{J} = \text{const} \equiv M. \tag{3.2}$$

Using (3.2), we rewrite the first equation of (1) in the form

$$\frac{dI}{dx} = gI(I - M). \tag{3.3}$$

Integrating (3.3) we get

$$Mgl = \ln \frac{I(0)[I(l) - M]}{I(l)[I(0) - M]}. \tag{3.4}$$

From (3.2) we have

$$M = I(l) - \mathcal{J}(l) = I(0) - \mathcal{J}(0). \tag{3.5}$$

Therefore (3.4) can be rewritten in the form

$$Mgl = \ln\left[\frac{I(0)\,\mathcal{J}(l)}{I(l)\,\mathcal{J}(0)}\right] = \ln\left[\frac{I(0)}{I(l)}K\right]. \tag{3.6}$$

From (3.5) and (3.1) we obtain

$$\mathcal{J}(l) = \frac{K}{K-1}[I(l) - I(0)]. \tag{3.7}$$

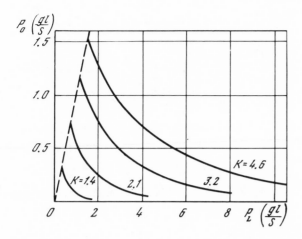

Fig. 7. Light power passing through the active medium of a SMBS laser vs. the power of the exciting light for different values of K.

Substituting (3.7) in the first equation of (3.5) we get

$$M = I(l) \frac{K[I(0)/I(l)] - 1}{K - 1} .\qquad(3.8)$$

Using (3.8), we get from (3.6)

$$I(l) = \frac{1}{gl} \frac{K - 1}{K[I(0)/I(l)] - 1} \ln\left[\frac{I(0)}{I(l)} K\right].\qquad(3.9)$$

This is the solution of our problem. For the numerical calculations it is useful to rewrite it in parametric form. We choose the parameter $z \equiv I(0)/I(l)$. Then

$$I(l) = \frac{1}{gl} \frac{K - 1}{Kz - 1} \ln kz \quad\text{and}\quad I(0) = z\, I(l).\qquad(3.10)$$

It is obvious that z varies in the range $0 < z \le 1$. Changing over to the corresponding powers, we obtain

$$P_L = \frac{K - 1}{Kz - 1} \ln Kz \quad\text{and}\quad P_0 = zP_L.\qquad(3.11)$$

Here P_L and P_0 is the pump power entering and leaving the active medium, respectively, the power being measured in units of S/gl, where S is the cross section area of the pump and lasing beams. As seen from (3.11), lasing is possible at $P_L \ge P_{thr} \equiv \ln K$. Figure 7 shows a plot of P_0 against P_L obtained from (3.11) for different values of K. The dashed line corresponds to $P_0 = P_L$.

If the laser operates in a stationary regime, the equation $P_0 = P_L - P_g$ is valid, where P_g is the power of the generated radiation. Using (3.11), we can easily obtain the pump conversion efficiency

$$\eta(P_L, K) \equiv \frac{P_g}{P_L} = \frac{P_L - P_0}{P_L} .$$

Figure 8 shows the dependence of the conversion efficiency on the pump radiation power at different attenuations in the resonator. We see that the efficiency can be close to 100%.

2. Single-Frequency SMBS Laser

Equations (1), and consequently also the theory developed in the preceding section, have been proved only for the case of monochromatic pump radiation. For a correct comparison

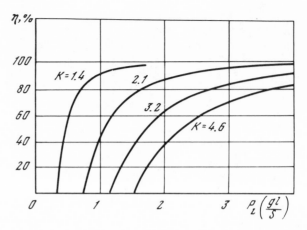

Fig. 8. Dependence of the laser efficiency on the pump radiation power.

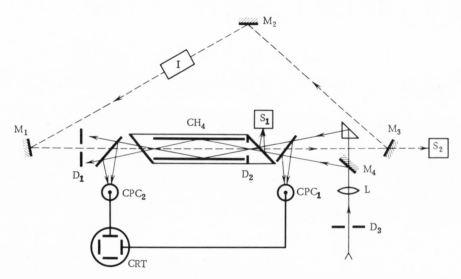

Fig. 9. Experimental setup for the investigation of a methane laser. M_1, M_2, M_3, and M_4, dielectric mirrors with transmission 2, 2, 40, and 50%, respectively (M_1, spherical, R = 160 m; M_2, M_3, M_4 plane); I, optical insulator and quartz plate that rotates the polarization plane through 45°; D_1 and D_2, diaphragms 5 × 5 mm; D_3, diaphragm 3 × 3 mm; L, lens focal length 37 cm; CPC_1 and CPC_2, FEK-09 coaxial photocells; CRT, cathode ray tube of I2-7 oscilloscope; S_1 and S_2, systems for the measurement of the pump and lasing radiation parameters.

of the experimental results with the theory it is therefore necessary to use in the experiments just this kind of radiation. Yet the pump light should, as we have explained, satisfy the conditions that the steady lasing state be reached. Starting from these requirements, we have developed a single-mode ruby laser, the diagram of which is given in Sec. 5 of the Appendix. Its radiation was used to pump the single-frequency SMBS laser (Fig. 9).

The active medium of this laser is methane at room temperature and at 130 atm pressure.* The resonator is made up of mirrors M_1, M_2, and M_3, and its optical length is 6.6 m.

———

*Under these conditions, as established in Chapter II, the gain due to the SMBS is ≈0.09 cm/MW, and the gain line width is ≈20 MHz.

The pump radiation from the ruby laser is split by mirror M_4 into two beams of equal intensity. These beams enter a hollow light pipe placed in the cell with the methane, are reflected several times from its walls, and then leaves the cell. The light pipe has a square cross section 6 × 6 mm and a length 94 cm, while the length of the cell is 96 cm. The light pipe is made of polished Plexiglas plates. The angles between the light-pipe walls and the pump beams equal approximately 0.5°. At these small angles, the coefficient of reflection from the glass is close to unity, so that the pump intensity remains practically constant over the entire length of the cell in the absence of lasing. The light pipe is so oriented that the plane of the pump polarization is parallel to two of its walls (and is perpendicular to the other two). The polarization of the pump radiation reflected from the walls remains therefore unchanged.

The presence of the light pipe in the resonator facilitates the development of the lasing on the angle modes; lasing from the cell windows is also possible. To avoid these effects, diaphragms D_1 and D_2 are placed in the resonator, the cell windows are inclined 45°, and the cell is painted black.

The lens L produces in the plane of diaphragm D_2 the image of diaphragm D_3. The dimension of the image is equal to the dimension of the diaphragm D_2. Consequently, the measuring apparatus S_1 registers only the radiation that enters the light pipe.

The pump radiation entering the cell has a maximum power ~0.6 MW at an approximate pulse duration at half height 200 nsec and at a spectrum width <15 MHz ($5 \cdot 10^{-4}$ cm^{-1}). The divergence of each beam is ~3.5 mrad. The pump parameters satisfy the conditions required to attain stationary lasing.

Two photocells register the pump light entering and leaving the cell, respectively. The signals from these photocells are fed simultaneously to the horizontal and vertical deflecting systems of the cathode ray tube. As a result, we see on the tube screen the dependence of the instantaneous radiation power passing through the cell (P_0) on the instantaneous pump power entering the cell (P_L).

In the absence of a resonator, no nonlinear processes develop in the methane, so that P_0 is a linear function of P_L (line 1 in Fig. 10). The same figure shows an oscillogram obtained in the presence of a resonator (the direction of motion of the electron beam is indicated by the arrow). The curve is two-valued because at the start of the pump pulse (up to the region marked by the asterisk) a transient lasing regime is observed, while at the end of the pulse (beyond this region) the regime is close to quasi-stationary.

In our laser we have K ≈ 7. For this value of K, using Eqs. (3.11), we plotted the dashed curve 3 of Fig. 10. The disparity with the experimental curve (past the asterisk) with the cal-

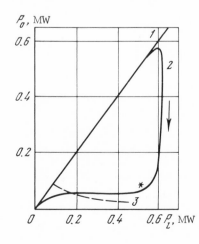

Fig. 10. Experimental oscillograms. 1) In the absence of a resonator; 2) with a resonator; 3) theoretical curve.

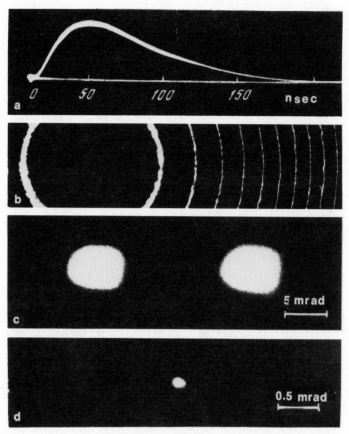

Fig. 11. Characteristics of SMBS laser emission and pump. a) Oscillogram of generated radiation; b) spectrogram of generated radiation (the free spectral range of the Fabry–Perot etalon is $3.33 \cdot 10^{-2}$ cm^{-1}); c) distribution of the pump radiation in the far zone (the two spots correspond to two pump beams); d) the same for the laser emission.

culated one apparently points to a certain deviation of the lasing from a quasi-stationary regime. Account must also be taken of the fact that Eqs. (1), and consequently also Eqs. (3.11), have been derived only for the case of plane pump and lasing waves.

It is seen from Fig. 10 that an appreciable fraction of the pump power is converted into the Brillouin-laser emission. It reaches 80%. At the same time, the radiation power at the laser output does not exceed ~35% of P_L, this being due to the large losses in the resonator. By decreasing the losses (a realistic possibility) one can, of course, increase the efficiency.

An oscillogram of the emission of the Brillouin laser is shown in Fig. 11. The laser operates on a single axial mode. The spectrum of the generated radiation reveals one line; no generation due to repeated scattering through 180° has been observed. Figures 11c and 11d show for comparison photographs of the distribution of the radiation in the far zone for the pump and for the lasing. The divergence of the generated light is practically always diffractive, ~0.12 mrad. The photographs were obtained by a procedure described in the appendix. By reducing these photographs it is easy to find the angle of distribution of the radiation energy. It turned out that half the energy is concentrated in a cone with an apex angle 0.15 mrad. We indicate for comparison that half the energy of each pump beam propagates in an angle of 5.1

mrad. Owing to so strong a difference in the divergences, the brightness of the radiation at the laser output is approximately 700 times larger than the pump brightness.

Let us estimate the possibilities of increasing the brightness with the aid of a Brillouin laser. As shown by theory and experiment, the conversion efficiency in such a laser can approach 100%. It is possible to obtain in it a converted-radiation divergence $\delta\theta_g \sim 10^{-4}$ rad. The influence of the pump radiation divergence ($\delta\theta$) on the laser operation has not been analyzed in detail, but it is clear that $\delta\theta$ can be quite large. Indeed, from expression (3)

$$\nu_B = \nu - \frac{2vn}{c}\,\nu\sin\frac{\theta}{2}\ .$$

It follows that the change of the scattering angle (θ) leads to a shift of the gain line. As seen from (3), this shift is equal to

$$\partial\nu_B = \Delta\nu\partial\left(\sin\frac{\theta}{2}\right),$$

where $\Delta\nu \equiv 2vn\nu/c$. At $\theta = 180° \pm \delta\theta/2$ we have

$$\partial\nu_B = \Delta\nu\,\frac{1}{2}\left(\frac{\delta\theta}{4}\right)^2.$$

(3.12)

It is clear that the laser efficiency will not become much worse so long as $\partial\nu_B$ is smaller than the width of the gain line ($\delta\nu_0$). For this purpose, as can be easily deduced from (3.12), it is necessary to have

$$\delta\theta \lesssim 4\sqrt{2}\ \sqrt{\delta\nu_0/\Delta\nu}\ \text{rad}\ .$$

(3.13)

The ratio $\delta\nu_0/\Delta\nu$ for compressed gases and liquids usually lies in the range 0.01–0.1. Taking for the sake of argument $\delta\nu_0/\Delta\nu = 0.01$, we obtain from (3.13) $\delta\theta \lesssim 0.57$ rad. Assuming $\eta = 100\%$ and that the cross-sectional areas of the laser and pump beams are equal we find from (1.16) that the increase of the brightness is $Q = (\delta\theta/\delta\theta_g)^2 \lesssim (0.57/10^{-4})^2 \approx 3\cdot10^7$. It is seen that SMBS lasers have potentially high capabilities.

We note in conclusion that under certain conditions ring lasers can operate effectively on one SMBS component and without a Faraday decoupler, so that their practical use is simplified. To prove this statement let us find the lasing threshold of the second harmonic under the assumption that the attenuation for this component in the resonator is the same as for the first (K). In this situation, lasing on the second component is possible if its gain, which is equal to $\exp\left(\int_0^l g\mathcal{J}dl\right)$, exceeds K. Thus, at threshold we have

$$\exp\left(\int_0^l g\mathcal{J}dl\right) = K.$$

(3.14)

On the other hand, from (3.14) and (1) we have

$$K = \exp\left(\int_0^l gIdl\right).$$

(3.15)

Taking (3.2) into account, it follows from (3.14) and (3.15) that at the threshold of the lasing of the second component we have $\mathcal{J} = I$. In this case we get from (1) $dI/dx = gI^2$. Integrating this equation and taking (3.1) into account, we obtain $glI(l) = K - 1$ or $P_L = K - 1$. Thus, lasing in a ring laser without an optical insulator takes place on a single component if the pump power lies in the range

$$\ln K \leqslant P_L \leqslant K - 1.$$

(3.16)

At $P_L = \ln K$, the conversion efficiency is $\eta = 0$. At $P_L = K - 1$ we have

$$\eta \equiv 1 - P_0/P_L = 1 - \mathscr{I}(0)/\mathscr{I}(l) = 1 - 1/K.$$

Thus, (3.16) corresponds to variation of the efficiency in the range

$$0 \leqslant \eta \leqslant 1 - 1/K.$$

(3.17)

We see that the dynamic range of pump power and the value of the maximum efficiency in the single-frequency regime increase with increasing K. In our experiments, as indicated, $K \approx 7$. Substituting by way of example this value in (3.16) and (3.17) we obtain $1.95 \leq P_L \leq 6$ and $0 \leq \eta \leq 86\%$.

CHAPTER IV

OPERATION OF SMBS AMPLIFIER IN THE SATURATION REGIME

1. Characteristics of SMBS Amplifier in the Stationary Regime

Amplification of light is presently used extensively in the construction of high-power laser systems. It has been established long ago that the SMBS light can be amplified. Up to now, however, only the case of the linear (without saturation) amplification of light in the scattering medium was investigated [10, 41, 57, 58]. At the same time, for practical applications greatest interest attaches to operation of an amplifier in the saturation regime, when a considerable fraction of the exciting light is scattered. In this chapter we investigate the influence of the saturation effect on the amplification of light in SMBS.

From the theoretical point of view, this question was first considered by Tang [15]. We shall attain the solution in a form more convenient for comparison with the experiment.

The problem of the interaction in the nonabsorbing medium of the amplifier, of the signal and pump radiations differs from that considered in the preceding chapter only in the boundary conditions. We can therefore use the already known expression (3.4)

$$Mgl = \ln \frac{I(0)[I(l) - M]}{I(l)[I(0) - M]},$$

where

$$M \equiv I(0) - \mathscr{I}(0) = I(l) - \mathscr{I}(l).$$

Hence

$$[I(0) - \mathscr{I}(0)]\,gl = \ln \frac{I(0)[I(l) - I(0) + \mathscr{I}(0)]}{I(l)\,\mathscr{I}(0)}.$$

(4.1)

We assume the ratio of the intensity of the input signal to the intensity of the pump to be given:

$$\mathcal{I}(0)/I(l) \equiv \chi. \tag{4.2}$$

We can now rewrite (4.1) in the form

$$I(l)\left[\frac{I(0)}{I(l)} - \chi\right]gl = \ln\left\{\frac{I(0)}{I(l)}\left[\frac{1}{\chi} - \frac{1}{\chi}\frac{I(0)}{I(l)} + 1\right]\right\}. \tag{4.3}$$

We present the solution (4.3) in parametric form. We choose the parameter $z \equiv I(0)/I(l)$. Then

$$I(l) = \frac{1}{gl}\frac{\ln[z + (z - z^2)/\chi]}{z - \chi} \quad \text{and} \quad I(0) = zI(l), \tag{4.4}$$

where $0 < z \leq 1$. Changing over to the corresponding powers, we obtain

$$P_L = \frac{\ln[z + (z - z^2)/\chi]}{z - \chi} \quad \text{and} \quad P_0 = zP_L. \tag{4.5}$$

Here, as before, P_L and P_0 denote the pump radiation power entering and leaving the amplifier, respectively, the power being measured in units of S/gl. Expression (4.2) is equivalent to

$$\chi \equiv P_s/P_L, \tag{4.6}$$

where P_s is the input-signal power.

We note that the gain of the signal in the absence of the saturation is equal to exp P_L, where P_L is measured in the indicated units. At a gain $\sim e^{25}$ it is necessary to take into account the spontaneous scattering (4), which was disregarded in the derivation of the preceding formulas. Expression (4.5) can therefore be used only if $P_L < 25$.

In an amplifier operating in the saturation regime, an appreciable fraction of the pump radiation is converted into signal radiation. Equations (4.5) make it possible to obtain the efficiency $\eta \equiv (P_L - P_0)/P_L$ of this conversion. Figure 12 shows a plot of the efficiency against the pump power for different χ. It is seen that the efficiency in the amplifier can be close to 100%.

The signal gain, equal to $\mathcal{I}(l)/\mathcal{I}(0)$, is very simply connected with the conversion efficiency. Indeed, from (4.5) we have

$$\frac{\mathcal{I}(l)}{\mathcal{I}(0)} = 1 + \frac{I(l)}{\mathcal{I}(0)}\frac{[I(l) - I(0)]}{I(l)} = 1 + \frac{\eta}{\chi}. \tag{4.7}$$

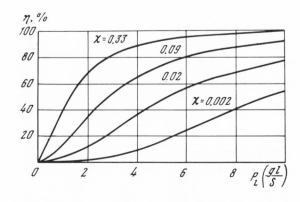

Fig. 12. Dependence of SMBS amplifier efficiency on the exciting-light power.

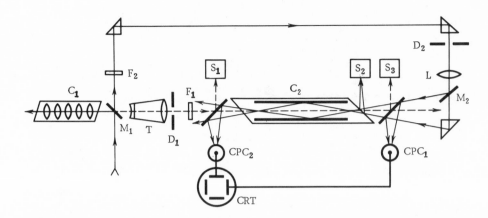

Fig. 13. Experimental setup. M_1 and M_2, dielectric mirrors with transmissions 65 and 50% respectively at $\lambda = 0.694\,\mu$m. C_1, cell with five lenses (multiple focusing is used to decrease the SMBS threshold), the focal length of the first lens is 54 mm and that of the remaining lenses 27 mm, the distance between lenses is 108 mm; T, telescope with fourfold magnifications; D_1, 5×5 mm diaphragm; D_2, 3×3 mm diaphragm. F_1 and F_2, filters; C_2, cell with hollow glass light pipe, length of cell 96 cm, length of light pipe 94 cm, cross section 6×6 mm, cell windows inclined 45°; L, lens with focal length 37 cm; CPC_1 and CPC_2, FÉK-09 coaxial photocells; CRT, cathode ray tube of oscilloscope I2-7; S_1, S_2, and S_3, systems for the measurement of the parameters of the input signal, the pump, and the amplified signal, respectively.

If the amplified light has a higher directivity than the pump light, then an amplifier with high conversion efficiency can be regarded as a device that increases the radiation brightness. The estimate given in the preceding chapter for the maximum possible brightness increase holds true for such an amplifier.

2. Experimental Investigation of Amplifier Operation in the Saturation Region

To investigate the operation of the amplifier we used an installation [59], a diagram of which is shown in Fig. 13. The radiation from a ruby laser was split by mirror M_1 into two beams. One of them was focused a number of times into a cell C_1 filled with methane gas. This gives rise to SMBS, and an appreciable fraction (up to 95%) of the exciting light is reflected. A telescope T and a diaphragm D are used to adduce from the reflected light a signal with almost plane wave front (the radiation passing through the diaphragm is subject to diffractive divergence).

The other beam is used to pump the amplifier. The two beams enter a hollow light pipe of square cross section, placed in a cell C_2 with methane.* The use of a light pipe ensures constancy of the gain over the amplifier cross section. A lens L at the entrance to the light pipe produces the image of the diaphragm D_2 so that the measuring apparatus S_2 registers only the pump radiation that enters the light pipe.

---

*The methane in both cells is at room temperature and under pressure 115 atm. Under these conditions, the gain is g ≈ 0.08 cm/MW and the gain line width is $\delta \nu_0 \approx 20$ MHz.

The optical paths are chosen such that the signal and the pump enter the amplifier simultaneously.

The pump source is the ruby laser described in Sec. 5 of the appendix. Its power at the entrance to the cell C_2 can reach ~1 MW. The divergence of each pump beam is ≈ 5 mrad; the angle spacing between them is 25 mrad. Owing to the use of the Faraday insulator (to decouple the laser from the cell C_1) the signal spectrum contains only one SMBS component. The spectral widths of the pump and of the input and amplified signals is < 20 MHz.

Insofar as we know, the question of the time τ_{st} to establish the steady-state gain in the presence of saturation has not been investigated before. In the absence of saturation we have $\tau_{st} \approx \tau_s P_L$, where τ_s is the damping time of the hypersound in the active medium of the amplifier [60, 61]. As noted above, P_L should be less than 25. We have chosen the pump-pulse duration to be ≈ 120 nsec at half-height, which is 15 times larger than the time $\tau_s \equiv 1/(2\pi \delta\nu_0)$. We note also that this pulse duration is 36 times larger than the time l/c needed for the light to pass through our amplifier.

We continue the description of the experimental setup. The coaxial photocells CPC_1 and CPC_2 register the pump light entering and leaving the cell C_2, respectively. The signals from these photocells are applied simultaneously to the horizontal and vertical deflecting systems of the cathode ray tube. Using the oscillogram, it is easy to obtain (just as in Chapter II) the dependence of the instantaneous power of the scattered light on the instantaneous pump power. When the amplifier operates in the quasi-stationary regime, these dependences should, of course, not be sensitive to the wave form of the pump pulse.

By placing the photocell CPC_2 in the position S_1 (Fig. 13), we registered the input-power signal (P_s) as a function of the pump power (Fig. 14). The lower branch of the curve corresponds to the rise and the upper to the descent of the pump pulse. It is seen that the ratio $P_s/P_L = \chi$ changes little in a large range of values of P_L. Thus, we are able to investigate in this scheme the operation of a SMBS amplifier at a signal/pump ratio which is practically constant during the pulse. The value of χ can be varied by varying the transmissions of the filters F_1 and F_2.

At $\chi = 0$ a very small fraction of the pump radiation is scattered in the amplifier. We therefore observe a linear dependence of P_0 on P_L (Fig. 15). At $\chi > 0$, starting with a definite pump power, this dependence becomes nonlinear, thus evidencing that saturation has been reached. The oscillogram branches corresponding to increasing and decreasing pump powers are similar to each other. This indicates that the amplification process is close to quasi-stationary. Figure 15 shows the averages of the two branches of these dependences, the deviations on the rising and descending parts of the pulse being designated as the measurement errors. The dashed curves are plots of Eqs. (4.5).

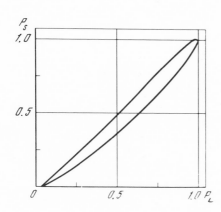

Fig. 14. Ratio of instantaneous signal power to instantaneous pump power. The measurement units are taken to be the maximum values of P_s and P_L.

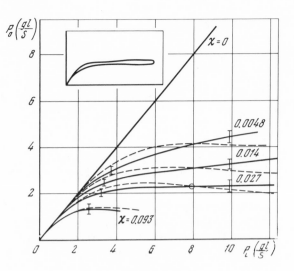

Fig. 15. Dependence of the instantaneous power of the radiation passing through the cell on the instantaneous pump power. Solid lines, experiment; dashed lines, calculation. The upper-left insert shows the experimental oscillogram.

The accuracy with which the absolute power was measured in these experiments was low (~30%). Therefore the scale of the experimental curves was chosen such that one of them coincided with the theoretical one (at the point marked by the circle).

We see that the theoretical and experimental results are in satisfactory agreement with one another. The reasons for some discrepancies between them are apparently that the theoretical analysis does not take into account the losses in the light pipe and pertains to the case of plane pump and signal waves, whereas in our experiments there were such losses and only the signal wave was nearly plane.

It is seen from Fig. 15 that a noticeable fraction of the pump radiation is transformed into signal radiation. Thus, at $\chi = 0.027$ and $P_L = 12$, the conversion efficiency is $\eta = 82\%$.* As shown by the measurements, the amplified signal radiation retains a diffraction divergence ~0.12 mrad. Consequently its brightness is approximately 1600 times larger than the pump brightness. Consequently, in addition to the lasers, SMBS amplifiers can also be used successfully for an effective conversion of laser radiation.

CHAPTER V

Q SWITCHING BY SMBS

1. Lasing Dynamics

So far we have investigated SMBS under conditions that exclude the influence of the scattered light on the operation of the exciting-radiation generators. It is of interest also to consider another situation, where there is a strong coupling between the scattering medium and the source of the exciting light (the laser). In this case the development of the SMBS can cause the laser to go over to the Q-switched regime. We have established that such a process can be observed in lasers assembled in accordance with the scheme shown in Fig. 16 [62]. In contrast to other studies of SMBS in laser resonators [28-33], in this scheme the radiation is not focused and there is no preliminary Q switching. This makes it possible to prevent heating and breakdown in the focal region [63], and facilitates the interpretation of the results.

To investigate the lasing dynamics we use a laser with the following characteristics: the amplifying medium A was a ruby crystal 23 cm long at a temperature 23°C; the medium B was

_____
*Using (4.7), we easily find that the gain under these conditions is ~31.

Fig. 16. Laser scheme. A, amplifying medium (ruby); B, medium in which the SMBS takes place (N_2, Xe, CS_2, ether, etc.); M_1 and M_2, dielectric mirrors.

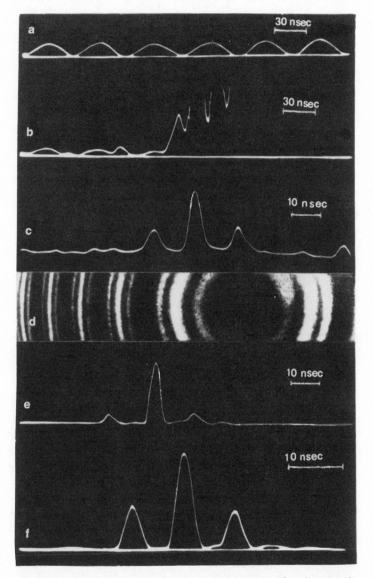

Fig. 17. Lasing dynamics. a) Structure of pulse in the free-running regime; b) excitation of SMBS; c) structure of pulse at wavelength 0.694 μm; d) emission spectrum in the region of 0.694 μm (Fabry–Perot etalon dispersion region 0.25 cm^{-1}); e) pulse at wavelength 0.828 μm; f) structure of pulse at wavelength 0.694 μm (modulation by scattering in ether); the time resolution of the recording apparatus is ~1.5 nsec.

nitrogen* compressed to 500 atm in a cell 1 m long; the optical length of the resonator was
L = 7.5 m; the distance from mirror M_2 to the cell where the nitrogen was L_1 = 2.5 m; the reflection coefficients of the dielectric mirrors M_1 and M_2 at a wavelength 0.69 μm were 99 and
60%, respectively; both mirrors were plane.

The diameter of the generated light beam at the exit from the ruby was $\approx$3 mm. The path
of the light to the mirror M_1 and back was 14 m. Taking into account the divergence of the
generated radiation ($\sim 2 \cdot 10^{-3}$ rad), the beam diameter increases over this path to 30 mm, and
consequently the radiation intensity is decreased by an approximate factor of 100.

The time evolution of the process is seen in Fig. 17. The first to occur is lasing by the
mirrors M_1 and M_2 at the frequency ν determined by the positions of the center of the gain
line in A. A microsecond pulse is generated with a regular structure and a period 2L/c (see
Fig. 17a), due to the beat of two axial modes. The peak power of the pulse is $\sim$10 kW.

The radiation passing through the cell with the nitrogen is partially reflected as a result
of Mandel'shtam−Brillouin scattering and goes over into its first Stokes component. This leads
to the return of additional spikes (Fig. 17b) besides the regular pulsations. The maxima of the
additional spikes are shifted in time by $2L_1/c$ relative to the maxima of the regular pulsations.
With increasing intensity of the first SMBS Stokes component the fraction of the light reflected
by the nitrogen increases, and this is equivalent to the action of a mirror having a growing reflection coefficient. Eventually the role of this mirror becomes decisive, and lasing develops
in the resonator which this mirror forms together with the mirror M_2. As a result, at the
exit from the laser, beyond mirror M_2, a light pulse appears with power $\sim$100 MW and duration
$\sim$4 nsec (Fig. 17c). The emission spectrum (Fig. 17d) reveals several lines, the spacing between which is equal to the Brillouin shift produced by 180° scattering in nitrogen (0.06 cm^{-1}),
and the intensities of which correspond to the temporal distribution of the radiation energy
(Fig. 17c).

During the course of the lasing, the light intensity in the resonator exceeds the threshold
for stimulated Raman scattering in nitrogen. Consequently, a light pulse of wavelength 0.828 μm,
energy up to 0.2 J, and duration $\sim$3 nsec, appears behind the mirror M_1 (Fig. 17e).

With increasing pump energy, the entire process described above is repeated. Several
high-power pulses having the same form as in Figs. 17c and 17e are then generated.

Besides compressed nitrogen, we used other substances for the Q switching, namely
xenon at 35 atm, liquid ether, acetone, carbon disulfide, carbon tetrachloride, and toluene in
water. Figure 17f shows by way of example an oscillogram of the lasing observed by using
ether as the scattering medium. Comparison of Figs. 17c and 17f shows that the temporal
characteristics of the radiation depend little on the type of active medium. They are likewise
not very sensitive to changes of the reflection coefficients of the resonator mirrors. Raman
scattering in xenon is practically impossible, so that when xenon was used the laser emitted
only at a wavelength 0.694 μm.

The experimental characteristics of the laser (see Fig. 18) were applied with the mirror
M_1 having a reflection coefficient $\sim$4%. It should be noted that the laser generates nanosecond
pulses having the same energy as in the free-running regime (and even higher near the threshold!).

A very important question is that of the limiting values of the pulse intensity and duration
that can be attained by Q-switching with the aid of SMBS. In our experiments the laser emis-

---

*At this pressure, the gain due to the SMBS is 0.13 cm/MW, and the gain line width is $\approx$6.5
MHz [49]. We have observed Q-switching at pressures 100-600 atm.

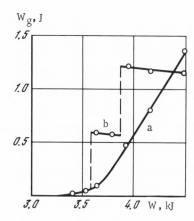

Fig. 18. Dependence of the emission energy (W_g) in different lasing regimes on the pump energy (W). a) Nitrogen pressure in cell 1 atm, free running; b) nitrogen pressure in cell 518 atm, Q-switching.

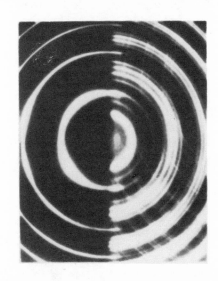

Fig. 19. Spectrograms of emission generated without the medium B (left) and in the presence of B (right). The free spectral range of the Fabry—Perot etalon is 0.833 cm^{-1}.

sion intensity at the end face of the ruby element was ~1 GW/cm^2 and was limited by the disintegration of the ruby and of the windows of the cells with the scattering medium. The question of how short the generated pulses can be calls for further study, but there are no grounds for counting on obtaining pulses much shorter than 1 nsec. The limit is determined here by the time of establishment of the hypersound in the scattering medium, which is usually ~1 nsec for condensed media [57].

Lasing first develops in the investigated regime at a frequency close to the center of the gain line in the medium A, and subsequently takes place on several Stokes components due to the successive scattering in the medium B. In our laser, this process ceases when the gain in the medium A decreases as a result of emission of an appreciable fraction of the energy stored in it. It is clear that if for some reason (for example, insufficiently broad A emission line) the process were to terminate earlier, the radiated energy would be less. It was therefore important to determine how many components take part in the operation of our laser.* Multiplying the number of components by the Brillouin shift we can determine the spectral interval needed for effective lasing in the Q-switching regime.

To determine the number of components we have compared the laser emission spectrum in the presence and in the absence of the scattering medium B (Fig. 19). The medium B was ether filling a cell 40 cm long; the mirrors M_1 and M_2 had reflections 4 and 50%, respectively; the resonator length was 2.8 m. The main error in these experiments is due to heating of the ruby by the pump radiation and to the corresponding drift of the gain line (at a rate ~0.13

_____

* The spectrogram in Fig. 17d does not permit this, for one cannot exclude the possibility of the presence of indiscernible weak components in it.

cm^{-1}/deg [64]). As shown by independent measurements, the heating was ~5°C within a time 0.7 msec. To decrease its influence, a thick glass plate was placed in the resonator from which the cell with the ether was removed; by rotating this plate we changed slightly the resonator losses and hence the lasing threshold. As a result of this, the instance of the start of lasing with and without the ether coincided within ±20 μsec at a fixed laser pump energy. Taking into account this accuracy, the accuracy with which the ruby temperature was kept constant (±0.02°C), and the stabilization of the pump energy (±1%), the total error did not exceed 0.03 cm^{-1}, which is much less than the Brillouin shift in ether ($\Delta\nu = 0.13$ cm^{-1}). It is seen from Fig. 19 that in Q switching by SMBS the number of Stokes components does not exceed 5.

If compressed gases are used as the scattering media, the value of $\Delta\nu$ can be ~0.01 cm^{-1} [50]. Therefore, taking our result into account, the Q-switching regime is perfectly feasible also for a rather narrow (~0.1 cm^{-1}) gain line in A. Examples of media with such gain line widths are $CaF_2 : Dy$ [65] and $Ar : C_3F_7I$ [66].

2. Conditions under Which Q Switching Is Possible

A mathematical description of the entire process in analytic form is difficult, but certain conditions for the realization of this regime become clear even from consideration of the initial lasing stage [67]. Let I be the emission intensity in the resonator at the frequency ν, and let I_1 be the emission intensity at the frequency $\nu_1 \equiv \nu - \Delta\nu$, where $\Delta\nu$ is the frequency shift produced in SMBS at 180°. If $I_1 < I$ and the coefficient of reflection from M_1 is small, then the equations for I and I_1 can be approximately written in the form

$$\frac{dI}{dt} = \frac{I}{\tau}(\alpha - 1), \tag{5.1}$$

$$\frac{dI_1}{dt} = \frac{I_1}{\tau_1}(\alpha_1 - 1) + \frac{c}{L}glII_1, \tag{5.2}$$

where τ is the time of the attenuation of the light in the resonator and α(t) is the ratio of the gain in A (per unit length) to its threshold value for the frequency ν, τ_1 and α_1(t) are the same for the frequency ν_1, c is the speed of light, L is the optical length of the resonator, g is the gain due to SMBS in the medium B, and l is the length of B. If the gain line in A is uniformly broadened, then

$$\alpha_1 = \alpha\frac{\sigma_1}{\sigma}\frac{\tau_1}{\tau}, \tag{5.3}$$

where σ and σ_1 are the amplification cross sections in A for the frequencies ν and ν_1, respectively. It follows from (5.3) that

$$\frac{d}{dt}\left(\ln\frac{I_1}{I}\right) = \frac{c}{L}glI - \frac{1}{\tau}\left[\alpha\left(1 - \frac{\sigma_1}{\sigma}\right) + \left(\frac{\tau}{\tau_1} - 1\right)\right]. \tag{5.4}$$

For the laser to operate in the regime described above it is necessary, at the very least, that the ratio I_1/I increase with time. As seen from (5.4), this is possible only if

$$I > I_{thr} \equiv \frac{L}{c\tau}\frac{1}{gl}\left[\alpha\left(1 - \frac{\sigma_1}{\sigma}\right) + \left(\frac{\tau}{\tau_1} - 1\right)\right]. \tag{5.5}$$

In addition, I_1 should be comparable with I at least at the end of the lasing pulse. It follows from (5.4) that $\frac{d}{dt}\left(\ln\frac{I_1}{I}\right) < \frac{c}{L}glI$. Hence

$$\ln\frac{I_1}{I} - \ln\frac{I_1(t_0)}{I_{thr}} < \frac{c}{L}gl\int_{t_0}^{t}Idt < \frac{c}{L}glE, \tag{5.6}$$

where E is the radiation energy density at the frequency ν and $I(t_0) \equiv I_{thr}$. From (5.6) it follows that to attain $I_1/I \sim 1$ we must have

$$E \gtrsim E_{thr} \equiv \frac{L}{cgl} \ln \frac{I_{thr}}{I_1(t_0)} . \tag{5.7}$$

If $\Delta\nu < \delta\nu_A$ where $\delta\nu_A$ is the width of the gain line in the medium A, and the gain line has a Lorentz shape, then $(1 - \sigma_1/\sigma)$ in Eq. (5.5) can be replaced by $[2\Delta\nu/\delta\nu_A]^2$. Taking this into account we obtain from (5) at $\tau_1 = \tau$ and $\alpha \sim 1$

$$I_{thr} \sim \frac{L}{c\tau} \frac{1}{gl} \left[\frac{2\Delta\nu}{\delta\nu_A}\right]^2 . \tag{5.8}$$

Substituting in (5.8) typical values $L/c\tau \sim 1$, $g \sim 0.01$ cm/MW, $\Delta\nu \sim 0.1$ cm^{-1}, and $\delta\nu_A \sim 10$ cm^{-1}, at a length $l \sim 10$ cm, we obtain

$$I_{thr} \sim 4 \text{ kW/cm}^2 . \tag{5.9}$$

This intensity is perfectly attainable in free-running laser operation.

As shown above, in addition to the condition $I > I_{thr}$, the condition $E > E_{thr}$ is also necessary. The quantity $I_1(t_0)$ in the expression for E_{thr} is difficult to determine accurately, but it is clear that it is not less than the level of the spontaneous noise from the medium A. To estimate the threshold energy density we can put $I_1(t_0) \sim 1$ W/cm^2. We then have from (5.7) and (5.9)

$$E_{thr} \approx \frac{L}{c} \frac{8}{gl} . \tag{5.10}$$

Putting $L \sim 100$ cm, $g \sim 0.01$ cm/MW, and $l \sim 10$ cm, we obtain from (5.10)

$$E_{thr} \sim 0.3 \text{ J/cm}^2 . \tag{5.11}$$

The free-running spike duration is usually ~ 1 μsec. As follows from (5.11), at such a duration it is necessary to have a light intensity $I \sim 0.3$ MW/cm^2, which exceeds (5.9) by two orders of magnitude.

Thus, if a condition (5.7) is satisfied and $\tau_1 = \tau$, then the condition (5.5) is certainly satisfied. The situation can change, however, if $\tau_1 < \tau$. In that case, according to (5.5), I_{thr} can increase significantly.

3. Experimental Verification of the Q-Switching

Conditions

To investigate the laser operation at $\tau_1 = \tau$ and $\tau_1 < \tau$, we have replaced the mirror M_1 by a Fabry−Perot etalon with variable gap. The reflection coefficients of the etalon mirrors were 4%; the mirror M_2 had a 50% reflection. The medium A was ruby crystal 23 cm long, and the medium B was liquid ether placed in a cell of 40 cm length. The optical length of the resonator (L = 2.8 m) was chosen to be large enough to make the distance between its modes (c/2L) less than the width of the Brillouin line in the medium B. The distance between the modes of the Fabry−Perot etalon was comparable with the Brillouin shift in the ether. Generation of high-power pulses was detected by the breakdown of the air in the focus of a lens ($f = 2.5$ cm) placed behind the mirror M_2. It was observed when the distance between the Fabry−Perot etalon modes was equal to $\Delta\nu/m$, where m is an integer (i.e., at $\tau_1 = \tau$), and was not observed when m was not an integer, in complete agreement with (5.5). Taking this result into account, all the remaining experiments described in this chapter were carried out at $\tau_1 = \tau$.

TABLE 4

Substance	l_{thr}, cm	g, cm/MW	$(gl)_{thr}$, cm²/MW
CCl_4	89.5±9.5	0.006 [57]	0.54±0.06
Acetone	26.7±2.7	0.022 [57]	0.59±0.06
Ether	22±2	0.024 [12]	0.53±0.05
CS_2	5.3±0.65	0.14 [57]	0.74±0.09

It follows from (5.10) that lasing in the Q-switching regime is possible only when the product gl exceeds the threshold value

$$(gl)_{thr} \approx 8L/cE. \tag{5.12}$$

In our laser the variation of E from experiment to experiment did not exceed 10%, this being attained by stabilizing the pump energy and the ruby temperature. By varying the length of B we determined the threshold lengths l_{thr} for different liquids, and then calculated $(gl)_{thr}$ by using the known values of g. In these experiments, the mirror M_1 had a reflection coefficient 19%; the remaining elements were the same as before. The results are listed in Table 4.

As expected, $(gl)_{thr}$ remains practically constant when g varies in a wide range. The small increase of its value for CS_2 is apparently due to the fact that free running in our experiment proceeds on two axial modes [43], the separation between which ~54 MHz is comparable with the width 75 MHz of the Brillouin line in CS_2 [57]. The experimentally determined $(gl)_{thr}$ is approximately double the value calculated from (5.12). This can be regarded as perfectly satisfactory agreement if it is recognized that it is difficult to determine the absolute value of the energy density.

We have thus ascertained the conditions necessary for Q switching by the SMBS in the resonator. We note in conclusion that a medium in which SMBS can take place is always present in lasers. It may be the liquid or gas cells, the material of the various windows, and also the amplifying medium.* Consequently, any laser is capable, in principle, of operating in the regime investigated by us. This, of course, calls for satisfaction of the conditions obtained above.

CHAPTER VI

INVERSION OF THE EXCITING-RADIATION WAVE FRONT IN SMBS

1. Comparison of the Wave Fronts of the Exciting and Scattered Light with the Aid of a Phase Plate

The divergence of the emission of the laser described in the preceding section was the same under Q switching as in the free running regime. It can therefore be concluded that the light scattered as a result of SMBS propagates backward in the same solid angle as the exciting radiation. This fact was noted earlier [69, 70], but it was never ascertained whether it is due only to the geometry of the experiment or has a more profound nature. To answer this question, we have compared the wave fronts of the reflected and exciting light [71].

---

*Q switching by SMBS in these elements was probably observed in [68].

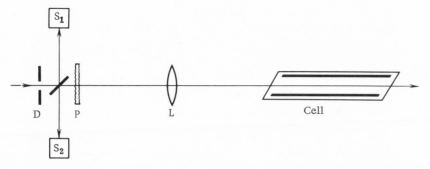

Fig. 20. Diagram of experimental setup. D) 6×6 mm diaphragm;
P) plate 1.3 thick, distance between plate and diaphragm 10 cm; L)
lens of 10 cm diameter with focal length 100 cm; length of cell 96
cm; length of light pipe 94 cm, cross section $4 \times .4$ mm, cell filled
with methane at pressure 125 atm; S_1 and S_2) systems for the measurement of the parameters of the laser light and the reflected light.

The experimental setup is shown in Fig. 20. The wave front of the ruby-laser radiation
is distorted with a phase plate P prepared by etching polished glass in fluoric acid. The rough-
nesses on the surface of the plate had dimensions ~150 μm and a depth ~1 μm (see [72] concerning the optical properties of such plates). The laser beam had a divergence 0.14×1.3
mrad. The divergence of the light passing through the plate was 3.5 mrad. This light entered
a hollow glass light pipe of square cross section, placed in the cell with the compressed methane.
To prevent lasing, the windows of the cell were inclined 45°.

The plate P was illuminated by a rectangular beam shaped by diaphragm D. With the aid
of a large-aperture lens L, the image of the illuminated region was projected on the entrance
to the light pipe, the size of the image being equal to the dimension of the input aperture of the
light pipe. Consequently the entire laser radiation registered by the measuring system S_1, after
passing through the plate P and the lens, entered the light pipe. The system S_2 registered the
reflected light, which also passed through the lens and the plate.

The ruby laser operates on a single axial mode and its emission on entering the cell has
a maximum power 1.3 MW at a pulse duration at half-height 110 nsec, and at a spectrum width
< 20 MHz. The decoupling between the laser and the cell was with the aid of a Faraday insula-
tor. A single line is observed in the spectrum of the reflected light (Fig. 21), and its displacement relative to the laser-emission line corresponds to scattering through 180°.

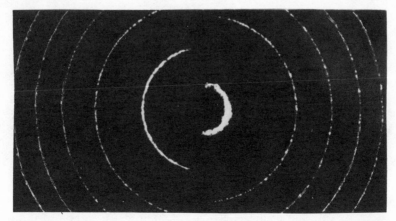

Fig. 21. Spectrograms of exciting (left) and scattered (right)
radiation. The Fabry−Perot etalon free spectral range is
$3.33 \cdot 10^{-2}$ cm^{-1}.

Photograph 22a shows the distribution of the laser radiation in the far zone. A photograph of the far zone of the reflected radiation is shown in Fig. 22b. We see that the scattered radiation passing through the plate P has practically the same divergence as the laser light. This is confirmed also by the agreement between the ratio of the intensities determined by the reduction of the negatives of Figs. 22a and 22b with the value of the reflection coefficient (~25%) obtained in calorimetric measurements. Equality of the divergences was observed also as a small <1% coefficient of reflection from the scattering medium.

A different picture is observed if the cell with the methane is replaced by a flat mirror (see Fig. 22c). In this case, the divergence of the reflected light greatly exceeds the divergence of the laser radiation and amounts to 6.5 mrad.

As the coherent light beam passing through the etched plate propagates, microinhomogeneities are produced in it as a result of interference between waves traveling in different directions [72]. To assess the influence of this exciting field radiation on the SMBS, we photographed the far zone of the reflected light in the absence of a plate (Fig. 22d). In this case the divergence of the scattered radiation greatly exceeds the divergence of the exciting light.

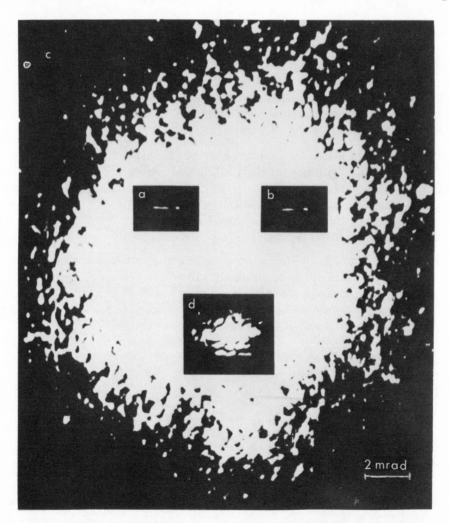

Fig. 22. Photographs showing the far-zone distributions of the laser emission (a), of the scattered radiation (b), of the light reflected by a flat mirror (c), and of the scattered light in the absence of the plate P (d).

2. Influence of the Structure of the Exciting Radiation

Field on the Shape of the Scattered-Light Front

From the experimentally observed fact that the wave front of the back-scattered radiation is "straightened out" by the same phase plate which has distorted the initial laser wave, it follows that the scattered field (signal) E (r, x) in the plane x = x$_0$ coincides (apart from a factor) with the complex-conjugate field of the exciting radiation E$_L^*$(r, x):

$$E_S\,(r,\,x_0) \approx \text{const}\; E_L^*\,(r,\,x_0).$$ (6.1)

The plane x = x$_0$ is perpendicular here to the average beam direction and is located near the plate on the side of the scattering medium, and r is the transverse coordinate. In other words, the wave front of the scattered light duplicates (with reversal of the sine) the front of the exciting radiation. At first glance this seems strange, since the spontaneous-scattering light and the gain g do not depend strongly on the angle.

We present theoretical arguments [71] favoring the satisfaction of relation (6.1). The dependence of the gain and of the reactive component of the nonlinear polarizability on the scattering angle θ in the range $0 \leq \pi - \theta \lesssim 3 \cdot 10^{-3}$ can be neglected. The propagation of the signal wave $E_S = e^{ik_S x}\varepsilon_S\,(r,\,x)$ in the direction (+x) will therefore be described by the parabolic equation

$$\frac{\partial \varepsilon_S}{\partial x} - \frac{i}{2k_S}\Delta_\perp \varepsilon_S - \frac{1}{2}\Gamma\,(r,\,x)\,\varepsilon_S = 0,$$ (6.2)

where the gain Γ(r, x), by virtue of the foregoing, is determined simply by the intensity of the exciting light, mainly $\Gamma = A|E_L\,(r,\,x)|^2$, where A is a certain constant. The most essential factor in the analysis is that the field $E_L = e^{ik_L}\varepsilon_L\,(r,\,x)$ satisfies (neglecting the term corresponding to the gain) an equation that is the complex conjugate of (6.2)

$$\frac{\partial \varepsilon_L}{\partial x} + \frac{i}{2k_L}\Delta_\perp \varepsilon_L = 0$$ (6.3)

(the small difference between the coefficients k$_L$ and k$_S$ of the transverse Laplacian can be neglected). We consider the system of functions f_i(r, x), i = 0, 1, 2, ..., which satisfy the orthogonality relation in the cross section x = x$_0$ and the same equation that describes the propagation of the complex-conjugate field of the exciting radiation;

$$\int f_i^*\,(r,\,x_0)\,f_j\,(r,\,x_0)\,dr = \delta_{ij}; \qquad \frac{\partial f_i}{\partial x} - \frac{i}{2k}\Delta_\perp f_i = 0.$$ (6.4)

Then the orthogonality relation holds in any cross section x = const. We choose a function f_0^* that coincides with the field of the exciting light: $\varepsilon_L = Bf_0^*$(r, x), where B is a certain constant and the remaining functions f_i^*, i = 1, 2, ..., are chosen arbitrarily, starting from the orthogonality condition (6.4). Representing the signal field in the form of the expansion

$$\varepsilon_S = \sum_{i=0}^{\infty} C_i\,(x)\,f_i,$$ (6.5)

we obtain for the coefficients C$_i$

$$\frac{dC_i}{dx} - \frac{1}{2}\sum_{j=0}^{\infty} \Gamma_{ij}(x)\,C_j = 0,$$ (6.6)

$$\Gamma_{ij} = AB^2 \int |f_0|^2 f_i^* f_j\,dr.$$ (6.7)

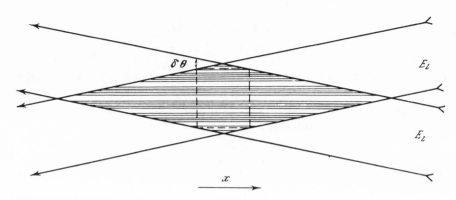

Fig. 23. Interference of two intersecting plane waves. The position of
the scattering medium is shown shaded.

We shall not investigate in detail the properties of the solutions of the system of equations
(6.6) and (6.7), and note only the following. In the case where the interference leads to appre-
ciable oscillations of the quantity $|f_0|^2$ over the cross section (this is precisely the situation
in the experiment), the diagonal element Γ_{00} (which can arbitrarily be called the gain of the
zero order-function) is ~2 times larger than the gains Γ_{ii} for the remaining functions and the
values of the off-diagonal coefficients $|\Gamma_{0i}|$, $|\Gamma_{ij}|$, i, j $\neq$ 0. The amplitude $C_0(x)$ can therefore
be expected to increase most rapidly, and this results in the required relation (6.1).

We note also that if the exciting radiation has an amplitude profile that is constant over
the cross section, $|f_0|^2$ = const, then the signal should not reproduce preferentially the complex-
conjugate front. This agrees qualitatively with the result of the experiment without the phase
plate.

The physical picture of the described phenomenon can be easily understood by considering
the spontaneous-emission gain in the field of two plane waves with identical wavelengths λ,
propagating in the direction $(-x)$ at a small angle $\delta\theta$ between them.* Interference fringes are
produced in the region of their intersection. Let the scattering medium be in this region
(Fig. 23). The spontaneous-scattering light can be represented in the form of a set
of pairs of plane waves with the same angle between them, propagating in arbitrary directions. The local maxima of the interference pattern for the pair propagating in the $(+x)$
direction coincide in space with the maxima of the fringes of the exciting field. Since the gain
is determined by the product of the intensities of the exciting and scattered light (averaged
over the cross section), the gain for this pair will be larger than for all others. Obviously, the
field of this pair is precisely the one described by expression (6.1). On the other hand, if the
scattering is in the field of one plane wave, then the gain for the signal waves (or wave pairs)
traveling in different directions will be practically the same. As a result, the entire set of
waves is effectively amplified and consequently the front of the exciting light is not reproduced.

Using the wave-pair model, let us estimate the conditions under which the reproduction
of the front is possible. Let a pair of waves of the amplified radiation be inclined at an angle
μ relative to the pair of pump waves. Then the corresponding interference fringes will make
the same angle. It is clear that the gain changes insignificantly if the spatial displacement of the
fringes over the length of the interaction region is smaller than the width of these fringes, i.e., at

$$\mu l < \lambda/2\delta\theta,$$

where l is the length of the region and $\lambda/2\delta\theta$ is the width of the fringes.

---

*A model of this type was used in [73], which is devoted to an investigation of SRS.

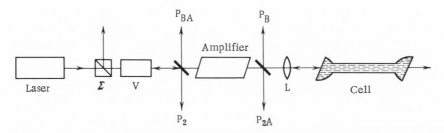

Fig. 24. Experimental setup. Σ, polarizer (Glan prism); V, Faraday cell rotating the polarization plane through 45°; amplifier, ruby crystal 24 cm long and 20 mm in diameter with end faces inclined at 4°; L, lens with focal length 25 cm; the cell is filled with carbon disulfide and has a diameter of 3 mm at a length of 1 m; it is placed 70 cm away from the amplifier.

Consequently a set of waves with an angle spread $\sim\lambda/2l\delta\theta$ will be effectively amplified. If this spread is smaller than the diffraction angle λ/d (here d is the transverse dimension of the interaction region), then it will have practically no effect on the accuracy with which the front is reproduced. We arrive thus at the condition

$$\lambda/2l\delta\theta < \lambda/d, \quad \text{or} \quad \delta\theta > d/2l. \tag{6.8}$$

Only when (6.8) is satisfied, i.e., only at a sufficiently large divergence of the exciting radiation, can we count on obtaining an inverted front. In our experiment with the phase plate, the condition (6.8) was satisfied.

3. Compensation for the Phase Distortions in an Amplifying Medium with the Aid of a "Brillouin Mirror"

Optical quantum amplifiers are presently used extensively to increase the power of laser radiation. The amplifiers can, however, while increasing the radiation power, produce as a rule a deterioration in its directivity, due to the static or dynamic (under the influence of the pump) inhomogeneities of the refractive index in the amplifying medium [74].

As shown above, the wave front of light reflected as a result of SMBS can duplicate the front of the exciting radiation. By using this effect we have succeeded in cancelling out the distortion in an amplifier, thus obtaining a considerable improvement in the divergence of the amplified radiation [74].

The experimental setup is shown schematically in Fig. 24. A light beam from a ruby laser with a divergence close to the diffraction value passes through an amplifier of poor optical quality, and then enters a glass cell with carbon disulfide, where the SMBS process develops. The cell "operates" as a mirror that converts the laser-radiation wave front into its complex conjugate. When the reflected light propagates, its wave front at any point should remain complex-conjugate to the laser-radiation front.* Consequently, the reflected light, after passing through the amplifier, should have the same front as the laser emission entering the amplifer.

In our scheme the backward-propagating light is spatially separated from the laser light with the aid of a Faraday cell and a polarizer, which operate as an optical decoupler.

---

* This statement is of course incorrect if the gain is not constant over the amplifier cross section, i.e., if the front is subject to amplitude distortion.

The amplifier is a ruby crystal with weak-signal gain that can be varied in the range 5–19. The laser radiation (P_L) entering the amplifier has a divergence ≈ 0.13 mrad at half-height, at a beam diameter 6 mm, a spectral width < 15 MHz, a maximum power ~0.1 MW and a pulse duration ~200 nsec.

The laser radiation passing through the amplifier is focused by lens L in front of a cell with carbon disulfide, entering the cell in the form of a diverging beam. Inasmuch as the refractive index of glass is smaller than that of CS_2, total internal reflection of the light takes place on their interface, and the light propagates in the cell as if in a waveguide with mirror walls. To prevent lasing, the ends of the cell are made broader and are painted black, and the cell windows are inclined at an angle of 60°.

Figure 25a shows the far-zone distribution of the unamplified laser radiation. Figure 25b shows, in the same scale, a photograph of the far zone of the amplified radiation (P_{LA}). The comparison of Figs. 25a and 25b shows that the amplifier distorts strongly the front of the light passing through it, increasing the divergence from 0.13 to ~2.5 mrad.

An appreciable fraction (~60% in energy) of the radiation entering the cell with the carbon disulfide is reflected backward as a result of the SMBS. The distribution of the reflected light (P_B) in the far zone duplicates in its details Fig. 25b.

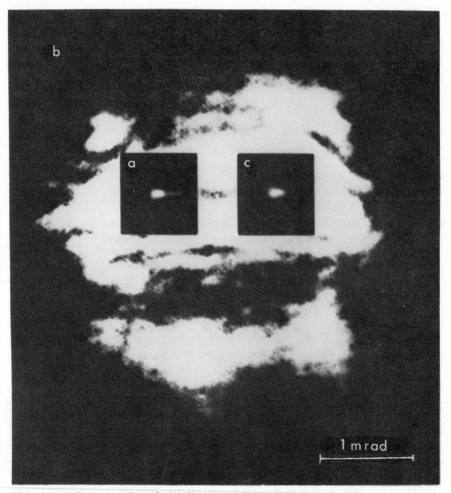

Fig. 25. Photographs of the far-zone distribution of the laser radiation (a), of the amplified laser radiation (b), and of the reflected light passing through the amplifier (c).

After the reflected light passes through the amplifier, its divergence is considerably decreased and becomes equal to ≈ 0.15 mrad. As seen in Fig. 25c, the far-zone distribution of the reflected and then amplified light (P_{BA}) is quite close to the distribution of the unamplified laser radiation. This effect is observed with the amplifier operating either in the linear or in the saturation regime (in the latter case the energy density P_{BA} in the beam reached ~ 4 J/cm^2).

It has thus been established that by using SMBS it is possible to compensate for the distortions in the amplified medium and to obtain an amplified light beam with a divergence close to the diffraction value. This method is not selective and can be used in laser systems operating at different wave lengths. The small buildup time of the SMBS ($\sim 10^{-9} - 10^{-8}$ sec) makes it possible to cancel out the dynamic inhomogeneities. By way of example, we indicate that these inhomogeneities can be cancelled out in liquid amplifying media.

CHAPTER VII

SMBS IN THE CASE OF EXCITING RADIATION WITH A BROAD SPECTRUM

In the experiments described in the preceding chapters, the SMBS was usually excited by radiation with a spectrum narrower than the spontaneous-scattering line ($\delta \nu_0$). The width of this line in compressed gases is frequently smaller than 10^{-3} cm^{-1} (30 MHz). Since the spectrum of high-power lasers usually subtends over a wider band it is important to ascertain in practice how the width of the pump section influences the stimulated scattering.

D'yakov [76] has shown in a theoretical paper that SMBS and stimulated Raman scattering (SRS) can be produced just as effectively by broadband pumping as by narrowband pumping. When radiation with a broad spectrum is used, effective forward SRS was observed [77-79], but for backward SRS the gain was practically zero [79]. It was also reported that the use of broadband radiation leads to a sharp decrease of the SMBS intensity [80, 81]. Thus, on the basis of all the published experimental results, an impression is gained that effective backward scattering (of both SMBS and SRS) is impossible with broad band pumping, although the theory [76] predicts the opposite.

To verify this theory, we have compared the SMBS threshold powers at different spectral widths of the exciting light and at different lengths of the scattering medium [82]. The experimental setup is shown in Fig. 26. The ruby-laser beam enters a cell with methane gas.* A light pipe placed in the cell ensures uniform illumination over the entire length of the cell. The uniform illumination over the cross section of the scattering region is obtained with the aid of an external light pipe. At the entrance of this light pipe, the lens L_1 produces the image of a diaphragm illuminated by the laser beam. The rays traveling at different distances from the optical axis of the system strike the walls of the light pipe in different places and at different angles. Therefore the output end of the light pipe turns out to be uniformly illuminated regardless of the distribution of the illumination at its entrance. The lens L_2 projects the image of the exit end on the entrance of the cell. Thus, the scattering develops in an exactly known uniformly illuminated volume. Consequently, the results of the experiments are not affected by the spatial distribution of the laser light, which inevitably changes in the course of the laser tuning which is needed to vary the width of the spectrum.

---

*Methane pressure 150 atm, approximate temperature 20°C. Under these conditions $\delta \nu_0 = 7 \cdot 10^{-4}$ cm^{-1} and the gain for narrow-band excitation is $g_0 = 0.094$ cm/MW.

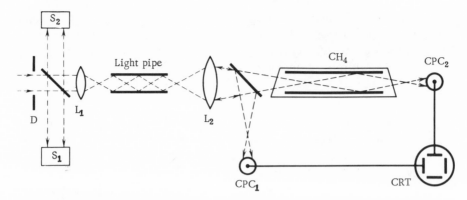

Fig. 26. Experimental setup. D) 4×4 mm diaphragm; L_1) lens with focal length 14 cm; the light pipe (length 23 cm, cross section 2×2 mm) is made up of Plexiglas plates; L_2) lens with focal length 30 or 50 cm for cell lengths 24 and 99 cm, respectively; the cell angles are inclined 11° in mutually perpendicular planes; the short cell contains a hollow glass light pipe 23 cm long with 2×2 mm cross section, while the long one contains a light pipe 97 cm long and 4×4 mm in cross section; the radiation loss in the light-pipe walls is ~14%; CPC_1 and CPC_2) FÉK-09 coaxial photocell; CRT) cathode-ray tube of I2-7 oscilloscope; time constant of measuring system ~10 nsec. S_1 and S_2, systems for the measurement of the parameters of the exciting and scattered light.

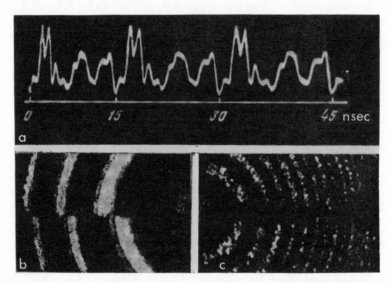

Fig. 27. Characteristics of exciting and scattered light in broad band pumping. a) Oscillogram of part of the laser pulse (resolution 0.7 nsec); b) spectrograms of exciting (top) and scattered (bottom) light, obtained simultaneously with the aid of a Fabry—Perot etalon with a dispersion region $1.667 \cdot 10^{-1}$ cm^{-1}; c) analogous spectrograms registered with an etalon with dispersion region $1.085 \cdot 10^{-2}$ cm^{-1}, the interval between the lines being equal to the distance between the axial modes of the laser ($2.17 \cdot 10^{-3}$ cm^{-1}).

The laser generates modes with transverse indices and is tuned from a single-mode to a multiple mode (with respect to the longitudinal index) regime. The width of its spectrum at half-height is $\delta\nu < 5 \cdot 10^{-4}$ cm^{-1} in the former case, i.e., $\delta\nu_0$, and $-\nu = (0.9\text{-}2.7) \cdot 10^{-2}$ cm^{-1} in the latter case, which is 13-40 times larger than $\delta\nu_0$. In multimode generation, the emission oscillogram (Fig. 27a) takes the form of periodically repeating fluctuations, the average duration of which agrees with the width of the spectrum. The individual laser modes registered in the spectrogram 27c indicate that the spectrum drifts little during the time of generation. The light intensity averaged over the period has a nearly Gaussian variation. The intensity variation in single-mode generation is similar. The pulse duration at half-height is $\approx$180 nsec. Owing to the use of a Faraday cell (to decouple the laser from the cell with the methane), a single SMBS component is observed in the spectrum of the back scattered light. Its structure details duplicate the spectral distribution of the excited radiation (Figs. 27b and c), and the frequency shift ($\sim$4.9 $\cdot 10^{-2}$ cm^{-1}) does not depend on the width of the pump spectrum.

In this scheme there is no external signal. Therefore the interaction of the radiation with the medium leads to amplification of the thermal-scattering light, which is usually $\sim$e^{25} times weaker than the exciting radiation. The amplification law is given by e^G, with $G = Ig_0l$ for the case of narrow-band excitation. Here I is the intensity of the exciting radiation and l is the interaction length. If the pump intensity exceeds a definite threshold level I_{thr}, with $G(I_{thr}) \approx 25$, then an appreciable fraction of the pump is converted into scattered light,* and the intensity of the light passing through the medium is stabilized at the indicated threshold level.† In our experiments the threshold was determined by two photocells which registered the power of the exciting radiation entering and leaving the cell. The signals from the photocells were fed simultaneously to the horizontal and vertical deflecting systems of a cathode ray tube, and the resultant oscillogram showed the dependence of the instantaneous radiation power passing through the scattering medium on the pump power. After measuring the threshold power (H_0) for narrow band excitation (Fig. 28) we calculated the threshold intensity $I_{0\,thr}$ and then found that in our case $G(I_{0thr})$ equals 24.

The results of the measurements of the SMBS threshold at the different widths of the pump spectrum ($\delta\nu$) and lengths of the scattering region (l) are shown in Fig. 28 as functions of the parameter $\delta\nu \cdot l$. This parameter has a simple physical meaning: $\delta\nu$ (cm^{-1}) $\cdot l \equiv l/l_{coh}$, where l_{coh} is the coherence length of the exciting light [83]. To compare the results with the theory, it is desirable to consider two situations: 1) $\delta\nu \cdot l \ll 1$; 2) $\delta\nu \cdot l \gtrsim 1$.

From the theory developed for the first case [76] (see also [84]) it follows that when the excitation is by stationary pumping with a broad spectrum the SMBS threshold is practically independent of the width of the spectrum. The spectrum of the scattered light duplicates in this case the shape of the pump spectrum. Our results confirm these conclusions fully.

There is no exact theory for the case $\delta\nu \cdot l \gtrsim 1$. There is, however, an estimate of the gain for broad band pumping [85]. According to this estimate

$$G(I) = \frac{1}{2} l [g_0I - 8\delta\nu - 8\delta\nu_0 + \sqrt{(g_0I - 8\delta\nu - 8\delta\nu_0)^2 + 32\delta\nu_0 g_0I}]. \tag{7.1}$$

An analysis of (7.1) shows that at $G \gg 1$ and $\delta\nu_0 \cdot l < 1$

$$G = Ig_0l - 8\,\delta\nu \cdot l. \tag{7.2}$$

---

* This is precisely the threshold that must usually be exceeded to observe stimulated scattering [53].

† In broadband pumping, as shown by the experiments with our installation, it is the average light intensity which is stabilized.

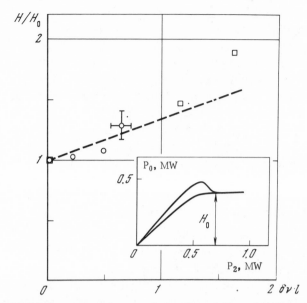

Fig. 28. Dependence of the SMBS threshold on the width of the pump spectrum and on the length of the scattering medium. Circles, experimental results at $l = 24$ cm; squares, at $l = 99$ cm. Dashed line, calculation by formula (7.3). H, threshold power at arbitrary spectral widths; H_0, threshold power. On the lower right is shown the experimental oscillogram: P_L, instantaneous pump power; P_0, instantaneous radiation power passing through the scattering medium. In a certain range of variation of the P_L branch, the oscillograms corresponding to the rise and descent of the pulse coincide, thus indicating that a quasi-stationary scattering regime has been attained.

Assuming $G(I_{thr}) = 24$, we obtain from (7.2)

$$I_{thr} = \frac{24}{g_0 l}\left(1 + \frac{1}{3}\delta \nu l\right).$$

Hence

$$I_{thr}/I_{0thr} \equiv H/H_0 = 1 + {}^1/_3\, \delta \nu \cdot l. \qquad (7.3)$$

As seen from Fig. 28, the experimental results are in satisfactory agreement with (7.3). Prior to the present paper, there were no published data* to compare quantitatively with formulas (7.1)-(7.3). We note, however, that results that are at first glance incompatible, namely those of [81], where a very low efficiency of SMBS excitation by broad band pumping was observed, and of [58], where this efficiency was high, can be attributed apparently to the fact that the parameter l/l_{coh} was different in the two cases.

*The relation (7.3) was subsequently confirmed in [86].

Using broadband excitation of SMBS in a short cell, we have also compared the wave fronts of the reflected and excited light by the procedure used in the experiments of the preceding chapter for the case of narrow-band pumping. The laser beam divergence was ~$2 \cdot 10^{-4}$ rad and close to the diffraction value. After passing through the lens L_1, the light pipe, and the lens L_2, the divergence of the exciting radiation increased by 150 times. At the same time, the angular distribution of the light scattered in the cell and passing through the same elements duplicates the far-zone distribution of the laser radiation. It follows therefore that the electric field of the scattered wave coincides (apart from a factor) with the complex-conjugate laser field.

We note in conclusion that a study of the singularities of the SMBS points to two possibilities of constructing Brillouin lasers that operate effectively under broad band pumping cone: 1) $\delta \nu L \ll 1$ and, consequently, $\delta \nu l \ll 1$. Here, as before, L is the optical length of the resonator (to be specific, a ring resonator). In this case the resonator length must be strictly adjusted, inasmuch as this is between its longitudinal modes $1/L \gg \delta \nu$; 2) $\delta \nu_0 L > 1$, $\delta \nu l < 1$ and the length L is equal to the length of the pump-laser resonator. In this case there is no need for the adjustment, since $1/L < \delta \nu_0$. The high scattering efficiency is ensured here by the fact that the repetition periods of the pump electric field (Fig. 27a) and the electric field of the scattered light are equal.

APPENDIX

EXPERIMENTAL TECHNIQUE

1. Divergence Measurement Procedure

A quantitative method of measuring the angular divergence of laser radiation was developed by Waynant [87]. In this method, photographs are taken of the field in the focal plane of the lens that focuses the laser radiation. The density curve of the photographic material is plotted with the aid of the investigated laser, which calls for stable operation of this laser for a long time. A self-calibration method based on obtaining simultaneously several images with different exposures was proposed later [88]. In this method, the laser radiation focused on a scattering screen is photographed with the aid of several lenses. To obtain different exposures, filters are placed in front of the lenses. To obtain a small number of beams with an intensity that decreases in accordance with known law, it was proposed in [89] to use successive reflections in a thick transparent plate of high optical quality, mounted at an angle to the laser beam.

In our study we used a simpler self-calibration method [90]. It is based on splitting one light beam into a number of spatially similar beams of different intensity with the aid of a mirror wedge. Figure 29 shows how this is realized. The light beam after passing through the wedge produces beam I; two reflections in the wedge produce a beam II that makes an angle 2φ with beam I. A number of beams making angles 2φ with one another and an angle of $2\varphi(N-1)$ with the first beam (where N is the number of the beam) is produced in similar fashion. It is easily seen that the intensity of the N-th beam is weaker by a factor $R_1 R_2$ than the intensity of the (N−1)-th beam; here R_1 and R_2 are the reflection coefficients of mirrors M_1 and M_2.

If the mirror wedge is placed past the lens, then the laser radiation will be focused in a number of points that lie on a circle with the radius equal to $f - q$ (here f is the focal length of the lens and q is the distance from the lens to the wedge).* The method of reducing the

---

* The mirror wedge can, of course, be used also to study other characteristics of laser light (for example, the distribution of the radiation in the mirror zone and of the spectrum).

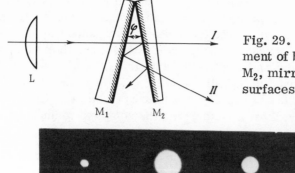

Fig. 29. Optical system for the measurement of beam divergence. L, lens; M_1 and M_2, mirrors; φ, angle between the mirror surfaces.

Fig. 30. Distribution of laser radiation in the far zone.

photography results in the self-calibration method is described in detail in [88] and we shall not dwell on it here.

Figure 30 shows photographs of the emission of an He−Ne laser, obtained by the described method. The wedge was made up of two dielectric mirrors with thin liner between them ($R_1 = R_2 = 70\%$, $\varphi = 19'$). The first beam is attenuated by a suitable filter to make its intensity equal to that of the 5th beam. The value obtained for the divergence angle was $4' \pm 0.4'$, in good agreement with the specifications of the laser.

Let us examine the capabilities of the method. The number of beams that can be obtained in principle is $90°/\varphi$. For dielectric mirrors with a maximum reflection at the laser-emission wavelength, the reflection and transmission coefficients remain practically unchanged at incidence angles from 0° to 30° [91]. Consequently, so long as the angle between the N-th beam and the first beam is less than 60°, we can regard the reflection coefficients to be the same for all the beams. This makes it possible to use the described method at divergence angles up to several degrees.

2. Cell for Optical Investigations of Compressed Gases

To work with compressed gases, we have constructed the cell shown in Fig. 31. We used in the construction the principle of the uncompensated area [92]. Thanks to the use of rubber

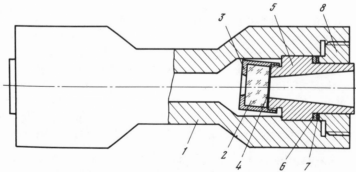

Fig. 31. Cell for optical investigations of compressed gases. 1) Body of cell; 2) quartz window 17 mm thick and 24 mm in diameter; 3) clamping nut; 4) polyethylene gasket 0.1 mm thick; 5) nipple; 6) rubber gasket; 7) steel gasket; 8) lock nut. The optical aperture of the cell is 14 mm.

gaskets, the forces necessary to produce the preliminary compression were negligible. The use of thin polyethylene gaskets made it possible to get along without polishing the nipple (the surface finish of all the metallic parts of the cell need not be better than grade $\Delta 5$).

We operated with cells of this design for seven years at pressures from 5 to 615 atm. Gas leakage was never observed in the course of operation.

3. Faraday Decoupler

To decouple the ruby laser from other elements of the experimental setup we used an optical decoupler based on a Faraday cell placed between two Glan prisms rotated 45° relative to each other. It is known that the transmission of such a device depends on the light propagation direction [93]. In the cell we used a rod of TF-7 glass 10 cm long with oblique ends. The pulse current source was an LC network (8 capacitors 140 μF each and 8 inductors 80 μH each), discharging into the solenoid of the cell through an adjustable discharge gap. The current pulse duration was ~ 2 msec and its wave form was close to rectangular. The transmission of our insulator in the forward direction was $\sim 50\%$, and in the backward direction it was 2500 times smaller. A number of structural elements of the insulator are described in [94].

4. Single-Mode Ruby Laser with Pulse Duration 60 nsec

To measure the gains in the stationary regime it is necessary that the exciting-radiation pulse duration greatly exceed the hypersound damping time in the investigated medium (~ 10 nsec for compressed gases). On the other hand, at the present time a complete theory of SMBS has been developed only for the case of monochromatic pump radiation. In the experiments of Chapter II we used a single-mode laser with emission pulse duration at half-height ~ 60 nsec. Its diagram is shown in Fig. 32.

In this laser, the selection of the longitudinal mode was effected by means of a Fox selector [95] (mirrors M_1, M_3, M_2) 0.3 m long, making possible stable laser operation on one axial mode. It emits light pulses of power ~ 0.5 MW at a spectrum width < 30 MHz (30 MHz is the apparatus function of the employed 15-cm Fabry–Perot etalon). The wave form of the pulse is very close to Gaussian.

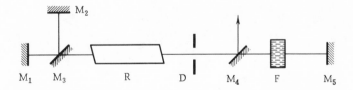

Fig. 32. Diagram of ruby laser. M_1, M_2, and M_5, dielectric mirrors with transmission $< 2\%$; M_3, mirror with transmission 50%; M_4, mirror for the extraction of the radiation with transmission 30%; R, ruby laser 23 cm long; D, diaphragm of 3 mm diameter; F, solution of vanadium phthalocyanine in nitrobenzene with initial transmission 40%; distance between mirrors M_1 and M_5 is 2.5 m; the pump system, which is not shown in the figure, consists of two IFP-5000 lamps and an elliptic reflector.

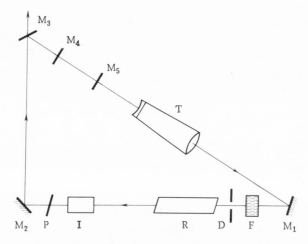

Fig. 33. Diagram of laser. M_1 and M_2, dielectric mirrors with transmission <2%; M_3, mirror for the extraction of the radiation, with transmission 40%; M_4 and M_5, mirrors perpendicular to the laser beam, with transmission 53% and separated by a distance ~35 cm; all the mirrors are flat; T, telescope with threshold magnification; F, saturable filter with initial transmission 30-60%; D, diaphragm of 7 mm diameter; R, ruby 23 cm long, with temperature maintained constant within 0.02°C; I, optical decoupler; P, quartz-crystal plate rotating the polarization plane of the generated light by 45°.

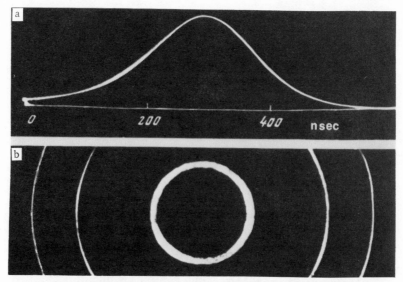

Fig. 34. Characteristics of pump radiation. a) Pulse wave form; b) spectrogram (Fabry–Perot etalon free spectral range $1.08 \cdot 10^{-2}$ cm^{-1}).

5. Single-Mode Ruby Laser with Pulse Duration 60-200 nsec

For a number of experiments (see Chapters III, IV, VI, VII) the pulse duration from the laser described above was insufficient, and the laser illustrated in Fig. 33 was developed.

The ring resonator has an optical length 460 cm. Owing to the Faraday decoupler, the radiation in the resonator can propagate only in one direction [96], thereby ensuring reliable decoupling of the laser from the light reflected in the investigated schemes. The plate P compensates for the rotation of the polarization plane in the cell. The longitudinal modes are selected by the interferometer M_4M_5, and the transverse one by the telescope T [97]. The Q switching is by bleaching of a solution of vanadium pthalocyanine in nitrobenzene. By varying its concentration we varied the laser pulse durations in the range 60-200 nsec. The peak power was $\lesssim 1$ MW. The oscillogram and spectrogram of the lasing is shown in Fig. 34. The lasing is on one mode. The width of the spectrum is < 15 MHz (15 MHz is registered on the spectrogram by using the Fabry−Perot etalon described below; the true width is of course smaller). In the absence of the interferometer, as a rule, 4-20 longitudinal modes are generated. The spectrum of each mode is in this case also narrower than 15 MHz.

6. Fabry − Perot Etalon with 46-cm Base

For a correct comparison of the experimental results with the theory, the spectral measurements must be carried out with a resolution comparable with the gain-line width. To attain a resolution better than 10^{-3} cm^{-1}, we have developed a Fabry−Perot interferometer with 46-cm base. Its construction is shown in Fig. 35. Each mirror of the etalon is supported on three steel balls pressed into a steel mount, which is fastened with epoxy resin to a quartz tube. Fine adjustment is by rotating the knobs (5). The force is transmitted thereby through the springs and the clamping ring with the balls to the mirror. The rough adjustment is carried out prior to the final assembly, by means of an autocollimator and an additional adjusting unit. In this design there is no need for high precision of the parts. To estimate the width of the apparatus function of our etalon, we plotted the spectrum of a single-mode He−Ne laser (we used lenses with $f = 1.6$-3 m). The lines recorded on the spectrogram corresponded in this case to a spectral width $\sim 4 \cdot 10^{-4}$ cm^{-1}. Thus, the apparatus function did not exceed $4 \cdot 10^{-4}$ cm^{-1} (12 MHz).

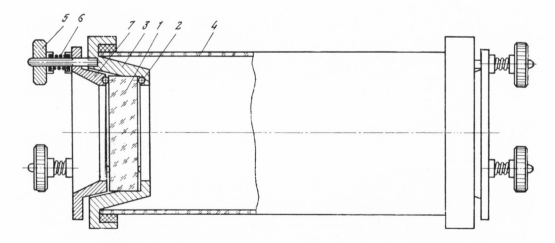

Fig. 35. Fabry−Perot etalon. 1) Mirror; 2) ball bearing; 3) mount; 4) quartz tube; 5) adjustment knob; 6) spring; 7) tightening ring.

LITERATURE CITED

1. E. J. Woodbury and W. K. Ng, Proc. IRE, 50:2347 (1962).
2. R. Y. Chiao, C. H. Townes, and B. P. Stoicheff, Phys. Rev. Lett., 12:592 (1964).
3. D. I. Mash, V. V. Morozov, V. S. Starunov, and I. L. Fabelinskii, Pis'ma Zh. Éksp. Teor. Phys., 1:41 (1965).
4. G. I. Zaitsev, Yu. I. Kyzylasov, V. S. Starunov, and I. L. Fabelinskii, Pis'ma Zh. Éksp. Teor. Fiz., 6:802 (1967).
5. D. H. Rank, C. W. Cho, N. D. Foltz, and T. A. Wiggins, Phys. Rev. Lett., 19:828 (1967).
6. D. Pohl, I. Reinhold, and W. Kaiser, Phys. Rev. Lett., 20:1141 (1968).
7. I. M. Aref'ev and V. V. Morozov, Pis'ma Zh. Éksp. Teor. Fiz., 9:448 (1969).
8. O. Rahn, M. Maier, and W. Kaiser, Opt. Commun., 1:109 (1969).
9. M. Maier, W. Rother, and W. Kaiser, Appl. Phys. Lett., 10:80 (1967).
10. M. Denariez and G. Bret, Phys. Rev., 171:160 (1968).
11. I. L. Fabelinskii, Molecular Scattering of Light, Plenum Press, New York (1968).
12. V. S. Starunov and I. L. Fabelinskii, Usp. Fiz. Nauk, 98:441 (1969).
13. N. Goldblatt, Appl. Opt., 8:1559 (1969).
14. W. Kaiser, J. Acoust. Soc. Am., 49:959 (1971).
15. C. L. Tang, J. Appl. Phys., 37:2945 (1966).
16. F. Barocchi and M. Zoppi, Opt. Commun., 3:335 (1971).
17. H. Takuma and D. A. Iennings, Appl. Phys. Lett., 5:239, 241 (1964).
18. T. Ito and H. Takuma, J. Phys. Soc. Jpn., 24:965 (1968).
19. T. Ito and H. Takuma, Jpn. J. Appl. Phys., 8:941 (1969).
20. E. B. Aleksandrov, A. M. Bonch-Bruevich, N. N. Kosmin, and V. A. Khodovoi, Zh. Éksp. Teor. Fiz., 49:1435 (1965).
21. A. Pine, Phys. Rev., 149:113 (1966).
22. T. Ito and H. Takuma, in: Physics of Quantum Electronics (P. L. Kelley, editor), New York (1966), p. 200.
23. A. Yariv, IEEE J. Quant. Electron., QE-1:28 (1965).
24. V. N. Lugovoi and V. N. Strel'tsov, Zh. Éksp. Teor. Phys., 62:1312 (1972).
25. A. J. Glass, IEEE J. Quant. Electron., QE-3:516 (1967).
26. B. A. Akanaev, S. A. Akhmanov, and R. V. Khokhlov, Pis'ma Zh. Éksp. Teor. Fiz., 1:4 (1965).
27. Yu. V. Dolgopolov, Diploma Work, Physics Institute, Academy of Sciences of the USSR, Moscow (1967).
28. E. Burlefinger and H. Puell, Phys. Lett., 15:313 (1965).
29. R. V. Wick and A. H. Guenter, Appl. Opt., 7:73 (1968).
30. A. J. Alcock and C. DeMichelis, Appl. Phys. Lett., 11:42 (1967).
31. A. J. Alcock and C. DeMichelis, Appl. Phys. Lett., 11:185 (1967).
32. D. Pohl, Phys. Lett., 24A:239 (1967).
33. E. A. Tikhonov and M. T. Shpak, Pis'ma Zh. Éksp. Teor. Fiz., 8:282 (1968).
34. F. Gires and B. Sver, Proc. IEEE, 56:1613 (1968).
35. A. Z. Grasyuk, V. I. Popovichev, V. V. Ragul'skii, and F. S. Faizullov, Kvant. Élektron., No. 1:70 (1971).
36. V. V. Ragul'skii, Candidate's Dissertation, Physics Institute, Academy of Sciences of the USSR, Moscow (1973).
37. A. Fox and T. Li, Proc. IRE, 48:1904 (1960).
38. E. Snitzer, in: Advances in Quantum Electronics, Am. Opt. Soc. (1961), p. 348
39. R. Y. Chiao and P. A. Fleury, Proc. Conf. on Phys. Quant. Electronics (1966), p. 241.
40. A. Z. Grasyuk, V. I. Popovich, V. V. Ragul'skii, and F. S. Faizullov, Pis'ma Zh. Éksp. Teor. Fiz., 12:286 (1970); Preprint No. 94, FIAN (1970).
41. D. Pohl, M. Maier, and W. Kaiser, Phys. Rev. Lett., 20:366 (1968).
42. A. Lauberau, W. Englisch, and W. Kaiser, IEEE J. Quant. Electron., QE-5:410 (1969).
43. V. I. Popovichev, V. V. Ragul'skii, and F. S. Faizullov, Kvant. Élektron., No. 1:135 (1971).

44. P. V. Avizonis, A. H. Guenter, T. A. Wiggins, P. V. Wick, and D. H. Rank, Appl. Phys. Lett., 9:309 (1966).

45. A. Z. Grasyuk, V. F. Efimkov, I. G. Zubarev, V. I. Mishin, and V. G. Smirnov, Pis'ma Zh. Éksp. Teor. Fiz., 8:474 (1968).

46. V. V. Bocharov, M. G. Gangardt, A. Z. Grasyuk, I. G. Zubarev, and E. A. Yukov, Zh. Éksp. Teor. Fiz., 57:1585 (1969).

47. F. J. McClung and D. H. Close, J. Appl. Phys., 40:3979 (1969).

48. S. R. J. Brueck and A. Mooradian, Appl. Phys. Lett., 18:229 (1971).

49. E. E. Hagenlocker, R. W. Minck, and W. G. Rado, Phys. Rev., 154:226 (1967); Appl. Phys. Lett., 7:236 (1965).

50. T. T. Saito, L. M. Peterson, D. H. Rank, and T. A. Wiggins, J. Opt. Soc. Am., 60:749 (1970).

51. V. I. Kovalev, V. I. Popovichev, V. V. Ragul'skii, and F. S. Faizullov, Kvant. Élektron., No. 7:78 (1972).

52. V. I. Borodulin, N. A. Ermakova, L. A. Rivlin, V. V. Tsvetkov, and V. S. Shil'dyarev, Zh. Éksp. Teor. Fiz., 49:1718 (1965).

53. M. Maier, Phys. Rev., 166:113 (1968).

54. M. Maier and C. Renner, Phys. Rev. Lett., 34A:299 (1971); Opt. Commun., 3:301 (1971).

55. V. A. Zagoruichenko and A. M. Zhuravlev, Thermophysical Properties of Gaseous and Liquid Methane [in Russian], Izd. Standartov, Moscow (1969).

56. V. I. Kovalev, V. I. Popovichev, V. V. Ragul'skii, and F. S. Faizullov, Pis'ma Zh. Éksp. Teor. Fiz., 14:503 (1971).

57. D. Pohl and W. Kaiser, Phys. Rev., B1:31 (1970).

58. M. A. Bol'shov, Candidate's Dissertation, Moscow State University (1971).

59. V. I. Kovalev, V. I. Popovichev, V. V. Ragul'skii, and F. S. Faizullov, Zh. Éksp. Teor. Fiz., 64:2028 (1973).

60. N. M. Kroll, J. Appl. Phys., 36:34 (1965).

61. B. Ya. Zel'dovich, Pis'ma Zh. Éksp. Teor. Fiz., 15:226 (1972).

62. A. Z. Grasyuk, V. V. Ragul'skii, and F. S. Faizullov, Pis'ma Zh. Éksp. Teor. Fiz., 9:11 (1969); Preprint No. 160, FIAN (1968).

63. V. V. Korobkin, D. I. Mash, V. V. Morozov, I. L. Fabelinskii, and M. Ya. Shchelev, Pis'ma Zh. Éksp. Teor. Fiz., 5:72 (1967).

64. J. D. Abella and H. Z. Cummins, J. Appl. Phys., 32:1177 (1961).

65. D. Birnbaum, Lasers [in Russian], Sov. Radio, Moscow (1967), p. 160.

66. V. S. Zuev, V. A. Katulin, V. Yu. Nosach, and O. Yu. Nosach, Zh. Éksp. Teor. Fiz., 62:1671 (1972).

67. V. I. Popovichev, V. V. Ragul'skii, and F. S. Faizullov, Kvant. Élektron., No. 5(11):126 (1972).

68. H. Samelson, A. Lempicki, and V. Brophy, J. Appl. Phys., 39:4029 (1969).

69. R. G. Brewer, Phys. Rev., 140:A800 (1965).

70. T. A. Wiggins, R. W. Wick, and D. H. Rank, Appl. Opt., 5:1069 (1966).

71. B. Ya. Zel'dovich, V. I. Popovichev, V. V. Ragul'skii, and F. S. Faizullov, Pis'ma Zh. Éksp. Teor. Fiz., 15:160 (1972).

72. T. A. Wiggins, T. T. Saito, L. M. Peterson, and D. H. Rank, Appl. Opt., 9:2177 (1970).

73. N. Bloembergen and Y. R. Shen, Phys. Rev. Lett., 13:720 (1964).

74. G. D. Baldwin and E. P. Riedel, J. Appl. Phys., 38:2720, 2726 (1967).

75. O. Yu. Nosach, V. I. Popovichev, V. V. Ragul'skii, and F. S. Faizullov, Pis'ma Zh. Éksp. Teor. Fiz., 16:617 (1972).

76. Yu. E. Dyakov, Pis'ma Zh. Éksp. Teor. Fiz., 11:362 (1970).

77. V. V. Bocharov, A. Z. Grasyuk, I. G. Zubarev, and V. F. Mulikov, Zh. Éksp. Teor. Fiz., 56:430 (1969).

78. J. Y. Beitz, G. W. Flynn, D. H. Turner, and N. Satin, J. Am. Chem. Soc., 92:4130 (1970).

79. A. Z. Grasyuk, I. G. Zubarev, and N. V. Suyazov, Pis'ma Zh. Éksp. Teor. Fiz., 16:237 (1972).

80. J. P. Budin, A. Donzel, J. Ernest, and J. Rapfy, Electron. Lett., 3:31 (1967).

81. R. H. Stolen, E. P. Ippen, and A. R. Tynpes, Appl. Phys. Lett., 20:62 (1972).

82. V. I. Popovichev, V. V. Ragul'skii, and F. S. Faizullov, Pis'ma Zh. Éksp. Teor. Fiz., 19:350 (1974).

83. M. Born and E. Wolf, Principles of Optics, Fifth Edition, Pergamon Press (1975).

84. Z. A. Baskakova, Candidate's Dissertation, Moscow State University (1973).

85. Yu. E. D'yakov, Kratk. Soobshch. Fiz., No. 4:23 (1973).

86. I. G. Zubarev and S. I. Mikhailov, Kvant. Élektron., 5:1150 (1974).

87. R. W. Waynant, Appl. Opt., 4:1648 (1965).

88. I. M. Winer, Appl. Opt., 5:1437 (1966).

89. P. V. Avizonis, T. T. Doss, and R. Heimlich, Rev. Sci. Instr., 38:331 (1967).

90. V. V. Ragul'skii, and F. S. Faizullov, Opt. Spektrosk., 27:707 (1969).

91. F. A. Korolev and A. Yu. Kliment'eva, Dokl. Akad. Nauk SSSR, 100:459 (1955).

92. D. S. Tsiklis, Handbook of Techniques in High-Pressure Research and Engineering, Plenum Press, New York (1968).

93. A. N. Zaidel', G. V. Ostrovskaya, and Yu. I. Ostrovskii, Techniques and Practice of Spectroscopy [in Russian], Nauka, Moscow (1972), p. 201.

94. V. I. Kovalev, V. I. Popovichev, V. V. Ragul'skii, and F. S. Faizullov, Prib. Tekh. Éksp., No. 5:199 (1972).

95. R. W. Smith, IEEE J. Quant. Electron., QE-1:343 (1965).

96. C. L. Tang, H. Statz, G. A. DeMars, and D. T. Wilson, Appl. Phys. Lett., 2:222 (1963); Phys. Rev., A136:1 (1964).

97. Yu. A. Anan'ev, Kvant. Élektron., No. 6:3 (1971).

COMPRESSED-GAS LASERS

V. A. Danilychev, O. M. Kerimov, and I. B. Kovsh

Results are presented of theoretical and experimental investigations of high-power compressed-gas lasers. The electroionization method of excitation is considered, as well as pumping by an intense electron beam. Experimental data are presented on the threshold and output characteristics, the efficiencies, and the gains of CO_2 and CO lasers operating without cooling the active medium. The prospects are discussed of utilizing the electroionization method to excite lasers on electronic transitions. Results are presented of the investigations of lasers operating with compressed xenon and $Ar:N_2$ mixtures, pumped with an electron beam, in which the lasing is on the electronic transitions of molecules Xe_2 ($\lambda \simeq 172$ nm) and N_2 ($\lambda \simeq 357$ nm). Prospects of using compressed-gas lasers in the thermonuclear fusion region, for selective stimulation of chemical reactions, and for applications in the material-finishing industry are discussed.

INTRODUCTION

Interest in compressed gases as active media for lasers is due to their high homogeneity, the possibility of attaining a high concentration of the active particles, and the existence of gas-laser systems with efficiencies up to 50%. In addition, at high pressures it is possible to tune the frequency smoothly, to generate ultrashort pulses, and to obtain high radiation power and coherence. These properties of compressed gases attract attention as possible active media for high-power lasers, the need for which is being particularly strongly felt of late in connection with the development of work on controlled thermonuclear fusion, task-oriented stimulation of chemical reactions, and other laser applications.

The problem of developing powerful compressed-gas lasers is fraught with two basic difficulties. First, the traditional method of exciting gas lasers, with a self-maintaining electric discharge, cannot be used to excite sufficiently large volumes of compressed gases, because of the instabilities that occur in the discharge and which lead to pinching of the discharge and to impossibility of volume excitation. It becomes therefore necessary to develop new methods for exciting compressed gases. Second, processes of population of laser levels in compressed gases can be accompanied by an appreciable increase in the role of quenching collisions of active particles with neutral molecules and electrons. In addition, impact broadening of laser levels, which is proportional to the gas pressure, leads to the need for an additional increase of the pumping rate with increasing pressure.

To excite lasers that use working media with large concentration of active particles, a method of pumping with electron beams was proposed in 1961 [1]. The application of this method has made it possible to obtain lasing in semiconductors [2, 3] as well as condensed and compressed gases [4, 5]. The working medium was excited with the aid of large-current nanosecond accelerators, the electron current of which reached 10^4 A at electron energies 1 MeV. The development of high-power accelerators for the excitation of lasers with large working-medium volumes is a complicated technical task. In addition, the average energy of the secondary electrons produced in the interaction between the electron beam and the laser active media amounts to ~10-30 eV, so that laser pumping with an electron beam is effective for the excitation of electron transitions and is ineffective for the excitation of vibrational−rotational transitions: the maxima of the excitation cross section of the latter lie in the region of 1-2 eV. However, the excitation of the vibrational levels is of great interest, for it is precisely the vibrational−rotational molecule transitions that make possible lasing with very high efficiencies, up to 30-50% [6].

In 1970, it was proposed [7, 8] to produce free electrons with the aid of an external ionization source, thereby raising the pressure and increasing the working volume of the gas lasers.

The first compressed-gas laser was developed in 1971 at the Lebedev Physics Institute, using a mixture of carbon dioxide and nitrogen at a total pressure of 25 atm [9, 10]. The radiation power per unit volume in this laser was approximately 10^6 times larger than in an ordinary CO_2 laser.

These studies have demonstrated the following:

(a) Triple quenching collisions have little effect on the lasing characteristics of compressed-gas lasers up to a pressure of 25 atm.

(b) The excitation method is characterized by a high degree of spatial homogeneity, and the discharge does not tend to contract.

(c) At a high degree of ionization, the internal field of the plasma exerts practically no screening action on the external field and does not lead to a nonuniform utilization of the working volume and to an abrupt reduction of the laser output energy, as suggested in [7]. These were named "electroionization" lasers.

The combined excitation of gas lasers by an electric discharge and ionizing radiation was investigated earlier in a number of studies at gas pressures and active-medium volumes typical of an ordinary low-pressure gas-discharge laser, under conditions when there is no pinching (contraction) of the autonomous glow discharge. Thus, by pumping an argon laser with a plasma-beam discharge, a plasma electron temperature ~100 eV was attained in [10], and lasing was produced at very low gas pressures, $~10^{-4}$ Torr. The effect of a proton beam on the generation of a gas laser based on CO_2 at pressures 1-10 Torr is the subject of the studies in [11]. In these studies, the possibility of going to higher pressures was not considered, and the experiments were performed under conditions when the pinching of the discharge did not take place, and the gas was not additionally ionized with an external source.

Attempts were made in [12] to raise the working-gas pressure. However, no appreciable increase of the pressure could be obtained in comparison with that in an ordinary gas-discharge laser: generation in a CO_2 laser with combined pumping by an electron beam and an electric discharge could not be obtained at pressures above 30 Torr.

By using various technical methods of stabilizing the glow discharge, the authors of [13, 14] succeeded in significantly raising (to ~1 atm) the pressure of the working gas in a CO_2 laser. In the lasers described in [13, 14], a transverse discharge was used (in place of the customarily employed longitudinal discharge), as well as specially shaped electrodes. These lasers were named TEA lasers. However, the use of a transverse discharge did not change in fundamental fashion the pumping mechanism, and did not make it possible to advance into the region of high pressures or to increase significantly the volume of the active medium.

Persson's work [15] on the stabilization of a glow discharge of low pressure by an electron beam served as the starting point of research aimed at finding methods of pumping atmospheric-pressure CO_2 lasers, performed at the Los Alamos laboratory in the USA. Members of this laboratory published in 1971 a communication reporting amplification in a mixture of carbon dioxide, nitrogen, and helium at atmospheric pressure, excited by a discharge stabilized with an electron beam [16]. The authors of that paper, however, were cautious in their estimates of the possibility of using the pumping method realized by them to excite an active medium at pressures higher than atmospheric.

At the present time, even though the electroionization principle of exciting dense gases is only about two years old, it is precisely with electroionization lasers that many outstanding accomplishments were made in laser technology:

1. Considerable volumes of dense gases were excited, amounting to dozens or even hundreds of liters [18].

2. Energy outputs of 50 J-liter^{-1}-atm^{-1} were realized at efficiencies 25-30% [17-20].

3. It was shown that such lasers can operate at high pressures ~100 atm [21].

4. Pulsed (including pulses shorter than a nanosecond), quasi-continuous, and continuous lasing regimes of high-pressure lasers were realized [17, 22].

5. Work on tuning the lasing frequency in a wide range of frequencies is being successfully pursued [23, 24].

In spite of all these accomplishments, many questions still await their solution and are connected both with the physical processes in the active medium and with the development of the individual laser elements.

A more complete investigation of the probabilities of various elementary processes is needed. The existing information concerning, for example, the CO_2 molecule alone, which is one of the most thoroughly investigated, is insufficient even for a rough estimate of the lasing power and of the gain attained between different highly excited levels. Yet it is just these transitions which make it possible to tune the frequency of a CO_2 laser in a wide range.

A very important question is that of the effective energy pickoff under conditions when ultrashort pulses are amplified. A closely related problem is that of the nonlinear optical phenomena that arise when a light pulse propagates in the active medium. At high pressures, an essential role can be played by effects that are not linear in the pressure, by triple quenching collisions, and by the onset of new optical transitions that are induced by pressure and by the strong electric field.

The chemical transformations that occur in the plasma produced by the electroionization method are of undisputed interest, since a feature of this plasma is the large difference between the vibrational and translational temperatures and the high density of the vibrationally excited molecules. At a translational temperature close to room temperature, the temperatures of the individual vibrational degrees of the molecule can reach thousands of degrees. Chemical reactions occurring under such thermodynamic disequilibrium conditions have very high rates, consume little energy, and produce a high yield of end products [25]. Some of the chemical reactions that occur in an electroionization plasma are of independent practical interest.

We have mentioned only some of the questions that are raised when more attention is paid to high-pressure gas lasers. It is seen, however, even from this short list that a large number of new physical problems must be solved if high-pressure gas lasers are to become practical. Some of them will be discussed in the present article.

CHAPTER I

ELECTROIONIZATION METHOD OF EXCITING
COMPRESSED-GAS LASERS

It is known that when electrons move through a molecular gas under the influence of an electric field, the electrons collide with the molecules. In inelastic collisions, the electrons transfer the energy accumulated between collisions to various degrees of freedom of the molecules. Figure 1 shows, in relative units, the effectiveness of various mechanisms whereby the electrons lose energy, as a function of the ratio of the electric field to the pressure in the case of molecular nitrogen [26]. Similar curves can be obtained also for other molecular gases. It is seen from Fig. 1 that at the parameter values $E/P < 10$ V-cm^{-1}-Torr^{-1} the principal electron energy loss mechanism is excitational of molecule vibrations. At the parameter values $E/P > 40$ V-cm^{-1}-Torr^{-1}, the main loss mechanism is excitation of the electron levels and ionization.

We see thus that for a laser operating on vibrational−rotational transitions and pumped with an electric current, the optimal range of electric fields lies much lower than the value of the field required to ignite the autonomous discharge.

This means that in the optimal scheme of an electric-discharge molecular gas laser it is necessary to forgo the autonomous electric discharge, to maintain the parameter E/P in the region of high effectiveness of the corresponding excitation mechanism, and to induce the conductivity of the gas by an independent method, for example, by applying ionizing radiation. In this case the concentration of the free electrons in the active medium does not depend on the electric field and is determined only by the intensity of the ionizing radiation. This is called the electroionization pumping method [9]. As a result of giving up the autonomous discharge, the flow of the electric current through the ionized gas becomes stable and the restrictions imposed in the case of the autonomous discharge on the pressure of the working gas and on the dimension of the system are lifted. It is this which is the most important feature of the electroionization method of supplying energy to the active medium of the laser, and distinguishes in principle the electroionization method from other combined-pumping methods.

The electroionization pumping method was developed as a result of research on semiconductor and condensed-gas lasers excited by electron beams [4].

Figure 2 shows schematically the excitation of an electroionization laser. The compressed gas is placed between two metallic electrodes to which the electric supply voltage is applied. The voltage is chosen to obtain the optimal value of E/P. To excite vibrational−rotational transitions in molecular gases, the optimal value of the parameter E/P is 5–10 times lower than the breakdown value. The ionization radiation makes the gas conducting, current

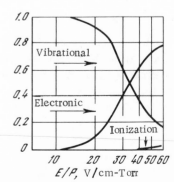

Fig. 1. Distribution of the pump power over the degrees of freedom of molecules in electric excitation of nitrogen [26].

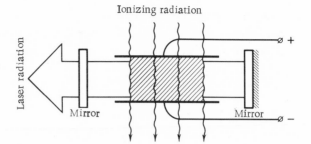

Fig. 2. Schematic illustration of the excitation of an electroionization CO_2 laser.

flows through the gas, and the energy released in the gas is

$$W = \sigma E^2 \tau_{\mathrm{p}},$$

(1)

where σ is the conductivity of the gas and is governed by the external ionization, E is the electric field, and τ_{p} is the duration of the electric-supply pulse. At optimal E/P and at an efficiency close to 100% this energy is transformed into vibrational energy of the molecules and then, as a result of the vibrational–translational relaxation it can be transformed into heat. By suitable choice of the gas composition it is possible to concentrate an appreciable fraction of the vibrational energy at the upper laser vibrational level and then remove it in the form of coherent radiation.

It is seen from Fig. 2 that the electroionization laser scheme does not differ in practice from that of the ionization chamber, the electric characteristics of which were investigated long ago by J. J. Thomson [27]. In the absence of recombination of the charged carriers and a high voltage on the electrodes, the current j_0 flowing through the ionization chamber reaches saturation [27]:

$$j_0 = j_s = qeL,$$

(2)

where j_s is the saturation-current density, q is the rate of production of the electron–ion pairs under the influence of the ionizing radiation, e is the electron charge, and L is the distance between the electrodes.

At high working-gas pressures and at a high ionization rate q, it is necessary to take into account the carrier recombination, and in this case the current does not reach saturation. According to [27], when recombination is taken into account, the following expressions are valid for the electric field E_e and for the potential V_e in the space-charge layer near the negative electrode:

$$E_e = 2 \left(\frac{\pi e}{R_i q} \right)^{1/2} \frac{j_0}{e}, \qquad V_e = \left(\frac{\pi e}{R_i q^3} \right)^{1/2} \frac{j_0^2}{e^2},$$

(3)

where e is the electron charge, R_i is the ion mobility, j_0 is the current density, and q is the rate of production of electron–ion pairs under the influence of the ionizing radiation per unit volume.

The electric field in the bulk of the gas, outside the space-charge layers, is given by

$$E_0 = \left(\frac{b}{q} \right)^{1/2} \frac{j_0}{e(R_i + R_e)}.$$

(4)

Here R_e is the electron mobility and b is the recombination coefficient.

Equations (2)–(4) were obtained under a number of simplifying assumptions for an electropositive gas in which the principal carrier annihilation mechanism is recombination. We shall not stop to analyze these simplifications (see the Appendix), but note that expressions (2)–(4)

and the expression for the dependence of the current on the voltage

$$j_0 = \frac{qLb^{1/2}e^{1/2}R_i^{1/2}}{2\pi^{1/2}R_e}\left[\left(1 + \frac{4\pi^{1/2}R_e^2e^2}{q^{1/2}R_i^{1/2}L^2b}V\right)^{1/2} - 1\right] \qquad (5)$$

(here L is the distance between electrodes and V is the voltage on the electrodes) describe well the experimental results [27-31].

It is seen from (5) that at high voltages between the electrodes the current through the gas is markedly limited by the space charge. Numerical estimates for typical laser mixtures show that in the case of a nonautonomous discharge described by the Thomson theory [Eq. (5)] the energy input to the gas from the external electric field, even under conditions of intense ionization, is insufficient to excite the laser.

This feature of the nonautonomous discharge was first pointed out in a 1970 paper [7] by A. V. Eletskii and B. M. Smirnov, who proposed to make an active CO_2-laser conducting by photoionization of alkali-metal vapors added to the working gas. They have assumed that the electric field must be weaker than the discharge-striking field. They expressed the natural fear that besides the difficulties in the realization of photoionization, the screening of the external field by the space-charge field near the electrodes will make the task of developing such a laser difficult. A paper published in the same year [8] contains a theory of the electroionization pumping method and reports experiments on the excitation of xenon, air, and nitrogen at pressures from 1 to 7 atm. The same paper reports measurements of the energy input to the gas, which has shown that this energy exceeds by many times the value predicted by Thomson's theory.

Thus, in spite of the large similarity between the electroionization laser and the ionization chamber, the processes that occur in the plasma of an electroionization laser differ substantially from those observed in an ionization chamber. The main difference is that the current flowing through the ionized gas is not limited by the space-charge field and consequently it becomes possible to pass through the gas currents that exceed by several orders of magnitude the currents in an ionization chamber.

1. Mechanism of Current Flow through the Active
Medium of an Electroionization Laser

In the experiment described in [8], the electric energy input to the gas was ~2 J/cm³. So large an energy input is impossible within the framework of Thomson's theory. Figure 3 shows the dependence of the electric energy input to N_2 on the voltage between the electrodes, obtained in experiments [21] and calculated in accordance with the theory of J. J. Thomson [27]. The result of this experiment can be explained in the following manner [9]: At a high degree of ionization, the gas column between the electrodes is a good conductor. The electric field in the cathode layer, from which the electrons have been depleted, can greatly exceed the field in the bulk of the gas, owing to the potential redistribution determined by the conductivity of the gas in the particular cross section, and can reach a value higher than the ionization threshold.

Electron multiplication takes place in the layer next to the cathode, and the limitation imposed by the space charge on the total current is cancelled out as a result of the increase of the ionic conductivity of the layer and of the increased emission from the cathode induced by the ion bombardment. Since the concentration of the neutral molecules in the active region of the electroionization laser is larger by 5-7 orders of magnitude than the electron concentration, in spite of the strong ionization of the gas, this cancellation can be easily effected. Thus, the cathode layer assumes in fact the role of an unlimited electron emitter. The emission is

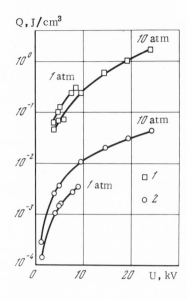

Fig. 3. Energy input to N_2 in the case of a strong-current nonautonomous discharge. 1) Experiment; 2) Thomson theory.

maintained automatically at a level governed by the conductivity of the bulk of the gas. No cascade discharge develops in the bulk of the gas if the field intensity is sufficiently lower than the breakdown value.

The theory of current stability under these conditions is given in [9, 32]. Calculations of the cathode potential drop, carried out in [4] (see also [21, 33-35]), are based on the approximate solution of the following nonlinear system of equations:

$$\frac{\partial n_e}{\partial t} = \frac{\partial}{\partial x}(n_e v_e) + \left(\frac{\partial n_e}{\partial t}\right)_i + \alpha n_e v_e - b n_e n_i^+ - \langle \sigma^- v_e \rangle n_e N^-,$$

$$\frac{\partial n_i^+}{\partial t} = -\frac{\partial}{\partial x}(n_i^+ v_i^+) + \left(\frac{\partial n_i^+}{\partial t}\right)_i + \alpha n_e v_e - b n_e n_i^+ - b_i n_i^+ n_i^-,$$

$$\frac{\partial n_i^-}{\partial t} = \frac{\partial}{\partial x}(n_i^- v_i^-) + \langle \sigma^- v_e \rangle n_e N^- - b_i n_i^+ n_i^-, \tag{6}$$

$$\frac{dE}{dx} = 4\pi e (n_i^+ - n_i^- - n_e).$$

This system describes the case of one-dimensional motion corresponding to large electrode dimensions and to a small distance L between them (see Fig. 2). The x axis is directed perpendicular to the plate; n_i^+, n_i^-, n_e, v_i^+, v_i^-, v_e are the concentrations and velocities of the positive and negative ions and electrons, respectively; b_i, b_e are the coefficients of the ion and electron−ion recombination, α is the first Townsend coefficient; $(\partial n_e/\partial t)_i = (\partial n_i^+/\partial t) = q$ is the rate of production of electron−electron pairs under the influence of the ionizing radiation; σ^- is the cross section for the sticking of the electrons to the electronegative molecules; N^- is the concentration of the electronegative molecules; E is the electric field; e is the electron charge; $\langle \ \rangle$ represents averaging over the electron velocities.

The first three equations are the charged-particle balance equations, while the last equation describes the space-charge field. These equations must be supplemented by the following boundary conditions:

$$[j_0 + \gamma_i j_i^+ + j_a]_{x=0} = [j_0]_{x=L}, \quad \int_0^L E \, dx = U. \tag{7}$$

Here j_0 is the electron-component current density, j_i^+ is the positive-ion current density, γ_i is the coefficient of electron emission from the cathode following bombardment by the ions,

j_a is the density of the field-emission current from the cathode, and U is the voltage between electrodes. In view of the low ion mobility we can neglect the contribution made to the total current by the ionic component of the current from the space charges next to the electrodes.

The quantity $(\partial n_e/\partial t)_i = (\partial n_i^+/\partial t)_i = q$ is determined completely by the flux density of the ionizing particles and by the number γ of the electron−ion pairs produced per centimeter of path of the ionizing particle. Figure 4 shows characteristic plots of γ against the electron-beam energy for several gases. Since the energy w_e consumed in the production of one electron−ion pair is practically independent of the type of gas ($w_e \simeq 25$-36 eV), the number of electron−ion pairs is determined essentially only by the gas density, by the initial energy of the fast particle, and by its charge and mass.

For an electron beam with current density j_e we have

$$q = \gamma j_e P/P_0 e. \tag{8}$$

Here γ is the number of electron−ion pairs produced in the given gas by one fast electron per centimeter of range at a pressure P_0, while P is the gas pressure.

Equations (6) and (7) and the equations that describe a low-pressure glow discharge [28, 29] are practically the same. They differ in that the right-hand sides of the first two equations in (6) have an additional term responsible for the production of the electron−ion pairs under the influence of the ionizing radiation. In the bulk of the gas, outside of the electrode space-charge layers, the field is lower than the critical discharge-striking value, so that the coefficient in this region is $\alpha \approx 0$, and the conductivity of the gas is due only to the ionizing radiation.

The bulk of the gas plays the role of the positive column of a glow discharge. In this volume, the stationary electron density n_e is determined by the expression

$$n_e = (q/b_e)^{1/2}, \tag{9}$$

which is valid for the recombination mechanism of electron annihilation. If sticking predominates, then

$$n_e = q/\langle \sigma^- V_e \rangle N^-. \tag{10}$$

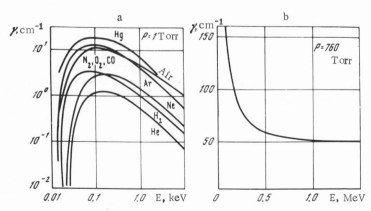

Fig. 4. Dependence of γ on the fast-electron energy. a) In different gases at a pressure 1 Torr [15]; b) in air at a pressure 1 atm.

We present the results of a calculation of the cathode potential drop V_e, of the thickness L_e of the space-charge layer, and of the parameter $E(0)/P$ in the cathode region for molecular nitrogen [21, 33] (see the Appendix):

$$V_e = \frac{(1 + \gamma_i)\, v_i(0)}{216\pi \cdot 10^{11} e} \left[\frac{P}{n_e k_e (E_0/P)} \right] \left\{ \frac{E(0)}{P} \right\}^2,$$

$$L_e = \frac{(1 + \gamma_i)\, v_i(0)}{108 \cdot 10^{11} e n_e k_e (E_0/P)} \left\{ \frac{E(0)}{P} \right\},$$

$$\frac{E(0)}{P} = \left[\ln \left(\frac{1 + \gamma_i}{\gamma_i} \right) \frac{108\pi \cdot 10^{11} e}{A v_i(0)(1 + \gamma_i)} \right]^{1/2} \left[\frac{k_e n_e}{P} \left(\frac{E_0}{P} \right) \right]^{1/2} + B. \tag{11}$$

The practical system of units is used in (11). The symbols E_0, n_e, k_e stand for the field intensity, stationary concentration, and mobility of the electrons in the bulk of the gas; $v_i(0)$ is the drift velocity of the ions in the cathode layer; $A = 1.17 \cdot 10^{-4}$ cm-Torr-V^{-2}, $B = 32.2$ V-cm^{-1}-Torr^{-1}. Equations (11) are valid for the range of the parameter $E(0)/P$

$$40 \text{ V/cm-Torr} < E(0)/P < 180 \text{ V/cm-Torr}, \tag{12}$$

in which the first Townsend coefficient can be represented in the form [36]:

$$\frac{\alpha}{P} = A \left\{ \frac{E(0)}{P} - B \right\}^2. \tag{13}$$

For the interval of the values of the parameter $E(0)/P$:

$$200 \text{ V/cm-Torr} < E(0)/P < 800 \text{ V/cm-Torr} \tag{14}$$

one uses a different empirical dependence of α/P on $E(0)/P$ [36]:

$$\frac{\alpha}{P} = A' \left\{ \frac{E(0)}{P} - B' \right\}, \tag{15}$$

where $A' = 1.25 \cdot 10^{-2}$ cm-Torr-V^{-2}; $B' = 50$ V-cm-Torr^{-1}.

For this region of values of the parameter $E(0)/P$, the following formulas are valid:

$$V_e = \frac{(1 + \gamma_i)\, v_i(0)}{216 \cdot 10^{11} e} \frac{P}{n_e k_e (E_0/P)} \left[\frac{E(0)}{P} \right]^2,$$

$$L_e = \frac{(1 + \gamma_i)\, v_i(0)}{108\pi \cdot 10^{11} e} \frac{1}{n_e k_e (E_0/P)} \left[\frac{E(0)}{P} \right],$$

$$\frac{E(0)}{P} = \left[\ln \left(\frac{1 + \gamma_i}{\gamma_i} \right) \frac{72\pi \cdot 10^{11} e}{A' v_i(0)(1 + \gamma_i)} \right]^{1/2} \left[\frac{k_e n_e}{P} \left(\frac{E_0}{P} \right)^{1/2} + B \right]. \tag{16}$$

Table 4 in the Appendix gives the values of the parameter $E(0)/P$, of the cathode potential drop V_e, of the thickness L_e of the cathode space-charge layer as functions of the electron density n_e in the main volume of the gas, and the values of the parameter E_0/P for molecular nitrogen at a pressure 13 atm. With decreasing electron density in the main volume of the gas, from $5 \cdot 10^{14}$ to $2 \cdot 10^{12}$ cm^{-3}, the cathode potential drop increases strongly. At an electron density $\sim 10^{10}$ cm^{-3} it reaches 25 kV and becomes comparable with the voltage between the electrodes. In this case the current decreases strongly and the laser cannot be excited. At high electron density, the thickness of the near-cathode layer is very small, $L_e \sim 10^{-4}$ cm, and the energy consumed in the formation of such a layer is a negligible fraction of the total energy released in the gas. In the absence of the influence of the space charge, the electric current depends

TABLE 1

Material	Absorption coefficient at wavelength $\lambda = 10.6\ \mu m$, cm^{-1}	Threshold energy flux, J/cm^2	Threshold power flux, W/cm^2
NaCl	$1.3 \cdot 10^{-3}$ [42]	76	$2 \cdot 10^8$
ZnSe	$5 \cdot 10^{-3}$ [42]	10	$2.5 \cdot 10^7$
BaF$_2$	0.1 [43]	4	10^7
Alloy Ge+Si+As (IKS-29)	$3.2 \cdot 10^{-2}$ [43]	15	$4 \cdot 10^7$
Alloy Ge+Se+As (M-8-4)	$1.4 \cdot 10^{-2}$	10	$2.5 \cdot 10^7$
Ge(M-4)	$7 \cdot 10^{-2}$	2.6	$6.1 \cdot 10^6$

on the voltage in accordance with Ohm's law:

$$j_0 = \sigma E_0;$$

(17)

here σ is the conductivity of the main volume of the gas and is governed by the external ionization, and E_0 is the electric field and is determined by the supply voltage V and by the distance L between the electrodes:

$$E_0 = V/L.$$

(18)

2. Experimental Technique

2.1. Construction of Laser Chambers

Investigations of the electroionization method of excitation of lasers operating on compressed molecular gases were carried out in two installations. In the first installation the gas was ionized with a beam of fast electrons of energy ~600 keV and pulse duration ~10^{-8} sec from an accelerator with a cold cathode [37]. The experiments were performed with this setup

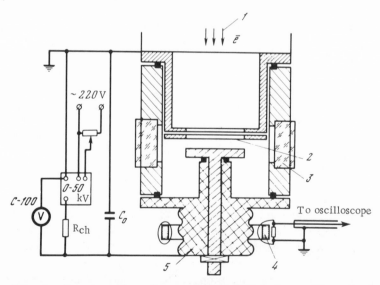

Fig. 5. Construction of experimental chamber with a discharge gap 4 cm long (chamber 1). Total volume of excitation region up to 5 cm^3. 1) Electron beam; 2) metallic foil; 3) mirror; 4) Rogowski loop; 5) insulator.

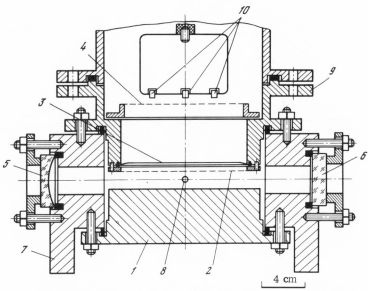

Fig. 6. Construction of experimental chamber with dielectric body (chamber 2). Discharge gap length 13 cm, total excitation volume up to 40 cm². 1) Lower electrode; 2) steel grid; 3) aluminum foil; 4) anode of electron gun; 5) total reflection mirror; 6) semitransparent mirror; 7) body of chamber; 8) opening for the entrance of the gas; 9) upper electrode; 10) cathodes of electron gun.

in three chambers of different volumes. The chamber constructions are illustrated in Figs. 5-7. Chambers 1 and 3 were of solid metal (1 of brass and 3 of Duralumin). Chamber 2 was made from a block of Plexiglas. Chambers 1 and 3 were designed for working-gas pressures up to 100 atm, while chamber 2 was rated 8 atm. Chamber 2, covered on the outside with a putty resistant to fluorine compounds, was used to investigate chemical electroionization lasers. In all three chambers, the upper current electrode was a steel grid welded to a metallic flange with a rectangular opening. This flange was secured with metallic screws to the exit flange of the accelerator. It was used simultaneously to clamp to the vacuum gasket a thin film that separated the vacuum volume of the accelerator from the chamber volume filled with compressed gas. The thin films used were foils of aluminum and beryllium alloy 50 μm thick, foils of niobium and molybdenum alloy 25-30 μm thick, and Mylar films 50 μm thick. The transparency of these films to an electron beam of energy 600-700 keV was high enough. The rectangular opening for the electron beam measured 10×40 mm in chamber 1, 20×100 mm in chamber 3, and 15×130 mm in chamber 2. The separator film withstood pressures higher than 5-10 atm only when a special holder was used. Figure 8 shows three types of holders for chamber 3, designed for pressures ~5 and 100 atm. The dimensions of the active region of the laser were determined for each chamber by measuring the cross section of the electron beam producing ionization of the working gas. Thus, the length of the active region was 4 cm for chamber 1, 13 cm for chamber 2, and 10 cm for chamber 3. The height of the active region was determined by the distance between the upper grounded electrode of the chamber (the grid) and the lower electrode (steel bar of the same area as the opening for the entrance of the electron beam).

A photograph and diagram of the second installation are shown in Fig. 9. The cross section and length of the active region were $20-100 \times 100$ mm² and 1000 mm, respectively. The generator for the electron gun was an Arkad'ev-Marx generator GIN-400-0.6/5 with output

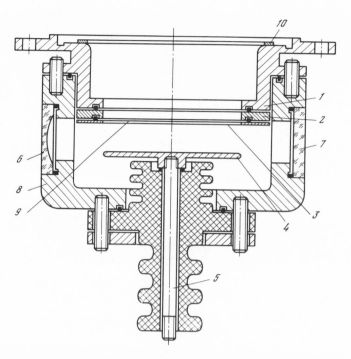

Fig. 7. Diagram of metallic experimental chamber with maximum working pressure higher than 100 atm (chamber 3). Length of discharge gap 10 cm; total excitation volume up to 40 cm^3. 1) Upper electrode; 2) support plate with openings; 3) steel grid; 4) lower electrode; 5) brass rod; 6) total-reflection mirror; 7) semitransparent mirror; 8) metallic chamber; 9) Mylar film; 10) anode of electron gun.

Fig. 8. Supporting plates with openings.

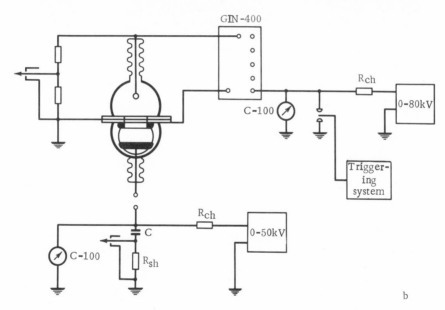

Fig. 9. Photograph (a) and schematic diagram (b) of electroionization laser
with active-region volume 10 liters (chamber 4).

capacitance 12 nF and output voltage up to 400 kV. We used either one GIN-400 generator, or
two connected in parallel. Figure 10 shows the construction of the vacuum chamber of the electron gun. The chamber was made of stainless steel and its dimensions with the flanges were
$\sim 1300 \times 400 \times 500$ mm. The cathode of the gun was fastened to a pyroceramic vacuum insulator [38]. The chamber was evacuated to a pressure 10^{-5} Torr by means of a turbomolecular
pump TMN-200. In the experiments we used both a cold pointed cathode (in which case the
electron-beam pulse duration was $\tau_e \sim 1$ μsec) and a thermionic cathode ($\tau_e \simeq 5$-30 μsec).
The emitting elements in the pointed cathode [37] were tubes of 10 mm diameter of thin (10-15
μm) tantalum foil (Fig. 10). The length of the pointed cathode was 1000 mm; the distance from
the anode to the pointed cathode was adjusted in the interval 30-100 mm. Figure 11

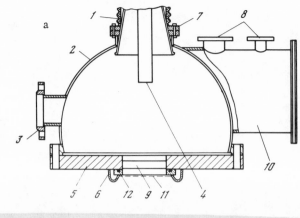

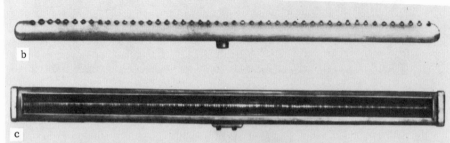

Fig. 10. Construction of electron gun of chamber 4 (a) and pointed cathodes: tubular (b) and knife-edge (c). a: 1) pyroceramic insulator; 2) chamber shell; 3) sighting window; 4) cathode lead; 5) separation plate; 6) screen; 7) insulator flange; 8) stubs for vacuum tubes; 9) supporting metal ribs for the foil; 10) flange for evacuation; 11) guard grid; 12) flange used to seal the foil.

shows the cathode unit of the heated cathode and emitter with dimensions 400 × 1000 mm, constructed with tungsten of tantalum wire. A stainless steel electrostatic screen was placed between the cathode unit and the chamber walls and used also to focus the electron beam onto the exit window. The thermionic cathodes were heated with a specially developed filament transformer of ~2 kW power with a high-voltage decoupling rated 200 kV between the primary and secondary windings.

The electron beam was introduced into the discharge gap of the laser chamber through a window measuring 10 × 100 cm in a plate that separated the vacuum chamber of the electron gun from the laser chamber (Fig. 12). The window was covered with an aluminum foil 25 μm thick or with a Lavsan polyester film 15-20 μm thick. The supporting element for the foil was a steel lattice with rectangular slots 10 × 100 mm. The supporting lattice served simultaneously as the anode for the electron gun. The foil was sealed with a rubber gasket and a rectangular flange to which a large-mesh steel grid was welded and served as the upper current electrode (cathode) of the laser chamber. The laser chamber was made of stainless steel. The dimensions of the chamber with the flanges and insulators were 1500 × 600 × 500 mm. To decrease the inductance of the discharge circuit the anode (Duralumin plate 20 × 130 × 1000 mm) was equipped with six parallel insulated leads and passed-through tubes of Plexiglas to connect to the high-voltage supply.

Prior to the admission of the working gas, the laser chamber was evacuated to 10^{-2} Torr with a forevacuum pump. Gases of commercial purity were used in the experiments in all the chambers.

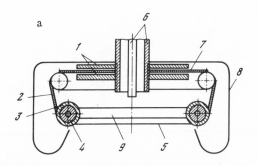

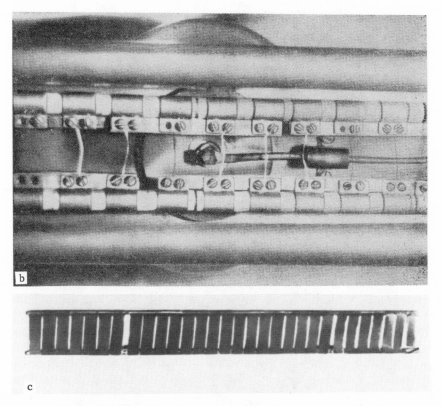

Fig. 11. Construction of cathode unit of heated cathode (a) and emitters based on tantalum wire (b) and Nb−La alloy foil (c). 1) Cathode holder; 2) suspension of cathode frame; 3) emitter contacts; 4) ceramic inserts; 5) emitter (foil); 6) filament leads; 7) supporting frame; 8) focusing screen; 9) cathode frame.

The end windows of the chamber had an optical diameter 140 mm and were designed to use mirrors of 160 mm diameter.

A detailed description of all the units of this setup and the design of the supply systems are given in [20, 39].

2.2. Optical Resonators

The laser optical resonators were made of spherical opaque glass and flat semitransparent glass. The total-reflection mirrors were made of copper, copper alloys, or U-8 steel; multilayer dielectric mirrors with reflectivity ~98% were also used. The semitransparent mirrors were sputtered on substrates of Ge, NaCl, BaF_2, and ZnSe. The first experiments were

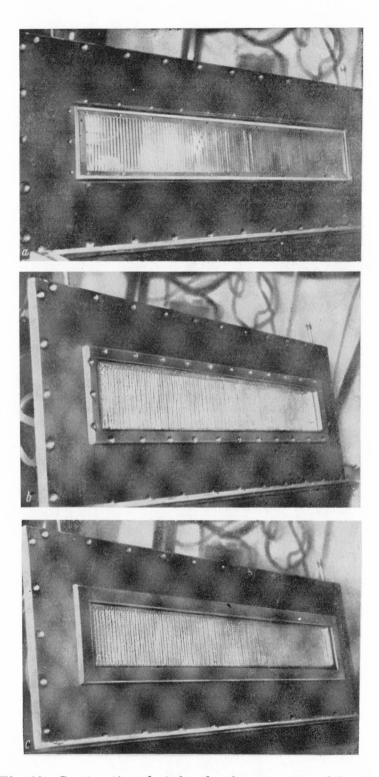

Fig. 12. Construction of window for the extraction of the electron beam in the discharge gap. a) Separating plate with supporting lattice; b) separating plate with sealed foil; c) the rectangular flange that seals the foil is covered with an electrostatic screen.

performed with gold and silver mirrors sputtered on substrates made of the above-mentioned materials in a relatively low vacuum $\sim 10^{-5}$ Torr. The strength of these coatings was low, not more than 0.03 J-cm^{-2}, and the mirrors "burned out" when the pump energy exceeded the threshold slightly. To increase the endurance of the mirror to the laser radiation at 10.6 μm wavelength, we prepared and investigated interference coatings, constituting two-component systems of alternating weakly absorbing dielectric layers with high and low refractive indices. The measurements of the radiation strength of the mirrors and of the materials were performed with the aid of an electroionization CO_2 laser, the radiation of which was focused with a NaCl lens (focal length 40 mm) on the investigated sample. The flux of the incident laser radiation was regulated by introducing attenuating plates and by smoothly varying the laser pump laser (the experiments were performed with the aid of chamber 2, at an output energy up to 1 J, radiation pulse duration at half height $\sim 4 \cdot 10^{-7}$ sec) [41]. In measurements of the radiation strength of the dielectric mirrors it was observed that the endurance of the coating depends to a considerable degree on the technology of depositing the coating on the substrate (the vacuum, the rate of sputtering, the substrate temperature, etc.). With increasing number of sputtered layers, the radiation endurance decreases. Tables 1 and 2 give the threshold values of the radiation flux required to disintegrate the transparent materials and the dielectric mirrors.

Measurements of the radiation endurance of the dielectric mirrors relative to unfocused radiation, carried out in chamber 4 in the regime of "short" ($T_p \simeq 10^{-7}$ sec) pulses, have led to similar results, but the absolute values of the threshold fluxes were much lower. The most enduring were single-layer dielectric coatings of As_2S_3 ($T_{10.6} \simeq 72\%$, sputtered on substrates of NaCl [$g_{thr} \simeq 2$ J-cm^{-2}]. In the "long" radiation-pulse regime ($T_p \simeq 2 \cdot 10^{-5}$ sec), the largest radiation endurance was possessed by the same single-layer coating of As_2S_3 on NaCl [$g_{thr} > 10$ J-cm^{-2}]. A three-layer mirror, for example, a dielectric $Sb_2S_3 + BaF_2$ mirror sputtered on NaCl, disintegrates already at $g \gtrsim 3$ J-cm^{-2}. Measurements of the endurance of polished metals relative to focused CO_2-laser radiation have shown that the largest radiation endurance is possessed by copper [$g_{thr} \gtrsim 300$ J-cm^{-2}].

2.3. Measurements of Laser Parameters

Diagrams of the electric power supplies for the laser chambers are shown in Figs. 5 and 13. The voltage was applied to the electrodes of the electroionization laser (chambers 1–3) from a bank of low-inductance ceramic capacitors K15-10 (~ 5 nF, U $\simeq 50$ kV). Depending on the volume of the working region, on the pressure, and on the gas composition, the capacitance of the bank ranged from 10 to 60 nF. To prevent breakdown of the capacitors on the surface, the K15-10 capacitors were potted in epoxy resin in dielectric guard rings or in a plate of Plexiglas. The circuit connecting the bank to chambers 1–3 had low inductance, and the total inductance of the discharge circuit was $< 10^{-7}$ H. To supply the laser with active-region volume

TABLE 2

Coating material	Substrate material	Mirror transmission coefficient at $\lambda = 10.6$ μm, %	Threshold energy flux, J/cm^2	Threshold power flux, W/cm^2
$Sb_2S_3 + BaF_2$	BaF_2	5	2	$5 \cdot 10^6$
»	K-8 glass	0	1.8	$4.5 \cdot 10^6$
»	NaCl	12	3	$7.5 \cdot 10^6$
»	»	30	4	10^7
$As_2S_3 + BaF_2$	»	22	5.1	$1.3 \cdot 10^7$
»	ZnSe	40	6.5	$1.6 \cdot 10^7$
IKS-A-6-45	NaCl	64	9.3	$2.3 \cdot 10^7$
$Sb_2S_3 + SrF_2$	Ge	5	10	$2.5 \cdot 10^7$
»	ZnSe	7	9	$2.3 \cdot 10^7$

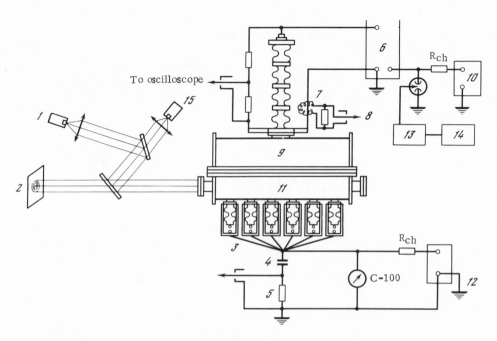

Fig. 13. Schematic diagram of the supply and control of the parameters of an electroionization laser with active volume 10 liters. 1) Calorimeter; 2) heat-sensitive screen; 3) connecting cables; 4) capacitor bank; 5) coaxial shunt; 6) voltage-pulse generator; 7) Rogowski loop; 8) triggering of oscilloscope; 9) electron gun; 10) rectifier; 11) laser chamber; 12) rectifier; 13) high-voltage pulse generator; 14) control-pulse generator; 15) photoreceiver.

10 liters (chamber 4), we used a bank of low-inductance capacitors IM-50/3 KM-50/4, or IK-100/0.25, the total capacitance of the different banks amounting to 1.5–16 μF. To decrease the inductance of the discharge circuit, the lead assembly consisted of a large number of parallel connected segments of RK-75 cable. The bank was connected to the anode of discharge chamber 4 through protective discharge gaps, so that no voltage could be applied to the discharge gap when the capacitors were charged, thereby avoiding leakage due to corona discharge between the anode and the chamber walls. The capacitor bank was charged with a rectifier delivering a voltage up to 100 kV. The bank was connected to the rectifier through a charging resistor 10^7–10^9 Ω.

The assembly of the measuring apparatus includes a number of pickups, with the aid of which the energy parameters of the insulation and the laser radiation characteristics are measured. The fast-electron beam current was measured with a Faraday cup or with a low-inductance shunt connected between the discharge-chamber anode (which in this case was the electron collector) and the ground. The oscillograms of the voltages on the cathode of the electron gun and the anode of the laser chamber 4 were obtained with the aid of resistive dividers; the discharge current was measured with a low-inductance coaxial shunt or with a Rogowski loop. The pulsed voltage from the Rogowski loop was used to trigger the oscilloscope sweep.

The parameters of the volume discharge initiated by the fast-electron beam were measured in the following manner. At a fixed pressure of the investigated gas mixture, the initial voltage U_0 on the capacitance was varied from 0 to $U_{st.br.}$, i.e., to the voltage at which autonomous breakdown of the gas gap took place. For each fixed voltage we measured the discharge-current amplitude, the wave form of the current pulse, and the voltage to which the capacitor was discharged. To eliminate errors connected with the instability of the operation of the electron guns, not less than 10 measurements were performed for each initial capacitor voltage, and the results were averaged. The capacitance C used for the measurement of the current−voltage characteristics was chosen such that the measured voltage across the capacitance as a result of the discharge was < 10% of the initial voltage. The energy delivered to the gas was determined by measuring the difference between the voltages across the capacitance C before (U_0) and after (U_∞) the discharge:

$$Q_p = C \frac{U_0^2 - U_\infty^2}{2}.$$ (19)

The measurement accuracy was determined mainly by the accuracy of the calibration of the pickups (the Rogowski loop, the shunt, the kilovoltmeter) and averaged ~20%.

The measurements thus yielded the parameters of the current pulse and of the energy input to the discharge gap as functions of the composition and pressure of the gas mixtures, of the intensity and duration of the ionization, and of the field in the discharge gap. In addition, photographs taken through the sighting windows of the working chamber made it possible to assess the spatial distribution of the current flowing through the gap.

The measurements of the output energy of the electroionization lasers were carried out with the aid of tungsten and graphite calorimeters [in the different experiments we used the instruments IEK-1, AKT-IM, IMO-2, "Scientech-302," and also a calorimeter with large (70 mm diameter) aperture developed in the optical design division of the Physics Institute of the Academy of Sciences of the USSR]. The energy measurement accuracy was determined by the accuracy of the calorimeter calibration and amounted to 10-15%, while the most accurate measurements were performed with the aid of the "Scientech-302" calorimeter (USA), the certified error of which is ~3%. In the measurement of the energy, the calorimeters were either placed directly at the exit mirror of the laser (chambers 1-3), or else part of the laser radiation was diverted with the aid of beam-splitting plates and focused on the calorimeters with the aid of NaCl lenses.

The wave form of the generated radiation pulse was registered with the aid of resistors of germanium doped with gold (Ge−Au), zinc (Ge−Zn), and antimony and zinc (Ge−Sb, Zn), and cooled with liquid nitrogen. The resolutions of the resistors were 10^{-7}, 10^{-9}, and 10^{-9} sec respectively for the gold-, zinc-, and zinc−antimony-doped resistors. We also used a "drag" detector with resolution ~$3 \cdot 10^{-9}$ sec. The Ge−Au photoresistor was connected to the oscilloscope through a cathode follower. Since the dark resistance of the high-speed photoresistors was low, the permissible supply voltage was also low, < 10 V. For this reason, the useful signal from the photoresistors connected directly to the cable turned out to be very small (against the background of the electric noise from the measuring apparatus). Figure 14 shows a circuit for pulsed supply of the photoresistors, which makes it possible to increase the useful signal in the cable by 10 times and to observe the wave form of the pulse on the oscilloscope electrodes. When a weak signal must be observed through an amplifier with small dynamic range, it is necessary to introduce into the electric circuit a differentiating network. To this end it suffices to apply the signal to the 75-Ω input of the amplifier through a decoupling resistor C_d, the value of which was chosen such as to satisfy the inequality

$$\tau_p < 75 C_d < \tau_{ref}.$$

where τ_p is the duration of the radiation pulse and τ_{ref} is the duration of the reference supply pulse.

The spatial distribution of the radiation field of the laser was investigated with the aid of a heat-sensitive screen consisting of carbon paper and a Mylar film or emery paper, at radiation energies of several hundered joules, and also with the aid of the "Radiovizor" in-

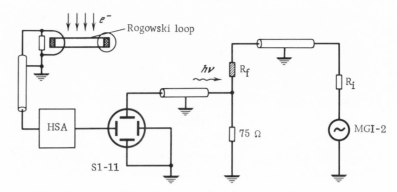

Fig. 14. Diagram of pulsed supply to photoresistor. HSA, horizontal sweep amplifier for S-1-11 oscilloscope; R_i, internal resistor for MGI-2 pulse generator.

strument, based on the quenching of the luminescence and the influence of infrared radiation. The procedure used to measure the laser beam divergence is described in Sec. 2 of Chapter II. The spectral measurement techniques are discussed in Sec. 3 of Chapter II.

3. Electric Characteristics of Active Medium

3.1. Calculation of the Characteristics of the Discharge Excited by the Electroionization Method

1. In accordance with the ideas concerning the charged transport processes in an ionized gas, which were discussed above, it can be assumed that the current excited by the electroinization method is due to the drift motion of the free electrons produced when the gas is ionized by the penetrating radiation in the volume of the gas gap. The concentration of the free electrons decreases with time because of electron−ion recombination and because of the sticking of the electrons to the electronegative molecules, and its dependence on the time is described by the equation

$$dn_e/dt = q + \alpha\, n_e v_e - bn_e n_i^+ - cn_e N_0.$$

(20)

The resistance of the gas gap is determined by the concentration and by mobility of the electrons (see Fig. 2)

$$R\,(t) = d[eSn_e\,(t)\,\mu_e\,(t)],$$

(21)

where S is the cross-sectional area of the discharge gap, the remaining symbols are the same as in the systems (6) and (11), and $\mu_e(t) = k_e/P$.

Calculation of the parameters of the nonautonomous discharge reduces in our case to a solution of the equation of the discharge circuit (C is the capacitance of the working capacitor bank, r is the resistance of the connecting busbars, L is the parasitic inductance of the circuit) simultaneously with Eqs. (20) and (21):

$$L\frac{dI}{dt} + [R\,(t) + r]\,I + \frac{\int I\,dt}{C} = 0.$$

(22)

Special measures were always taken in the experiment to decrease the energy loss in the discharge circuit, so that the connection from the capacitor bank to the discharge gap was so constructed as to satisfy the inequality

$$R(t) \gg r; \qquad R(t) \gg R_{cr} = 2\sqrt{L/C}. \tag{23}$$

When the last inequality of (23) is satisfied, as is known from the analysis of the tank circuit, the voltage drop across the tank-circuit inductance is small, so the first term of Eq. (22) can be neglected.

2. Inasmuch as Eq. (22) cannot be solved analytically in general form, and the rates of the generation of electron−ion pairs, recombination, and sticking depend to a strong degree on the actual composition of the investigated gas mixture, it is advantageous to consider limiting cases, when the rate of decrease of the electron concentration is determined only by recombination or only by sticking [21, 44].

a. Recombination mechanism. This case corresponds, for example, to a discharge in pure nitrogen or in a mixture of nitrogen with helium, in which there is no sticking of electrons to the molecules or the atoms. The Townsend multiplication of the electrons in the investigated field region that is typical of the operation of an electroionization laser based on vibrational−rotational molecule transitions can be neglected [36, 45] so that we obtain the following system of equations:

$$\frac{dn_e}{dt} = q - bn_e^2, \qquad I\frac{d}{eS\mu_e n_e} + \frac{\int I\,dt}{C} = 0, \qquad n_e(0) = 0; \qquad U_C(0) = U_0. \tag{24}$$

The working capacitance is chosen as a rule large enough to prevent the field in the discharge gap from decreasing noticeably during the discharge time. This is a necessary condition in laser pumping, for otherwise the rate of excitation of the upper laser level decreases abruptly during the pump pulse (see Fig. 1b). Therefore the change of the values of μ_e and b, which generally speaking depend on the field intensity, during the discharge time can be neglected in comparison with the change of the electron concentration. Assuming a rectangular ionizing-radiation pulse with duration T_{beam} and with amplitude q_0 we obtain the following expressions for the time dependences of the electron concentration, the discharge-current density, the voltage on the discharging capacitor, and the per-unit energy input:

$$n_e(t) = \begin{cases} n_{eq}\tanh(n_{eq}bt), & t \leqslant T_{beam} \\ \dfrac{1}{b(t-T_{beam}) + 1/n_{eq}\tanh(n_{eq}bT_{beam})}, & t \geqslant T_{beam} \end{cases} \tag{25}$$

$$j(t) = \begin{cases} \dfrac{eU_0\mu_e n_{eq}}{d}\tanh(n_{eq}bt)\,[\cosh(n_{eq}bt)]^{-\frac{e\mu_e S}{bCd}}, & t \leqslant T_{beam} \\ \dfrac{eU_0\mu_e}{d}[n_{eq}bT_{beam}\sinh(n_{eq}bT_{beam})]^{-\frac{e\mu_e S}{bCd}}\left[\dfrac{t-T_{beam}}{T_{beam}} + \dfrac{1}{n_{eq}bT_{beam}\tanh(n_{eq}bT_{beam})}\right]^{-\frac{e\mu_e S}{bCd}}, & t \geqslant T_{beam} \end{cases} \tag{26}$$

$$U_R(t) = \begin{cases} U_0\,[\cosh(n_{eq}bt)]^{-\frac{e\mu_e S}{bCd}}, & t \leqslant T_{beam} \\ U_0\left\{n_{eq}bT_{beam}\sinh(n_{eq}bT_{beam})\left[\dfrac{t-T_{beam}}{T_{beam}} + \dfrac{1}{n_{eq}bT_{beam}\tanh n_{eq}bT_{beam}}\right]\right\}^{-\frac{e\mu_e S}{bCd}}, & t \geqslant T_{beam} \end{cases} \tag{27}$$

$$Q = \frac{CU_0^2}{2Sd}\left\{1 - \left[\left(\frac{T_c - T_{beam}}{T_{beam}} + \frac{1}{n_{eq}bT_{beam}\tanh(n_{eq}bT_{beam})}\right)n_{eq}bT_{beam}\sinh(n_{eq}bT_{beam})\right]^{-\frac{2e\mu_e S}{bCd}}\right\}. \tag{28}$$

Here $n_{eq} = (q_0/b)^{1/2}$ is the equilibrium concentration of the free electrons, corresponding to a sufficiently long ionization-pulse duration; T_c is the time during which the flow of the current through the gas gap practically stops because the electron concentration decreases to the critical value n_{cr} (see Chapter I, Sec. 1).

b. Decrease of the electron concentration as a result of sticking (this mechanism can prevail in excited mixtures with large contents of electronegative gases such as O_2, CO_2, SF_6, etc.). The dependence of the discharge parameters on the time is determined in this case by the system of equations

$$\frac{dn_e}{dt} = q - cn_e N_0, \qquad I\frac{d}{eS\mu_e n_e} + \frac{\int I\,dt}{C} = 0, \qquad n_e(0) = 0, \qquad U_C(0) = U_0. \tag{29}$$

In the same approximation, in which a rectangular ionization pulse is assumed, we obtain the solution in the form

$$n_e(t) = \begin{cases} \dfrac{q_0}{cN_0}\,[1 - \exp(-cN_0 t)], & t \leqslant T_{beam} \\[2ex] \dfrac{q_0}{cN_0}\,\{[1 - \exp(-cN_0 T_{beam})]\exp[-cN_0(t - T_{beam})]\}, & t \geqslant T_{beam} \end{cases} \tag{30}$$

$$j(t) = \begin{cases} \dfrac{CU_0}{S}\,\{acN_0[1 - \exp(-cN_0 t)]\}\exp\{-a[cN_0 t + \exp(-cN_0 t) - 1]\}, & t \leqslant T_{beam} \\[2ex] \dfrac{CU_0}{S}\,\{acN_0[1 - \exp(-cN_0 T_{beam})]\}\exp[-cN_0(t - T_{beam})]\times \\[1ex] \qquad \times \exp\{-a[cN_0 T_{beam} - (\exp cN_0 T_{beam} - 1)\exp(-cN_0 t)]\}, & t \geqslant T_{beam} \end{cases} \tag{31}$$

$$U_C(t) = \begin{cases} U_0 \exp\{-a[cN_0 t + \exp(-cN_0 t) - 1]\}, & t \leqslant T_{beam} & (32) \\[2ex] U_0 \exp\{-a[cN_0 T_{beam} - (\exp cN_0 T_{beam} - 1)]\exp(-cN_0 t)\}, & t \geqslant T_{beam} & (33) \end{cases}$$

$$Q = \frac{CU_0^2}{2Sd}\,\{\exp\{-2a[cN_0 T_{beam} + (\exp cN_0 T_{beam} - 1)\exp(-cN_0 T_c)]\} - 1\}. $$

Here $a = eS\mu_e q_0/cdN_0^2 c^2$. For mixtures in which electron−ion recombination and electron sticking take place simultaneously, we obtain analytic expressions analogous to (25)-(28) and (30)-(33), but only in the case when the effects of sticking and recombination respectively can be regarded as small corrections. For example, in the case of weak sticking, when $cN_0 \ll bn_e$, we have

$$n_e(t) = \begin{cases} \sqrt{q_0/b}\,\tanh\left(\sqrt{q_0 b}\,t + \dfrac{cN_0}{2\sqrt{q_0 b}}\right), & t \leqslant T_{beam} \\[3ex] \left\{\left[\dfrac{b}{cN_0} + \dfrac{1}{\sqrt{q_0/b}\,\tanh\left(\sqrt{q_0 b}\,T_{beam} + \dfrac{cN_0}{2\sqrt{q_0 b}}\right)}\right]^{-1} - c^2 N_0^2(t - T_{beam})\right\}^{-1} - \dfrac{b}{cN_0}\right\}^{-1} & t \geqslant T_{beam} \end{cases} \tag{34}$$

3. The equations derived enable us to calculate all the parameters of the discharge excited in the gas mixture by the electroionization method, if we know the characteristics of the ionizing radiation, the value of the operating capacitance, and the geometry of the discharge gap. Conversely, we can calculate the ionization intensity, the field intensity, etc., which are needed to obtain given values of the electron concentration, the discharge current, and the per-unit energy input.

Let us estimate, for example, the maximum per-unit energy input that can be obtained if the working capacitance is increased without limit. For the case when the principal mechanism decreasing the electron concentration is electron−ion recombination, we easily obtain from (28)

$$Q\vert_{c\to\infty} = Q_{\text{lim}} \simeq \begin{cases} \dfrac{e\mu_e U_0^2}{bd^2}\left(n_{\text{eq}}\, bT_{\text{beam}} + \ln\dfrac{n_{\text{eq}}}{2n_{\text{cr}}}\right), & n_{\text{eq}}\, bT_{\text{beam}} \gtrsim 1, \\[3mm] \dfrac{e\mu_e U_0^2}{bd^2}\ln\dfrac{n_{\text{eq}}^2\, bT_{\text{beam}}}{n_{\text{cr}}}, & n_{\text{eq}}\, bT_{\text{beam}} \ll 1. \end{cases} \tag{35}$$

[In the derivation of (35) it was assumed that $b \simeq 10^{-6}$-10^{-8} cm^3-sec^{-1} [45] and $n_{\text{eq}} \gg n_{\text{cr}} = n_e(T_c)$.]

As follows from a comparison of (28) and (35), the growth of the per-unit energy input with increasing working capacitance practically stops at $C \gtrsim 2\mu eS/bd \log A$, where

$$A = \begin{cases} \dfrac{n_{\text{eq}}}{2n_{\text{cr}}}\, e^{n_{\text{eq}}\, bT_{\text{beam}}}, & n_{\text{eq}}\, bT_{\text{beam}} \gtrsim 1, \\[3mm] \dfrac{n^2\, bT_{\text{beam}}}{n_{\text{cr}}}, & n_{\text{eq}}\, bT_{\text{beam}} \ll 1. \end{cases}$$

We note that to obtain a large energy input in this case it is more convenient to use "long" pulses, in which, even at a relatively low intensity of the ionizing radiation ($q_0 \simeq 10^{17}$ cm^{-3}-sec^{-1} for $b \simeq 10^{-7}$ cm^3-sec^{-1} and $T_{\text{beam}} \simeq 10^{-5}$ sec) the condition $n_{\text{eq}}\, bT_{\text{beam}} \ll 1$ is satisfied and the limiting energy input increases in proportion to the product of the equilibrium electron concentration by the ionization pulse duration

$$Q_{\text{lim}} \sim n_{\text{eq}} T_{\text{beam}} \sim \sqrt{j_{\text{beam}}}\, T_{\text{beam}}.$$

In the "short" pulse regime, when $n_{\text{eq}} bT_{\text{beam}} \ll 1$, the possibility of increasing the energy input at the expense of the ionizing radiation is strongly limited by the fact that the $Q_{\text{lim}}(n_{\text{eq}}, T_{\text{beam}})$ dependence is much weaker (logarithmic).

For the case when the electron concentration in the ionized gas decreases only as a result of sticking, we obtain from (33)

$$Q\vert_{c\to\infty} = Q_{\text{lim}} = \frac{e\mu_e E_0^2}{cN_0}\, q_0 T_{\text{beam}}. \tag{36}$$

The dependence of the limiting energy input on the electron−ion pair generation rate is stronger (linear).

We present also expressions for the power input per unit volume of gas in the case of continuous electroionization excitation. For the recombination mechanism we have

$$W = e\sqrt{q/b}\,\mu_e P^2 (E/P)^2, \tag{37}$$

and for the sticking

$$W = e(q/cN_0)\mu_e P^2 (E/P)^2. \tag{38}$$

Since the kinetic coefficients are usually cited in the literature for pure gases, in calculations for mixtures we used in the present study "weighted" values of the coefficients, averaged with allowance for the composition of the mixtures with the aid of the relations

$$\frac{1}{\mu} = \sum_i \frac{1}{\mu_i}\frac{P_i}{P_{\text{tot}}}; \qquad b = \sum_i b_i \frac{P_i}{P_{\text{tot}}}; \qquad c = \sum_i c_i \frac{P_i}{P_{\text{tot}}}, \tag{39}$$

where μ_i, b_i, and c_i are respectively the electron mobility, the electron–ion recombination coefficient, and the sticking coefficient for the i-th gas, the partial pressure of which in the mixture is P_i.

To calculate the rate of generation of electron–ion pairs it is necessary to know the ionizing ability of the employed ionizing radiation. For a beam of fast electrons, for example, we have

$$q(x) = \frac{i_{beam}}{ew} \cdot \frac{dU_e(x)}{dx}, \tag{40}$$

where j_{beam} is the current density in the beam, e is the electron charge, dU_e/dx is the ionization energy loss, and w is the energy required to produce an electron–ion pair. The quantity w does not depend on the energy of the fast electrons and amounts usually to 20-40 eV [46]:

Ionized gas	He	Ne	Ar	Kr	Xe	N_2	O_2	Air	CO_2
w, eV	42.3	36.6	26.4	24.1	21.9	34.9	30.8	33.9	32.8

The distribution of the ionization losses for a beam of fast electrons in a medium is calculated presently by the Monte Carlo method on the basis of the beta theory [47]. Examples of such a distribution are shown in Fig. 15a. For estimates it is customary to use the averaged ionization loss $dE/dx \simeq 0.5 (U_0/R_E)$, where R_E is the range of the fast electron with initial energy U_0. In the considered energy region $U_0 \lesssim 1$ MeV, the quantity R_E is well approximated by the expression [48]:

$$R_E = 0.274 \frac{A}{Z\rho} \frac{U_0^2}{(U_0 + m_e c^2) m_e c^2} \tag{41}$$

(A is the atomic weight of the medium that decelerates the beam, Z is the charge of the nucleus, ρ is the density in g/cm³, m_e is the electron rest mass, and c is the speed of light).

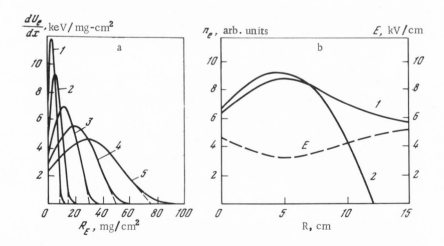

Fig. 15. Rate of electron–ion pair generation in a gas ionized by a beam of fast electrons vs. the penetration depth. a) Universal curves for different initial fast-electron energies: 1) $U_0 = 75$ kV, $R_E = 8.8$ mg/cm²; 2) $U_0 = 100$ kV, $R_E = 15$ mg/cm²; 3) $U_0 = 150$ kV, $R_E = 30$ mg/cm²; 4) $U_0 = 200$ kV, $R_E = 50$ mg/cm²; 5) $U_0 = 250$ kV, $R_E = 75$ mg/cm²; b) comparison of the relations calculated with (1) and without (2) the applied external electric beam taken into account.

It must be noted that in electroionization excitation of a gas, when the electrons that ionize the gas move in the field that is applied to the discharge gap, the distribution of the ionization losses changes somewhat in comparison with the usual conditions, when there is no field. This circumstance was first pointed out in [49]. The author of the same paper calculated the distribution of the ionization loss for electrons of energy $U_e \simeq 100$ keV for ionization of air placed in a field ~ 4 kV-cm^{-1}-atm^{-1} (Fig. 15b). It turned out that the effective range of the fast electrons, and accordingly the homogeneity of the distribution of the concentration of the secondary electrons, increase appreciably in such a field.

We note in conclusion that the problem of calculating the energy parameters of a discharge excited by the electroionization method can be inverted, namely, the values of the kinetic coefficients of the gas can be calculated from the experimentally measured functions $j(t)$ and $U_C(t)$ as well as from the value of Q_{input}. This is exactly the method used by us here to measure the recombination and sticking coefficients for a large number of gas mixtures. The simplicity and reliability of this method of measuring the energy coefficients of gases make it quite convenient in applications [50].

3.2. Experimental Investigation of a Nonautonomous Discharge Initiated in a Compressed Gas by an Intense Electron Beam − Discussion of Results

1. a. The glow of the gas gap in the case of a nonautonomous discharge initiated by an electron beam is a volume effect and is practically homoegeneous (Fig. 16a). With increasing relative field intensity E/P, the intensity of the volume glow increases somewhat, but it always remains much weaker than the glow due to autonomous breakdown, which develops when the field increases to a value E/P$_{st.br.}$ (Fig. 16b), although the energy inputs to the discharge remain of the same order. Consequently, the electron temperature in the volume discharge under

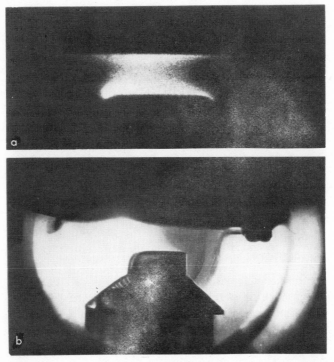

Fig. 16. Gas-gap glow excited by the electroionization method (a) and glow of autonomous-breakdown spark in the same gap (b).

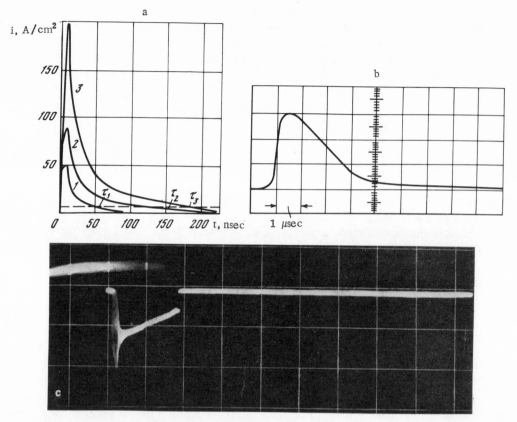

Fig. 17. Typical oscillograms of discharge-current pulses produced by electro-ionization excitation of compressed gas. a) $T_{beam} \simeq 10^{-8}$ sec, P = 5 atm, E/P = 10 V-cm^{-1}-Torr^{-1}: 1) CO_2; 2) CO_2; $N_2 = 1:1$; 3) N_2; b) $T_{beam} \simeq 10^{-6}$ sec; c) $T_{beam} \simeq 4 \cdot 10^{-5}$ sec.

consideration is much lower than in the breakdown channel. It should be noted that the breakdown of an ionized gas develops already in fields E/P < $(E/P)_{st.\,br.}$, and the time delay between the "volume" current pulse and the breakdown decreases rapidly with increasing initial field. For example, for the mixture $CO_2:N_2 = 1:1$ at a pressure P = 5 atm in chamber 1, the delay time decreases from ~$2 \cdot 10^{-3}$ to ~10^{-8} sec when E is increased from 0.9 $(E)_{st.br.}$ to 0.96 $(E)_{st.br.}$. The field region in which two alternating discharge forms exist — volume and spark — turns out to be larger the larger the working capacitance and the partial CO_2 pressure in the investigated $CO_2:N_2:He$ mixture.

We investigated also the flow of current initiated under "overvoltage" conditions. The working capacitance was connected in this case with the gas gap in a chamber similar in construction to chamber 1 through an additional discharge gap, which was triggered by the fast-electron beam itself ($\tau_{work} \simeq$ 1-2 nsec). This made it possible to increase the initial field in the gap between the electrodes to a value $E_0 \simeq 5E_{st.br.}$ at a pressure P = 1 atm. Under these conditions the discharge was not of the volume type. Breakdown developed already during the time of the ionization pulse, and several streamers were produced simultaneously and their number increased with increasing overvoltage. For example, one to two streamers are produced per square centimeter in nitrogen at $E_0/E_{st.br.} \simeq 2.5$.

b. Figure 17 shows typical oscillograms of discharge-current pulses obtained when the discharge is initiated by electron beams of various durations and intensities.

It must be noted that when the electron density in the discharge gap is decreased to a certain threshold value, the discharge current decreases jumpwise, in accordance with the

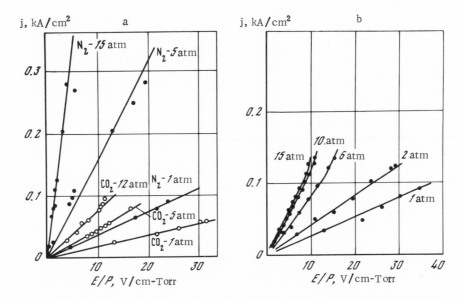

Fig. 18. Current−voltage characteristics of discharge excited by the electroionization method in CO_2 and N_2 (a) and in the $CO_2:N_2 = 1:1$ mixture (b) ($T_{exc} \simeq 10^{-7}$ sec) at various pressures.

conclusions of the theory considered above, practically to zero, i.e., the "strong current" volume discharge vanishes. This can be clearly seen when the discharge is initiated by a weak electron beam having a density that decreases with time. When the ionization intensity weakens to a certain limiting level, the density of the discharge current drops abruptly, even though the ionization of the gas by the electron beam still continues (Fig. 17c). The total discharge duration depends on the rate of decrease of the electron concentration after the ionization is stopped. In chamber 1, for example, the discharge in nitrogen stopped approximately after 10-20 μsec, while in CO_2 it stopped after 20-100 nsec. Estimates show that the electron concentration at which the discharge stops in the investigated $CO_2:N_2:$ He mixtures amounted under the experimental conditions to 10^{10}-$5 \cdot 10^{11}$ cm^{-3}.

Plots of the maximum amplitude of the discharge-current density against the relative field intensity are shown for different $CO_2:N_2$ mixtures in Fig. 18 (the measurements were performed in chamber 1 at an average current density in the fast-electron beam entering the discharge gap $j_{beam} \simeq 25$ A/cm^2 and at an ionization pulse duration $T_{beam} \simeq 10^{-8}$ sec). In all the mixtures the current increases practically linearly with increasing field, and only at fields close to the breakdown value ($E_0 \gtrsim 0.9$ $E_{st. br.}$) does the growth rate increase, the increase being faster the larger the CO_2 content in the mixture. The discharge-current density increases with increasing gas mixture pressure at a fixed value of E/P, and with increasing nitrogen content in the mixture at constant pressure (Fig. 19).

The function j^{ampl} (P, E/P) can be expressed by the following empirical formula:

$$j^{ampl}(P, E/P) \sim P^k (F/P)^m. \tag{42}$$

For carbon dioxide we have $k_{CO_2} = 0.5$, $m_{CO_2} = 1.1$; for nitrogen $k_{N_2} = 1.0$, $m_{N_2} = 0.9$ and when the composition of the $CO_2:N_2$ mixture is varied the parameters k and m vary smoothly between the corresponding values for the pure gases (for example, $k_{1:1} \simeq 0.7$; $m_{1:1} \simeq 1.05$).

The obtained dependence of the discharge current on the field intensity indicates that no cascade multiplication of the electrons takes place in the volume of the gas. The increase of

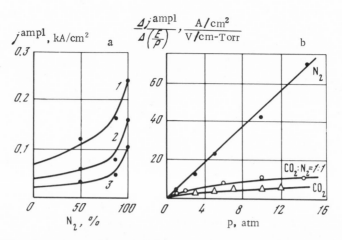

Fig. 19. Dependence of the maximum amplitude of the discharge current on the composition (a) for different pressures E/P at $P = 10$ atm and on the pressure (b) of the $CO_2:N_2$ mixture ($T_{exc} \simeq 10^{-7}$ sec). 1) 10; 2) 5; 3) 3 V/cm-Torr.

the current density with increasing field is due only to the corresponding increase of the electron drift velocity, and the effect of gas amplification begins to come into play only in the region preceding the breakdown.

From the general formulas (26) and (31) we easily obtain an expression for the maximum amplitude of the discharge-current density under the experimental conditions in chamber 1 ($C \simeq 10^4$ pF; $d = 0.3$ cm; $S = 4$ cm^2; $T_{beam} \simeq 10^{-8}$ sec; $J_{beam} \simeq 25$ A/cm^2; $E_e \simeq 600$ keV). For example, in the case of pure nitrogen we have

$$j^{ampl}_{(for\ N_2)} = en_p V_e \frac{\tanh (n_{eq}bT_{beam})}{[\cosh (n_{eq}bT_{beam})]^{\frac{e\mu_e S}{bcd}}} \simeq 6.3 \cdot 10^{-3} P \left(\frac{E_0}{P}\right)^{0,9}. \tag{43}$$

Here P is in Torr and E/P is in units of V/cm-Torr. For the $CO_2:N_2$ mixture with carbon dioxide content $\delta = P_{CO_2}/P_{tot}$ at $\delta \gtrsim 0.2$, when $N_0 \gg bn_e$, we have

$$\frac{j^{ampl}_{(\delta)}}{j^{ampl}_{(for\ CO_2)}} \simeq 1 + \frac{\rho(\delta)}{\rho_{CO_2}} \frac{w_{CO_2}}{w(\delta)} \frac{1-\delta}{\delta} \tag{44}$$

(ρ is the density of the ionized gas and w is the energy for the production of an electron−ion pair). Relations (43) and (44) are in good agreement with the experimental data (Fig. 18 and 19).

A comparison of the current oscillograms with the calculated functions $j^{ampl}(P, E/P)$ − Eqs. (26) and (31) − enables us to determine the electron−ion recombination coefficient. Under the experimental conditions, for $CO_2:N_2$: He laser mixtures in the field range $E/P = 1-5$ V/cm-Torr and at $b \simeq 10^{-7}$ cm^3/sec we have for pure nitrogen $b \simeq 3 \cdot 10^{-7}$ cm^3/sec, the recombination coefficient being practically independent of the pressure at $P = 1-10$ atm.

c. The energy input to the gas in the case of the considered nonautonomous discharge consists of the energy Q_e drawn from the fast beam electrons and the energy Q_C coming from the discharging working capacitance. Estimates show that for all the ionization intensities employed in the experiments we have $Q_e \lesssim 4 \cdot 10^{-4}$ J/cm^3-atm, and a value $Q_C \gtrsim 10^{-2}$ J/cm^3-atm is reached already in fields $E/P \gtrsim 2$ V/cm-Torr, so that the total energy input is $Q = Q_e + Q_C \simeq Q_C$.

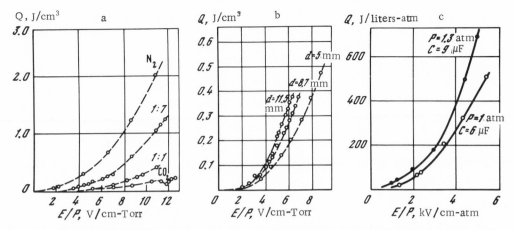

Fig. 20. Dependence of the energy input on the relative field intensity in the discharge gap at various durations and intensities of the gas ionization. a) Chamber 1, P = 10 atm; j_{beam} = 30 A/cm^2, T_{beam} = 10^{-8} sec, d $\simeq$ 0.3 cm, C = 10,000 pF; b) chamber 3, mixture CO_2:N_2 = 1:2, j_{beam} = 5 A/cm^2, P = 3 atm, T_{beam} $\approx$ 10^{-8} sec, C = 60,000 pF; c) chamber 4, mixture CO_2:N_2:He = 1:2:4, j_{beam} $\approx$ 0.1 A/cm^2, T_{beam} $\approx$ 3 $\cdot$ 10^{-5} sec.

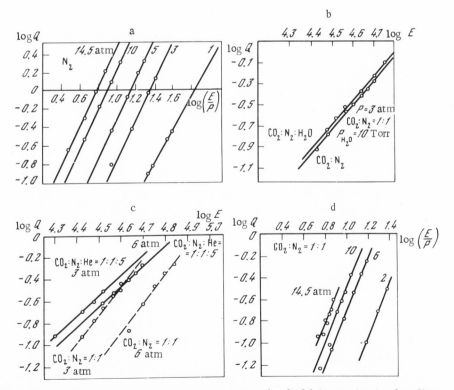

Fig. 21. Plots of the energy input against the field intensity in the discharge gap of chamber 1 for CO_2:N_2:He mixtures with varying compositions.

The energy input per unit increases rapidly with increasing initial field in the discharge gap: $Q \sim (E/P)^r$, the exponent r being determined only by the composition of the gas mixture and varying in the range $2 \lesssim r \lesssim 3$ (Figs. 20 and 21):

Composition of mixture $CO_2:N_2:He$	1:0:0	1:1:0	1:2:0	1:3:0	1:4:0	0:1:0	1:1:5
r	2.5	2.5	3.0	2.65	2.5	2	2

If we trace the dependence of the energy input on the composition of the $CO_2:N_2$ mixture, then it turns out that the maximum energy, at a fixed pressure and at an initial field E_0, can be delivered to pure nitrogen. With increasing partial CO_2 pressure (at a constant total pressure) the energy input decreases rapidly until the proportion of the $CO_2:N_2$ mixture reaches approximately 1:1. The decrease of the per-unit energy input then stops practically (Fig. 22a). At a constant value of the relative field intensity E/P, the per-unit energy input increases with increasing pressure in the discharge gap (Fig. 22b). For all the $CO_2:N_2$ mixtures, starting with $E/P \simeq 10$ V/cm-Torr, the growth of the per-unit energy input with increasing relative field intensity ceases to be monotonic. At pressures $P \gtrsim 5$ atm, descending sections appear on the $Q(E/P)$ curves. This effect is more noticeable the larger the partial pressure of the CO_2 in the gas mixture (Fig. 23). The decrease of the energy input is due to the decrease of the current on the descending part of the pulse, since there are no analogous descending parts on the plots of $j^{ampl}(E/P)$. It is most probable that this phenomenon is due to the rapid decrease of the concentration of the free electrons after the termination of the ionization pulse, owing to their sticking to the CO_2 molecules [45].

The known fact that the growth of the quantity E flattens out with increasing pressure [47] results in the existence of an optimal pressure for the per-unit energy input for each mixture. Under the experimental conditions, the maximum energy that can be delivered to a nonautonomous discharge without violating its homogeneity is 2-3 J/cm³ for $CO_2:N_2$ mixtures and 0.6-1 J/cm³ for $CO_2:N_2:He$ mixtures.

The measured dependences of the energy input on the composition and on the pressure of the gas mixture, and also on the field in the discharge gap, agree well with the calculated values (Fig. 24). The calculations were carried out with the aid of the equations derived in Sec. 3.

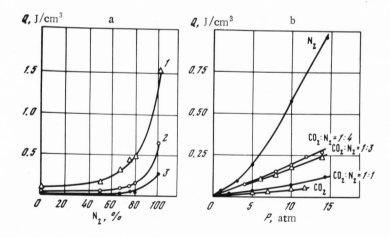

Fig. 22. Dependence of the per-unit energy input on the composition (a) and on the pressure (b) of the $CO_2:N_2$ mixture (the conditions of chamber 1). a) P = 10 atm, d = 0.3 cm. 1) E/P = 26 V/cm-Torr; 2) E/P = 18 V/cm-Torr; 3) E/P = 12 V/cm-Torr; b) E/P = 6 V/cm-Torr.

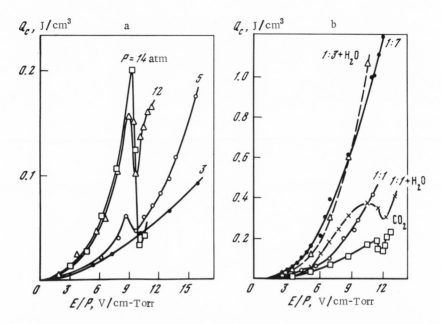

Fig. 23. Dependence of the per-unit energy input on the relative field intensity for pure CO_2 (a) and for $CO_2:N_2$ mixtures (b) (conditions of chamber 1). a) For different values of P; b) d = 0.3 cm; $P_{H_2O} \approx 10$ Torr, P = 10 atm.

The energy input depends significantly on the intensity and duration of the ionization, on the geometry of the discharge gap, and on the value of the working capacitance. An increase of the energy of the gas-ionizing electrons leads to a rapid growth of Q, but this growth slows down already at electron energies $U_{e0} \simeq 150$ keV (Fig. 25a).

An increase in the current density of the ionizing-electron beam (due to using more effective cathodes or to using a foil that is more transparent to the fast electrons) also leads to a noticeable increase of the energy input (Fig. 25b). The Q(j_{beam}, U_e) dependence can be easily explained on the basis of the known relations for the energy loss in the foil ΔU_f and to the

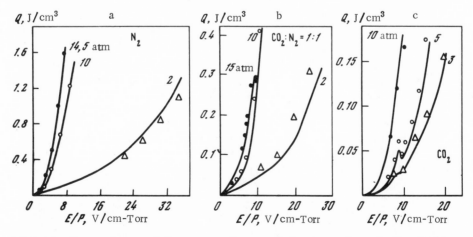

Fig. 24. Dependence of the per-unit energy input on the relative field intensity in a discharge gap for pure N_2 (a) and CO_2 (c) and for the $CO_2:N_2$ mixture (b) under conditions of chamber 1 at various pressures. Points, experimental value; curves, calculated data.

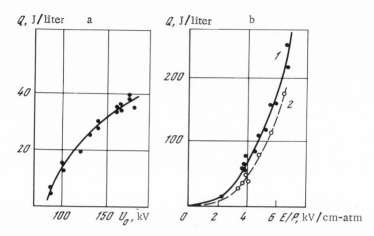

Fig. 25. Dependence of the per–unit energy input on the energy of the fast electrons ionizing the gas at a constant value of the relative field intensity $E/P \simeq 2$ kV-cm^{-1}-atm^{-1} (a) and on the relative field intensity at different current densities in the fast-electron beam (b). Ionization-pulse duration $\sim 10^{-6}$ sec. a) $CO_2:N_2:He = 1:1:4$, $P = 1$ atm, $C = 3$ μF, $V = 10$ liters; b) $CO_2:N_2 = 1:1$, $P = 1$ atm, $V = 2.1$ liters: 1) $j_{beam} = 0.4$ A/cm^2, 2) $j_{beam} = 0.3$ A/cm^2.

gas ionization ΔU_g, and also the dependence of the foil transparency T_f to fast electrons on the electron energy (see, e.g., [47]), which can be approximated in the energy region of interest to us in the following manner:

$$\Delta U_f = U_{e0} \exp\left[-k_f \frac{U_{e0} - U_f}{U_f}\right],$$

$$\Delta U_g = (U_e - \Delta U_f) \exp\left[-k_g \frac{U_{e0} - \Delta U_f - U_g}{U_g}\right], \tag{45}$$

$$T_f = \frac{i_{beam}}{i_{beam}^0} \simeq 1 - \exp\left[-k_f' \frac{U_{e0} - U_f}{U_f}\right]. \tag{46}$$

Here U_{e0}, j_{beam}^0, U_e and j_{beam} are the energy of the electrons and density of the current in the beam in front and behind the foil; U_f and U_g are the minimal energies necessary to pass through the foil and the discharge gap, respectively; k_f, k_f', and k_g are coefficients that depend on the foil material.

If the variation of the field intensity in the discharge gap during the discharge time τ_{dis} can be neglected, then the per–unit energy input is $Q \simeq j_{dis} E_{dis} \tau_{dis}$; $j_{dis} \sim n_e \sim \sqrt{q}$, and for the pair-generation rate we have (see Sec. 3.1):

$$q \simeq \frac{i_{beam}}{e} \frac{U_e}{R} \frac{1}{w} \sim j_{beam}^0 T_f \frac{U_{e0} - \Delta U_f}{R(U_e)}.$$

Thus, the per–unit energy input increases monotonically with increasing current density in the fast-electron beam. The fast growth of Q with increasing U_{e0} at low energies ($U_{e0} \simeq U_f + U_g$), due mainly to the increase of the transparency of the foil, reaches saturations already at $U_{e0} \simeq (2-3)(U_f + U_g)$

$$Q \sim \sqrt{q} \sim \sqrt{ j_{beam}^0 \frac{1 - \exp\left(-k_f' \frac{U_{e0} - U_f}{U_f}\right)}{U_{e0}\left[1 - \exp\left(-k_f \frac{U_{e0} - U_f}{U_0}\right)\right]} } . \tag{47}$$

Recognizing that for an electron gun with a point cathode we have $j_{beam}^0 \sim U_{e0}^{3/2}$ [51, 52], we obtain good qualitative agreement between the calculated $Q(U_{e0})$ dependence and the experimental dependence (Fig. 25a).

Under the experimental conditions at a constant value of E/P, the per-unit energy input increases with increasing working capacitance C and (at small C) with increasing height of the gap d (Fig. 26). Indeed, an increase of d at a fixed field E_0 leads to an increase of the charging voltage across the capacitance $U_C = E_0 D$ and accordingly to an increase of the total energy input $Q_{tot} \sim U_0^2 \sim d^2$ and of the per-unit energy input $Q = Q_{tot}/V \sim d$, as is indeed observed in experiment (Fig. 26a). With increasing capacitance, the energy input first increases (inasmuch as at small values of C the capacitors become charged practically fully during the discharge time and $Q \sim C$), after which the growth of Q slows down because the time constant RC of the discharge becomes too large (Fig. 26b). The observed Q(d, C) dependences agree fully with the calculated ones [see expressions (28), (33), (35)]. Experiment confirms also the conclusions drawn in the analysis of the dependence of the energy input on the ionization intensity. For example, an increase of the electron-beam current density from ~5 to ~30 A/cm² at a fixed fast-electron energy 600 keV and at an ionization pulse duration $T_{beam} = 10^{-8}$ sec (changeover from chamber 3 to chamber 1) leads to an increase of the per-unit energy input by a factor 2-3 (see Fig. 20), inasmuch as under the considered conditions we have for the CO_2:N_2 mixtures $n_{eq} b T_{beam} \gtrsim 1$ and $Q \sim n_{eq} b T_{beam} \sim (j_{beam})^{1/2}$.

A comparison of the results of the calculations with experiment has made it possible to determine the rate of sticking of the electrons to the CO_2 molecules in the absence of external ionization. It turns out that the sticking coefficient after the end of the ionization $(c_2) = \langle \sigma_2 v \rangle$ is much smaller than during the time of the fast-electron beam pulse (the value of a_1 was calculated from the experimental values of the maximum amplitudes of the current

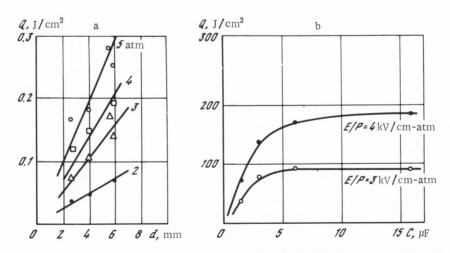

Fig. 26. Per-unit energy input vs. height of the discharge gap (a) and the working capacitance (b). a) Conditions of chamber 1, mixture CO_2:N_2 = 1:1.5, E/P = 10 V/cm-Torr; b) conditions of chamber 4, mixture CO_2:N_2:He = 1.5:1:4, V = 10 liters, $T_{beam} \simeq 10^{-6}$ sec.

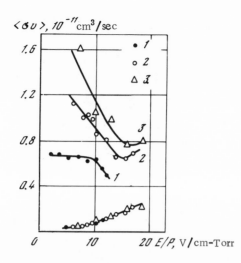

Fig. 27. Dependence of the coefficient of sticking of electrons to CO_2 molecules on the relative field intensity in the discharge gap in the presence (the three upper curves) and in the absence (lower curve) of external ionization. 1) $P = 10$ atm; 2) 5 atm; 3) 3 atm.

density) (Fig. 27). It is possible that the observed decrease of the coefficient c_1 with increasing pressure and field in the discharge gap is due to the fact that it was calculated without allowance for recombination, and that at high concentrations of the free electrons and positive ions, which are present in the discharge gap during the time of the ionization pulse, it is necessary to take into account, generally speaking, also the contribution of the recombination processes to the rate of decrease of n_e. (The sticking coefficient measured in the absence of ionization practically coincides with the published value [45].) We note that the descending sections on the $Q(E/P)$ curves appear at values of E/P ($\simeq 10$ V/cm-Torr) such that the fraction of the electrons with $\varepsilon \gtrsim 4$ eV, together with the rate of sticking, increases abruptly [45, 53].

Results similar to those considered above were obtained also in an investigation of the characteristics of a discharge excited by the electroionization method in other molecular gases and their mixtures (Fig. 28).

2. Thus, our experiments have shown that a nonautonomous volume discharge with an ohmic current-voltage characteristic develops in a compressed gas excited by the electroionization method if the ionization intensity is high enough. The qualitative and quantitative dependences of the energy parameters of this discharge on the characteristics of the ionizer

Fig. 28. Dependence of the energy input on the relative field intensity in the discharge gap for different gases and their mixtures. Conditions of chamber 4, $P = 1$ atm, $V = 10$ liter, $T_{beam} \approx 10^{-6}$ sec, $j_{beam} \approx 0.5$ A/cm^2.

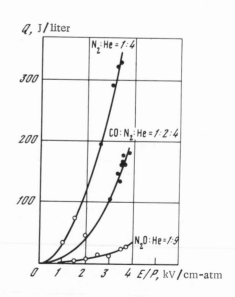

and of the supply system, as well as the structure of the discharge, remain constant when the volume of the discharge gap is varied in a wide range (the range $1-10^4$ cm^3 was investigated).

Of great importance from the point of view of using such a discharge to pump gas lasers is the question of the maximum energy-input values for the investigated discharge. The value of Q, as shown above, at a given composition of the working gas mixture and at a given pressure, is determined by the ionization intensity Q, by the duration T_{beam} of the ionization pulse, and by the field intensity E in the discharge gap.

According to estimates made with the aid of the equations in Sec. 3.1, if the discharge circuit parameters are properly chosen, the per-unit energy contribution for the CO_2:N_2:He mixtures, at reasonable ionization intensities, can reach in fields $E < E_{st.br.}$ values of the order of several J-cm^{-3}-atm^{-1}, and when a gas with high dielectric strength, such as nitrogen, is excited, the ionization intensity can exceed 10 J-cm^{-3}-atm^{-1}. These values, however, are not realized in experiments because of the development of transverse instabilities and because the space charge becomes localized and turns into a streamer breakdown. In the case of a relatively high degree of gas ionization and low discharge duration (the conditions of chamber 1-3), when the per-unit energy input under electroionization excitation of the mixtures CO_2:N_2:He amounts to $\lesssim 0.3-0.5$ J-cm^{-3}-atm^{-1}, the streamer breakdown develops with increasing field intensity in the discharge gap up to prebreakdown values $E/P \gtrsim (0.8-0.9)(E/P)_{st.br.}$. The delay of the breakdown relative to the start of the ionization varies in this case in a very wide range — from several milliseconds to several nanoseconds — with an insignificant variation of the field, something impossible in the pure streamer or pure Townsend breakdown mechanism [54]. Most readily the breakdown of the ionized gas takes place in "weak" fields ($\lesssim 0.9(E/P)_{st.br.}$ as a result of a strong distortion of the field by the space charge after the electron concentration decreases to a value $n_e \lesssim n_{cr}$, and streamer breakdown takes place in strong fields $E \gtrsim 0.95(E)_{st.br.}$; the streamer development is facilitated in this case by the presence of a high concentration of electrons and ions ($n_{e,i} \simeq 10^{12}-10^{14}$ cm^{-3}) [54].

In the case of relatively weak ionization of the gas or a long duration of the discharge, the transverse instability can develop also in fields much weaker than the static-breakdown field.

One of the main mechanisms of development of such an instability was analyzed theoretically by Suchkov and Belenov [9, 21, 44]. It consists qualitatively in the following: Assume that at the start of the discharge the gas temperature in a certain region of the interelectrode space exceeds the temperature in the neighboring regions. At one and the same gas pressure, a temperature rise in the entire discharge gap means a proportional decrease of the particle density and an increase in the electron drift velocity. Further, since the rate of production of the free electrons in the gas depends only on the total number of particles N and does not depend on the volume, while the rate of their annihilation by recombination decreases with increasing volume, higher values will obtain in the region with the lower density not only for the electron burst velocity but also for the degree of ionization n_e/N. Accordingly, in the region of increased temperature, the power released per particle $W \sim n_e(v_e/N)$ will also be higher. Thus, the fluctuations of the temperature during the course of the discharge will increase, and the particle density in this region will decrease, while the average particle density in the discharge will remain constant. When E/N reaches the breakdown value in this region, the volume discharge changes into a streamer discharge. The calculations of the dynamics of this instability, carried out in [21, 44], with allowance for the fluctuations of the voltage drop near the cathode, have shown that in fields with $E/P \simeq (0.5-0.8)(E/P)_{st.br.}$ the breakdown should develop at a per-unit energy input of the order of 0.8-1 J-cm^{-3}-atm^{-1}.

3. The nonautonomous volume discharge considered in the present chapter has all the properties needed to serve as a pump for high-pressure gas lasers, namely, high homo-

geneity when large gas volumes are excited, the possibility of attaining a large energy input, the possibility of "tuning" to the excitation of individual vibrational levels of the molecules by smooth variation of the average energy of the free electrons, and finally the possibility of independently varying the concentration and the energy of the free electrons and consequently stability at high pressures and volumes of the gas.

A comparison of the experimental data with the calculated ones shows that the expressions obtained in the analysis of the discharge circuit make it possible to calculate with sufficient accuracy, first all the parameters of the discharge excited by the electroionization method, and second, the characteristics of the supply system needed to reach the required values of the discharge current, the per-unit energy input, etc.

CHAPTER II

ELECTROIONIZATION CO_2 HIGH-PRESSURE LASER

Let us examine the well-known working-level scheme of carbon dioxide lasers (Fig. 29). This scheme was first proposed in the U.S.A. by Patel in 1964 [55]. The mechanisms whereby the CO_2 laser levels are excited was investigated by many workers, and a detailed discussion of this mechanism is contained in a number of reviews [55-59] and original articles (see, e.g., [26]).

The nitrogen molecules go over under the influence of the inelastic impact by the electrons from the lower level V = 0 to the upper vibrational levels. The vibrationally excited nitrogen molecules, which have a zero dipole moment, have a very long lifetime and the only mechanism of drawing their vibrational energy is in fact collisions with the molecules of the carbon dioxide. The levels 00^01 of CO_2 and V = 1 of nitrogen practically coincide, so that when an unexcited CO_2 molecule collides with an excited nitrogen molecule the excitation is transferred, with high probability, to the CO_2 molecule. The radiative transitions from the 00^01 level to the 10^00 and 02^00 levels produce the laser radiation. The levels 10^00 and 02^00 then decay as a result of collisions with the CO_2 molecules in the ground state, and the level 01^10 is produced. The transition of the CO_2 molecule from the level 01^10 to the ground state is the result of collisions of the CO_2 molecules at this level with the molecules of nitrogen, carbon dioxide, or helium, if the latter is present in the mixture. The rate of decay of the lower laser levels, in the case of correctly chosen mixtures, turns out to be higher than the rate of

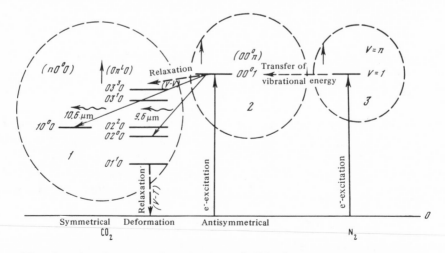

Fig. 29. Mechanism of excitation of CO_2-laser working levels.

excitation of the upper laser level even at a very high rate of excitation [55-59]. Inasmuch as the processes of excitation of the upper laser level and the decay of the lower laser levels are collisional, in the case of a proportional change of the number of particles with increasing pressure the ratio of the rates of excitation of the upper level and of the decay of the lower level turns out to be independent of the pressure, if the corresponding collision cross sections do not depend on the pressure.

Let us examine the dependence of the radiation power on the gas pressure. The rate of excitation of the nitrogen molecules dN/dt is proportional to the electron density n_e, to the molecule density N, and to the cross section σ for the excitation of the vibrational levels of the nitrogen molecule averaged over the electron velocities v_e. In the quasi-stationary regime, the rate of excitation of the upper laser level coincides with the rate of excitation of the nitrogen molecules, since, as already mentioned, the only channel whereby the vibrational energy can be drawn from the nitrogen molecules are the collisions with the CO_2 molecules. Generally speaking, a noticeable contribution to the rate of excitation of the upper laser level is made by direct excitation of the CO_2 molecules by electron impact, but allowance for this contribution does not lead to a change of the qualitative dependence of the radiation power on the pressure. In the case of radiative decay of the upper laser level, the radiation power per unit volume W is equal to the product of the radiation quantum energy $\hbar\omega$ by the rate of excitation of the upper laser level dN/dt:

$$dN/dt = n_e N < \sigma v_e >, \tag{48}$$
$$W = \hbar\omega n_e N < \sigma v_e >. \tag{49}$$

If the ratio of the electron and molecule densities does not depend on the gas pressure,

$$n_e/N = \text{const}, \tag{50}$$

the radiation power is proportional to the square of the pressure:

$$W \sim P^2. \tag{51}$$

These estimates are valid for the case of pulsed lasing of a radiation-pulse energy not exceeding the maximum attainable for carbon dioxide, $\lesssim 0.3$ J-cm^{-3}-atm^{-1}. This limiting value is determined by the maximum gas temperature at which inversion between the laser levels is possible at all. For a cw laser, the foregoing estimates are valid if the gas flow is fast enough and ensures a low working temperature of the gas in the active region.

The results of the experiment are in good agreement with this estimate. Figure 30 shows the calculated dependence of the radiation power of 1 cm^3 of active medium of a CO_2 laser on the pressure. The lower two points, 0.1 and 10 W/cm^3, correspond to the values obtained for

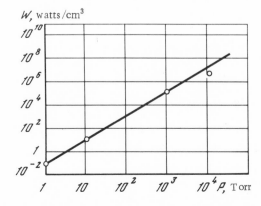

Fig. 30. Pressure dependence of the radiation power per unit volume of the active medium of a CO_2 laser.

ordinary gas-discharge lasers of low pressure [56], while the upper points are the experimental data obtained for an electroionization CO_2 laser [9].

1. Kinetics of Population of Working Levels;
Gain of Active Medium of Electroionization CO_2 Laser

The processes of excitation and relaxation of the vibrational levels in molecular CO_2 laser at low pressures have been investigated in sufficient detail. Therefore, using the results of the known experimental and theoretical studies, we can point out the most probable mechanism whereby energy is transformed in a gas consisting of CO_2 and N_2 molecules. The energy transfer mechanism is illustrated in Fig. 29.

The electrons excite directly the vibrational N_2 and CO_2 levels. This is followed by resonant exchange of energy between the vibrational levels of the N_2 molecules and the levels of antisymmetrical type of oscillations of the CO_2 molecules $(00^0 n)$ followed by the transformation, as a result of molecular collisions, from the antisymmetrical oscillation modes into the deformation and symmetrical modes. With further relaxation, the deformation mode goes over, by collisions with the molecules CO_2, N_2, He, H_2O, into the energy of translational motion. The direct transition of the vibrational energy into translational energy from the laser levels, as well as the radiative processes, can be neglected. For each normal mode of the CO_2 or N_2 molecule we can introduce a vibrational temperature. This is made possible by the fact that the processes of exchange of vibrational quanta within a given type of oscillations takes place within times that are much shorter than the time of the transition of the energy into the translational motion of the molecules and its exchange between different types of oscillations. These vibrational temperatures will differ in general from one another and from the gas temperature. The amount of energy per unit volume for each mode of the CO_2 or N_2 molecule is given by the equations [60]

$$E_i = h\nu_i \frac{x_i}{(1-x_i)^2} N_{CO_2}, \qquad E_2 = h\nu_2 \frac{2x_2}{(1-x_2)^2} N_{CO_2},$$

$$E_{N_2} = h\nu_{N_2} \frac{x_{N_2}}{(1-x_{N_2})^2} N_2, \qquad i = 1, 2, 3,$$

where $h\nu$ is the energy of the vibrational quantum; N_{CO_2} and N_{N_2} are the numbers of the CO_2 and N_2 molecules, respectively; $x_i = \exp(-h\nu_i/kT_i)$ (t_i is the vibrational temperature); the subscripts i = 1, 2, 3 pertain respectively to the antisymmetrical, deformation, and symmetrical modes of the CO_2 molecules. By virtue of the resonant interaction the vibrational temperatures of the symmetrical and deformation modes are close to each other: $T_2 \simeq T_3$. Therefore, inasmuch as $h\nu_3 = 2h\nu_2$ (Fig. 29), we have

$$x_3 = x_2^2.$$

Thus, the vibrational levels of the CO_2 and N_2 molecules can be broken up into four groups (Fig. 29): the ground-state levels of the CO_2 and N_2 molecules; levels of the antisymmetrical modes of the CO_2 molecules (group 2); the levels of the symmetrical and deformational modes (group 1); the vibrational levels of the N_2 molecules (group 3). At low vibrational temperatures ($kT_i < h\nu_i$, i = 1, 2, 3, 4), i.e., $x_i \ll 1$, it is the lowest vibrational levels that are mainly populated in each group.

In the case of the electroionization method of exciting the mixtures CO_2:N_2:He [E/P < (E/P)$_{st.br}$], almost 100 percent of the electron energy goes to the excitation of the vibrational levels of the N_2 molecules (group 3) and the levels of the antisymmetrical mode of the CO_2 molecules (group 2), while the population of the first group of levels by electron impact can be neglected.

We shall also assume from now on that increasing the pressure of the laser mixture to several dozen atmospheres does not lead to qualitative changes of the elementary excitation processes and relaxation of the working levels. This assumption is in fact not obvious, inasmuch as at high gas pressures there can appear new types of quenching processes, particularly those due to triple collisions, and was never investigated before the experimental results described below were obtained.

According to the mechanism described above for the excitation and relaxation, the kinetic equations for the populations of the upper laser level and the first excited vibrational levels of the N_2 molecules ($V = 1$) in the absence of induced transitions take the form (in the approximation $N_1 \ll N_{CO_2}$, $N_{N_2}^* \ll N_{N_2}$):

$$dN_1/dt = \gamma_1 W_p /h\nu + \alpha_{10} N_{CO_2} N_{N_2}^* - (\alpha_{01} N_{N_2} + 1/\tau_{CO_2}) N_1,$$
$$dN_{N_2}^*/dt = \gamma_2 W_p /h\nu + \alpha_{01} N_{N_2} N_1 - (\alpha_{10} N_{CO_2} + 1/\tau_{N_2}) N_{N_2}^*,$$
$$N = N_{CO_2} + N_{N_2}; \qquad N_{N_2}/N_{CO_2} = \delta,$$

(52)

where N_{CO_2}, N_{N_2} are the numbers of CO_2 and N_2 molecules in the mixture; $N_{N_2}^*$ is the number of excited molecules $N_2 (V = 1)$; $h\nu$ is the energy of the first vibrational level of N_2; N_1 is the number of CO_2 molecules on the upper level; W_p is the pump power; $\gamma_1 W_p$ and $\gamma_2 W_p$ are the fractions of the pump power required to excite by electron impact the level $CO_2 (00^0 1)$ and the vibrational levels of the nitrogen molecules; $\alpha_{10} \simeq \alpha_{01}$ is the rate constant in inelastic collisions between the molecules CO_2 and $N_2 (V = 1)$; τ_{CO_2}, τ_{N_2} are the relaxation times of the levels $(00^0 1) CO_2$ and $(V = 1) N_2$. The cross sections for the molecule−molecule collisions that lead to conversion of the vibrational quanta into translational energy is very small, 10^{-23} cm^2, so that the quantity $1/\tau_{N_2}$ in (53) can be neglected. The solution of the system (52) at a pump pulse duration $T_p < 1/\alpha_{01} N_{CO_2}$; $1/\alpha_{01} N_{N_2}$ is given by

$$\Delta N \simeq N_1 = \frac{\alpha N (\gamma_1 + \gamma_2) Q_p}{2\psi (1 + \delta) h\nu} (e^{\psi t} - e^{-\psi t}) e^{-\frac{\varphi}{2} t},$$

(53)

where $Q_p = W_p T_p$ is the per−unit pump energy:

$$\psi = \left[\frac{\varphi}{4} - \frac{\alpha_{01} N}{(1 + \delta) \tau_{CO_2}} \right]^{1/2}; \qquad \varphi = \alpha_{01} N + 1/\tau_{CO_2}.$$

Equation (52) in the approximation $1/[\tau(1 + \delta)] < \alpha N$, which is valid for the customarily employed laser mixtures, determines the maximum inversion (from the condition $dN_1/dt = 0$):

$$\Delta N_{max} \simeq N_1 \simeq \frac{Q_p (\gamma_1 + \gamma_2)}{(1 + \delta) h\nu} \left[1 - \frac{t^*}{(1 + \delta) \tau_{CO_2}} \right],$$

(54)

where

$$t^* = \frac{1}{2\varphi} \ln \frac{\varphi + 2\psi}{\varphi - 2\psi}.$$

In the regime when long pump pulses are used ($T_p > 1/\alpha_{01} N_{CO_2}$; $1/\alpha_{01} N_{N_2}$) the solution of (52) takes the form ($W_p = const$ at $0 \le t \le T_p$):

$$N_1 = \begin{cases} \dfrac{W_p (\gamma_1 + \gamma_2)}{h\nu} \tau_{CO_2} \left[1 - \exp \left(-\dfrac{t}{\tau_{CO_2}} \dfrac{1}{1+\delta} \right) \right], & t \le T_p, \\[3mm] \dfrac{W_p (\gamma_1 + \gamma_2)}{h\nu} \tau_{CO_2} \left[\exp \left(-\dfrac{t - T_p}{\tau_{CO_2}} \dfrac{1}{1+\delta} \right) - \exp \left(-\dfrac{t}{\tau_{CO_2}} \dfrac{1}{1+\delta} \right) \right], & t \ge T_p. \end{cases}$$

(55)

At low pump energies, when the thermal population of the lower laser level can be neglected, Eqs. (53) and (55) determine the time dependence of the population inversion, while the maximum inversion in the long pump pulse regime is reached at the end of the excitation pulse

$$\Delta N_{max} \simeq N_1 = \frac{Q_p(\gamma_1 + \gamma_2)}{h\nu} \frac{\tau_{CO_2}}{T_p} \left[1 - \exp\left(-\frac{T_p}{\tau_{CO_2}} \frac{1}{1+\delta} \right) \right]. \tag{56}$$

The half-width of the line of the vibrational−rotational transition in a mixtures of carbon dioxide with nitrogen and helium is determined by the formula [61]

$$\Delta\nu_J = \gamma_J (P_{CO_2} + \beta P_{N_2} + \alpha P_{He}), \tag{57}$$

where γ_J = 0.08−0.12 cm^{-1}-atm^{-1} is the coefficient of impact broadening in pure CO_2; J is the rotational quantum number of the level $00^0 1$; P_{CO_2}, P_{N_2}, and P_{He} are the partial pressures of CO_2, N_2, and He; β and α are coefficients that take into account the efficiency of the broadening of the rotational lines of carbon dioxide molecules in collisions with nitrogen molecules and helium atoms; $\beta \simeq 0.8$−0.9; $\alpha \simeq 0.5$−0.6.

It follows from (57) that at a working-mixture pressure of several atmospheres the value of $\Delta\nu_J$ becomes comparable with the distance between the rotational lines ($\simeq 2$ cm^{-1}) and the contribution of the neighboring lines to the gain of the line at the frequency ν becomes appreciable. Therefore at high pressures the expression for the gain [57] is supplemented by terms that take into account the contributions of the neighboring lines:

$$K(\nu) = \frac{c^2}{8\pi^2\nu^2} \frac{hcB_1}{kT} \sum_{J'} \frac{(2J'+1)\Delta\nu_{J'} A_{12}^{J'} \chi(\nu - \nu_{J'})}{(\Delta\nu_{J'})^2 + (\nu - \nu_{J'})^2} \times$$
$$\times \left\{ N_1 - N_2 \frac{B_2}{B_1} \exp\frac{hc}{kT} [F(J' \pm 1) - F(J')] \right\} \exp\left[-F(J')\frac{hc}{kT} \right], \tag{58}$$

where J' is the rotational quantum number of the upper laser level $10^0 0$; $F(J') = J'(J+1)$; $\Delta\nu_{J'}$ is the half-width of the vibrational−rotational lines; B_1 and B_2 are the rotational constants for the levels $00^0 1$ and $10^0 0$, respectively; and $A_{12}^{J'}$ is the probability of the transition $J' \rightarrow J'' \pm 1$ ($A_{12} = 0.21 \pm 0.03$ sec^{-1} for $J'' = 18$−22 [57], which is the rotational quantum number of the lower laser level).

The parameter $\varkappa(\nu - \nu_{J'})$ was introduced in Eq. (58) because the line shape of the rotational−vibrational transition differs somewhat from Lorentzian. The equation for the line shape takes the form

$$f(\Delta\nu_{J'}, \nu - \nu_{J'}) = \frac{1}{\pi} \frac{\Delta\nu_{J'} \chi(\nu - \nu_{J'})}{(\Delta\nu_{J'})^2 + (\nu - \nu_{J'})^2},$$

where the parameter $\chi(\nu - \nu_{J'})$ is close to unity when $(\nu - \nu_{J'}) \lesssim \Delta\nu_{J'}$, and is somewhat smaller than unity for $(\nu - \nu_{J'}) > \Delta\nu_{J'}$.

Since $B_1 \simeq B_2$, it follows that

$$\frac{B_2}{B_1} \exp\frac{hc}{kT} [F(J' \pm 1) - F(J')] \simeq 1.$$

We can then write (58) in the form

$$K(\nu) = \frac{c^2}{8\pi^2\nu^2} \frac{hcB}{kT} (N_1 - N_2) \sum \frac{(2J'+1)\Delta\nu_{J'} A_{12}^{J'}}{(\Delta\nu_{J'})^2 + (\nu_{J'} - \nu)^2} \exp\left[-F(J')\frac{hc}{kT} \right]. \tag{59}$$

2. Threshold Characteristics, Output Energy, Power, and Efficiency of Laser; Divergence of the Radiation

The results of the investigations of the threshold characteristics of CO_2 lasers at high pressure are shown in Figs. 31–35. The threshold parameter E/P, i.e., the ratio of the field intensity at which the generation pulse appears to the total pressure of the mixture, depends on the composition of the mixture, on the distance between the electrodes, on the working capacitance, and on the parameters of the employed resonator. At all mixture compositions, the field threshold $(E/P)_{thr}$ decreases with increasing pressure. A pressure larger than 4 atmospheres, at any mixture composition, corresponds to a slow decrease of $(E/P)_{thr}$ with pressure. At lower pressures, the measured value of $(E/P)_{thr}$ increases strongly. The dependence of $(E/P)_{thr}$ on the partial pressure of the nitrogen at constant pressure has a minimum (Figs. 33 and 34). It is seen from Fig. 33 that at a nitrogen content larger than 70% the threshold field intensity increases at all mixtures pressures. Addition of water vapor to the CO_2:N_2 mixture increases the threshold at all employed laser-mixture pressures. Addition of helium lowers the lasing threshold at a fixed mixture pressure, but the spark-breakdown voltage is decreased at the same time.

Figure 35 shows the dependence of the threshold pump energy on the pressure for CO_2:N_2 mixtures of various compositions. The threshold energy Q_{thr} increases with increasing pressure for all compositions of the CO_2:N_2 mixture. When the pressure is raised from 1 to 3–5 atm the value of Q_{thr} decreases linearly with pressure. An increase of the percentage of nitrogen in the mixture increases the threshold pump energy.

The linear dependence of the inversion on the pump energy [see Eqs. (54) and (56)] explains the dependence of the threshold characteristics on the pressure, mixture composition, and resonator parameters.

The equation

$$K(J_0) = K_{thr} = 1/2L + \ln(R_1 R_2)^{-1} + K_0 \qquad (60)$$

(L is the length of the active region; R_1 and R_2 are the mirror reflection coefficients; K_0 are the internal losses; K_{thr} is the threshold gain; $K(J_0) = K_{max}$ (J) is the maximum value of the gain) determines the threshold characteristics of the laser.

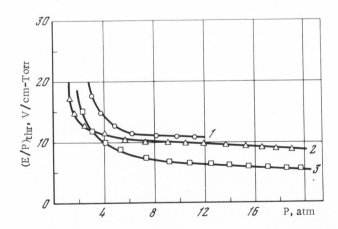

Fig. 31. Dependence of the threshold value of the parameter E/P on the pressure (L = 4 cm, chamber 1) for mixtures CO_2:N_2:He with various compositions. 1) 1:3:0; 2) 1:1:0; 3) 1:2:5.

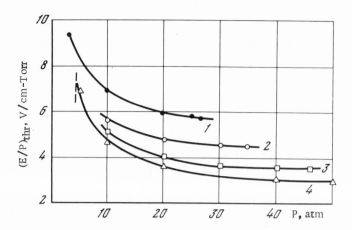

Fig. 32. Dependence of threshold value of the parameter E/P on the pressure for the mixtures CO_2:N_2:He (L = 10 cm, chamber 3, dielectric mirrors) of various compositions. 1) 1:2:4; 2) 1:2:8; 3) 1:2:16; 4) 1:2:32. The dashed line marks the breakdown threshold.

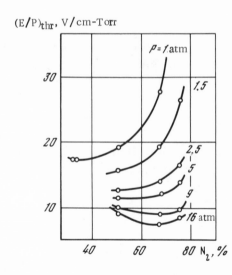

Fig. 33. Dependence of the threshold value of the parameter E/P on the nitrogen content in the mixture CO_2:N_2 at different pressures (L = 4 cm, chamber 1).

Fig. 34. Dependence of the threshold value of the parameter E/P on the nitrogen content in the mixture CO_2:N_2 (L = 12 cm, chamber 2), P = 2 atm.

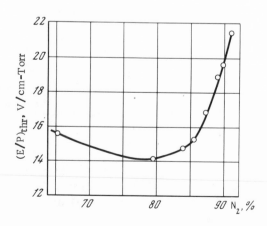

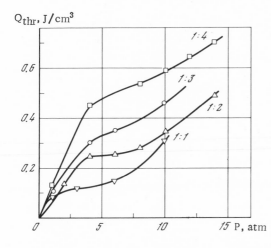

Fig. 35. Dependence of the specific threshold pump energy on the pressure (L = 4 cm, chamber 1, dielectric mirrors) for $CO_2:N_2$ mixtures of various compositions. $T_1 = 5\%$, $T_2 = 0\%$, $\lambda = 10.6$ μm.

At laser-mixture pressures lower than 3–5 atm, when the contribution to the gain of the vibrational–rotational line from the wings of the neighboring lines is negligible ($\Delta \nu_{J'} \ll \nu_{J'} - \nu_{J'-1}$), the gain is proportional to the ratio of the inversion of the populations to the line width of the vibrational–rotational transition [see Eqs. (54), (56), and (59)]:

$$K \sim \frac{\Delta N_{max}}{\Delta \nu} = \frac{Q_p (\gamma_1 + \gamma_2)}{P} \varkappa (\delta, \varphi), \tag{61}$$

where P is the pressure of the mixture (atm); φ is the partial pressure of He in the mixture; $\varkappa(\delta, \varphi)$ is a function that depends on the mixture composition. In this case Eq. (61) can be represented in the form

$$\varkappa' (\delta, \varphi) \frac{Q_{thr}}{P} = K_{thr}. \tag{62}$$

At pressures P > 3–5 atm (depending on the mixture composition), the contribution made to the gain by the wings of the neighboring lines becomes substantial, and the increase of the pump energy with increasing pressure is slower than proportional. Equation (62) describes well also the result of the measurements of the threshold parameter E/P. Indeed, the pump energy can be well approximated [44] by the empirical formula

$$Q_p = M(\delta, \varphi)P^{1/2}(E/P)^{K(\delta, \varphi)}, \tag{63}$$

where $M(\delta, \varphi)$ and $K(\delta, \varphi)$ are functions that depend only on the composition of the mixture [$K(\delta, \varphi) \simeq 2$]. Substituting (63) in (62) we obtain

$$(E/P)_{thr} = F (\delta, \varphi)/P^{1/4}.$$

At large pressures (P > 8–10 atm) the individual vibrational–rotational transitions overlap and the gain line width becomes independent of pressure. In this case $\Delta N_{thr} = const(P)$, $Q_{thr} = const/(\gamma_1 + \gamma_2)$, while $(E/P)_{thr}$ continues to decrease with increasing pressure. An appreciable decrease of E/P, however, causes the quantity $(\gamma_1 + \gamma_2)$ to decrease and an important role is assumed (see Fig. 15b) by the electron-impact population of the lower laser level. Consequently Q_{thr} again begins to increase with increasing pressure, and the quantity $(E/P)_{thr}$ practically ceases to decrease with increasing pressure.

Addition of helium to the working mixture increases the rate of excitation of the upper laser level on account of the increase of the electron temperature in the discharge [26, 39], the decrease of the rate of relaxation of this level, and the decrease of the width of the vibrational–rotational transition. Therefore, the quantities Q_{thr}, and most significantly $(E/P)_{thr}$, decrease with increasing partial pressure of the helium in the investigated mixture.

The shape of the generated radiation pulse is determined primarily by the excitation regime of the active medium. When the gas is ionized by a nanosecond beam of fast electrons, the duration of the radiation pulse decreases approximately linearly with increasing laser-mixture pressure.

When the laser is pumped with pulses of durations larger than or comparable with the relaxation time of the upper laser level (10-20 μsec-atm, see Sec. 4), and at a relatively low power, the radiation pulse practically duplicates the excitation pulse (see Figs. 36c and 17c), and remains unchanged when the composition and pressure of the gas mixture are changed. On the other hand, if the pump power is high enough, then a characteristic spike of short duration ($\sim 10^{-8}$-10^{-7} sec) appears at the start of the lasing pulse, and is accompanied by long "tail" (Fig. 36a). This effect, which was observed in the investigation of TEA and electroionization lasers [9, 21, 62], is explained by the fact that at high rates of population of the upper laser level, a large inversion ($\Delta N \gg \Delta N_{thr}$) is reached before the laser light field is produced inside the resonator — this is the so-called gain switching regime [63]. The radiation energy in the first spike of the generation pulse is determined mainly by the energy stored directly

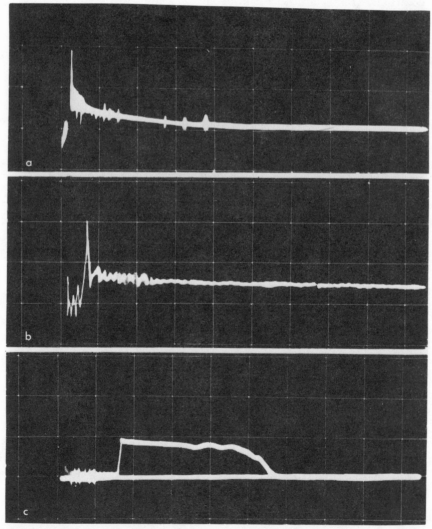

Fig. 36. Typical shapes of radiation pulses of electroionization CO_2 laser. a) Mixture $CO_2:N_2 = 1:2$, $T_{exc} \simeq 10^{-7}$ sec; b) mixture $CO_2:N_2 = 1:2$, $T_{exc} \simeq 10^{-6}$ sec; c) mixture $CO_2:N_2:He = 1:2:4$, $T_{exc} \simeq 4 \cdot 10^{-5}$ sec.

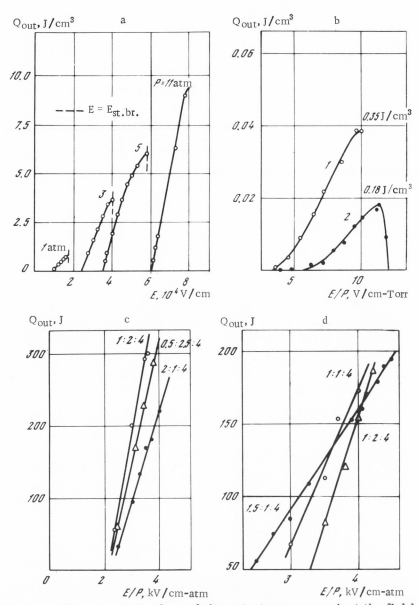

Fig. 37. Characteristic plots of the output energy against the field in the active region. a) Conditions of chamber 1, mixture CO_2:N_2 = 1:2; b) conditions of chamber 3 (curve 2) and chamber 4 (curve 1) at $T_{beam} \simeq 10^{-6}$ sec, mixture CO_2:N_2 = 1:2; c) conditions of chamber 4 at $T_{beam} \simeq 3 \cdot 10^{-5}$ sec, excitation volume 10 liters, mixtures CO_2:N_2:He of various compositions, P = 1 atm; d) conditions of chamber 4 at $T_{beam} \simeq 10^{-6}$ sec, excitation volume 10 liters, mixtures CO_2:N_2:He with different compositions, P = 1 atm.

on the level 00^01 CO_2, and only on the slope of the trailing edge is it determined by the energy obtained from the first vibrational level of N_2. Therefore with increasing content of the CO_2 in the working mixture the fraction of the radiation energy contained in the first spike of the lasing pulse increases (Fig. 36a). The total duration of the lasing pulse increases noticeably with the increasing nitrogen and helium content in the working mixture. The lasing duration increases also when the reflection coefficient of the output mirror is increased, owing to the lowering of the lasing threshold.

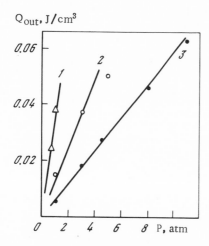

Fig. 38. Dependence of the specific energy yield on the pressure of the working gas mixture $CO_2:N_2 = 1:2$ ($E/P \simeq 10$ V/cm-Torr). 1) Chamber 4, $j_{beam} = 0.5$ A/cm^2, $T_{beam} \approx 10^{-6}$ sec; 2) chamber 3, $j_{beam} \approx 5$ A/cm^2, $T_{beam} \simeq 10^{-8}$ sec; 3) chamber 1, $j_{beam} \simeq 25$ A/cm^2, $T_{beam} \simeq 10^{-8}$ sec.

The output energy and the efficiency of an electroionization CO_2 laser at a given composition of the output mixture are determined mainly by the pump energy. The plot of $Q_{out}(E/P)$ can be approximated with good accuracy by a straight line (Fig. 37). The rate of increase of Q_{out} with the field increases with the gas pressure, and the fastest rate is observed when $CO_2:N_2:He$ mixtures are used in which the concentration of the carbon dioxide and the nitrogen are in a ratio $1:2$, i.e., $P_{CO_2}/P_{N_2} \simeq 0.5$. These mixtures turned out to be optimal for obtaining high radiation energies at all excitation regimes. At a fixed value of the relative field intensity E/P, the laser-radiation energy increases practically linearly with the pressure of the working gas mixture (Fig. 38), and the radiation power increases as the square of the pressure.

The increase of the power with increasing pressure of the working mixture results in damage of the resonator mirrors (photographs of the radiation-damaged mirrors are shown in Fig. 39). Therefore the specific energy and lasing power attained in the present study were limited not by the capabilities of the active medium but by the beam endurance of the resonator mirrors. The maximum output energy was more than 1000 J in the regime of "long" pulses and 200 J in the regime of "short" pump pulses (Fig. 37). The specific energy output in these regimes was ~70 J-liter^{-1}-atm^{-1} and ~40 J-liter^{-1}-atm^{-1}, respectively, while the specific

Fig. 39. Output mirror of chamber 4, damaged by laser radiation.

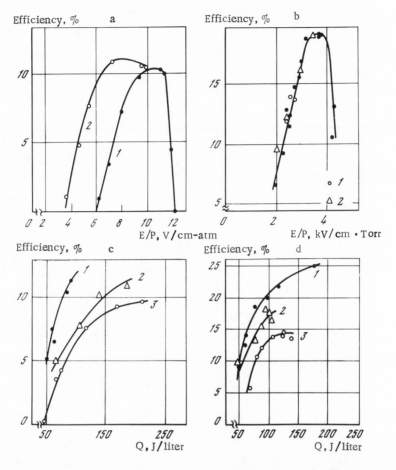

Fig. 40. Characteristic plots of the efficiency of an electroionization CO_2 laser against the field in the active region (a, b) and
against the pump-pulse energy (c, d).   a) Mixture $CO_2:N_2$ =
1:2, P = 1 atm, 1) $j_{beam} \simeq 5$ A/cm^2, $T_{beam} \simeq 10^{-8}$ sec;   2)
$j_{beam} \simeq 0.5$ A/cm^2, $T_{beam} \simeq 10^{-6}$ sec; b) chamber 4, mixture $CO_2:N_2:He$ = 1:2:4, P = 1 atm, $T_{beam} \simeq 3 \cdot 10^{-5}$ sec,
$j_{beam} \simeq 0.1$ A/cm^2: 1) C = 6 μF; 2) C = 9 μF; c) mixture
$CO_2:N_2$ = 1:1 (1), 1:2 (2), 1:3 (3), P = 1 atm; $T_{beam} \simeq 10^{-6}$ sec;
d) mixture $CO_2:N_2:He$ = 1:2:4 (1); 1:2:8 (2); 1:5:16 (3), P = 1
atm, $T_{beam} \simeq 3 \cdot 10^{-5}$ sec.

lasing power was 10^6 W/cm^3 when the medium was ionized by a nanosecond beam of fast electrons of high density.

Figure 40 shows characteristic plots of the dependence of the efficiency of an electroionization CO_2 laser on the relative field intensity (E/P) for different excitation regimes. The
maximum measured laser efficiency reaches 25%. The laser efficiency was limited by the fact
that in the experiments the choice of the reflection coefficient of the output mirror was determined not by the requirement that the resonator be made optimal, but by the requirement that
the output mirror have the strongest beam endurance. For example, in an electroionization
laser with an active region 100 cm long (chamber 4) we used dielectric mirrors with not more
than 1 or 3 evaporated reflecting layers ($R_2 \simeq 0.3$ and 0.5 respectively).

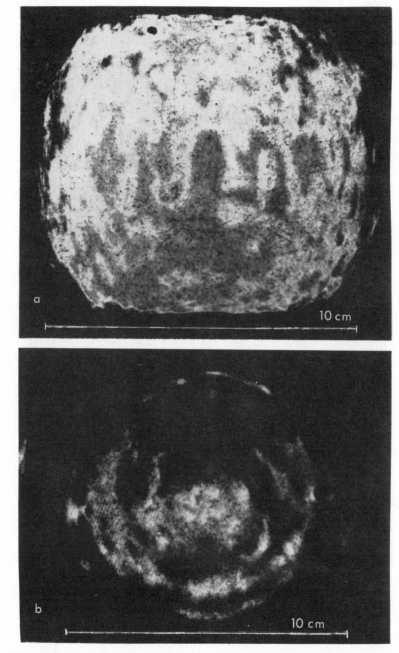

Fig. 41. Distribution of radiation field of high-power electro-ionization CO_2 laser in the near zone. Photograph of emery paper damaged by laser radiation at $Q_{out} \simeq 500$ J (a) and 100 J (b).

To measure the divergence of the output radiation of the laser, we used the scheme proposed in [64].

The beam divergence was measured at radiation energies ~100 J for a hemispherical (r = 5 m) and for a flat resonator. The total-reflection mirrors were made of gold coated on glass. The reflection coefficient of the output dielectric NaCl mirror was ~30%. To prevent parasitic lasing, which increases the beam divergence, an antireflection-coating was deposited on the laser anode (thin loosely woven cloth). The beam divergence at the 1/2 level for the hemispherical resonator was $\theta \simeq 2.5 \cdot 10^{-2}$ rad, and for the plane parallel resonator $\theta \simeq 6 \cdot 10^{-4}$ rad. To decrease the divergence further and reduce it to the diffraction limit $(\theta_{\text{dif}} = \lambda/D \simeq 10^{-4}$ rad) it is necessary to use unstable resonators, the use of which is particularly effective in the case of lasers with large volumes [65].

The distribution of the radiation field in the near zone is shown in Fig. 41. The characteristic annular structure was observed for all employed resonators, so that this annular structure cannot be attributed to interference effects that are produced when the laser radiation is reflected from the two faces of the semitransparent output mirror [66]. It appears that it is due to the interaction of lower modes with modes that differ both in the transverse and in the axial indices, but have a close frequency; the possibility of this phenomenon in a laser with large gain was indicated in [67].

3. Gain Spectrum of Electroionization CO_2 Laser

To obtain the emission spectra of a tunable electroionization CO_2 laser, one of the resonator mirrors was replaced by a diffraction grating (150 lines/mm, blaze angle 52°) (Fig. 42). The operation of the reflecting grating is described by the relation $m\lambda = d(\sin r + \sin i)$, which connects the diffraction order m, the grating period d, and the incidence and diffraction angles i and r. In the laser resonator, the grating operates in an anticollimation setting, with $i = r = \varphi$ (φ is the grating blaze angle). The energy from the resonator was directed by the grating in the direction of the zeroth order (m = 0), and also through the resonator mirror. The experiments have shown that without a diaphragm inside the resonator the lasing spectrum has a complicated structure and the individual vibrational−rotational lines are poorly resolved. The reason is that during the time of scanning of the grating the laser is abruptly switched from one transverse mode to another, and this affects the form of the spectrum. By introducing a diaphragm of 3 mm diameter it is possible to suppress the transverse modes and to separate the fundamental TEM_{00} mode. Figure 43 shows the lasing spectrum recorded with a 3-mm diaphragm (laser transition 00^01–10^00). The contours of the spectral lines no longer have a complicated structure.

Figure 43 shows the emission spectrum of an electroionization CO_2 laser for the mixture $CO_2 : N_2 = 1 : 1$ at various pressures. These spectra were obtained at the largest pump energy for each pressure $[Q_p^{max} = Q_p(E_{st.\,br.})]$. For mixtures with other ratios of CO_2 and N_2, the spectra are similar. The maximum tuning was obtaining with 15 rotational lines of the P branch [P(4)−P(32)] and 10 rotational lines of the R branch [R(10)−R(28)], corresponding to a tuning range from 930 to 985 cm^{-1}. Smaller values of the pump energy corresponded to fewer lines in the emission spectrum of the tunable CO_2 laser. At a pump energy close to threshold, lasing was observed only on one vibrational−rotational transition. This means that the gain for the other vibrational−rotational transitions is smaller than the threshold gain

$$K_{\text{thr}} = \tfrac{1}{2L} \ln (R_{\text{gr}} R_2)^{-1} \simeq 0.04 \ \text{cm}^{-1},$$

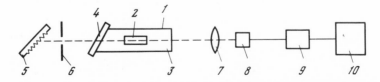

Fig. 42. Experimental setup for tuning the lasing frequency of an electroionization CO_2 laser. 1) Working chamber; 2) active region; 3) resonator mirror; 4) NaCl plate mounted at the Brewster angle; 5) diffraction grating (replica); 6) diaphragm; 7) NaCl lens; 8) radiation receiver; 9) cathode follower; 10) oscilloscope.

I, rel. units

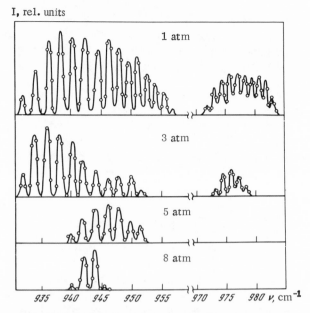

Fig. 43. Emission spectrum of electroionization CO_2 laser for the mixture CO_2:N_2
(1:1) (L = 4 cm, chamber 1) at various pressures.

where $R_p \simeq 0.8$ is the reflection coefficient of the grating at the frequency to which the grating is tuned, and $R_2 \simeq 0.9$ is a coefficient accounting for the reflection of the mirror and for other losses in the resonator. In this case the lasing is on a transition from a rotational sublevel of the 00^01 level, corresponding to the maximum of the expression (59) $J_{opt} \simeq 0.95\sqrt{T} - 1/2$.

The growth of the pump energy with increasing field intensity is limited by the static breakdown of the gas. To increase the breakdown threshold, the working gas was rid of the water vapor by passing the prepared mixture through a trap cooled to $-4°C$. The gain of the outermost lines observed in the emission spectrum of the tunable CO_2 laser is approximately equal to the threshold gain. Consequently, using Eq. (59) and the value of the threshold gain, it is possible to determine, from the obtained emission spectra, the gain at the center of the vibrational−rotational line at the frequency ν (J_{opt}) corresponding to the maximum of (59). Figures 44 and 45 show plots of the maximum gain [at $Q_p = Q_p (E \simeq E_{st. br.})$] against the pressure and the percentage content of the nitrogen in the CO_2 : N_2 gas mixture. It is seen that there exists an optimal nitrogen content in the mixture, approximately 45-65%, at which the gain is maximal. At a pressure P > 3 atm, the mixture CO_2 : N_2 = 1 : 2 has the the largest gain. For each mixture there exists an optimal pressure, and for the mixtures CO_2 : N_2 with less than 70% nitrogen the optimal pressure is less than 1 atm. For mixtures with large nitrogen content, the maximum shifts towards higher pressures, and amounts to 3 atm for the 1 : 4 mixture. When helium is added to the working mixture, at the same field intensity, the pump energy increases. This leads to a corresponding increase of the gain. The maximum attainable gain in mixtures containing helium, however, is lower than the gain of mixtures without helium. The reason for this experimental fact is that the spark breakdown field for mixtures containing helium is lower than for mixtures without helium.

We have also calculated the gain of an electroionization CO_2 laser for different mixtures and pressures. The results of the calculations of the maximum gain by means of Eqs. (54)

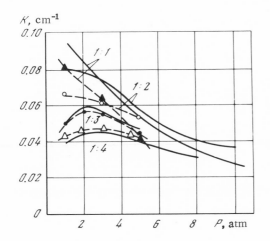

Fig. 44. Dependence of the gain of the active medium of an electroionization CO_2 laser on the pressure of the mixture $CO_2:N_2$ ($l = 4$ cm, chamber 1) at different mixture compositions. Solid curves, calculation; dashed curves, experimental data.

and (59), starting from the experimental values of the pump energy, are shown in Figs. 44 and 45. The agreement between experiment and calculation is good. The small quantitative discrepancy between the calculated and experimental values (~40%) is observed mainly at large gains, when the approximation $N_{00^01} \ll N_{CO_2}$ is not quite valid [for example, at $K_{thr} \simeq 0.08$ cm^{-1}, $P = 1$ atm, $N_{00^01} \simeq {}^1/_6 N_{CO_2}$ and Eq. (54) is not exact]. Another cause of the discrepancy is the inaccuracy in the values of the constants (A_{12}^J, $\Delta\nu_J$) used to calculate the gain. An increase of the pump energy and a change in the widths of the vibrational−rotational transition lines following the addition of helium turn out to be such that a calculation of the change in the gain, based on Eqs. (54) and (59), yields results that agree with the experimental data. It can thus apparently be assumed that the role of the helium at high pressures reduces mainly to an increase of the gain at a fixed electric pump field, owing to the increase of the pump energy and the decrease of the line width of the vibrational−rotational transition. The results of the calculation of the maximum gain at high pressures (up to 50 atm) in mixtures containing helium, on the basis of the threshold characteristics (Fig. 32) and the pump energy by formulas [54] and (59), is shown in Fig. 46. From a comparison of Figs. 44 and 46 it is seen that in mixtures with helium, maximum gains are attained at higher pressures than in mixtures without helium and the maximum shifts towards lower pressures with decreasing partial pressure of the helium.

Let us examine the question of the spectral resolution of the grating. It is known that $R = mN$, where N is the total number of the active lines of the grating. In our experiments the diaphragm diameter was 3 mm, corresponding to 600 active lines, so that the total resolution in first order was ~600. Actually, however, the resolution is higher.

Fig. 45. Dependence of the gain of the active medium of the electroionization CO_2 laser on the nitrogen concentration in the mixture $CO_2:N_2$ ($l = 4$ cm, chamber 1) for different pressures. Solid curves, calculation; dashed curves, experimental data.

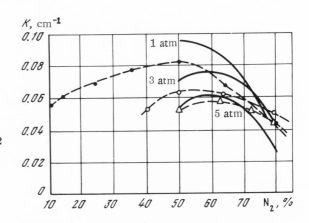

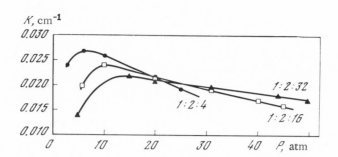

Fig. 46. Dependence of the gain on the pressure of
the working mixture $CO_2:N_2:He$ ($l = 10$ cm, chamber
3) at different mixture compositions.

This is caused by the specifics of the use of a grating in a laser, since the presence of
an active amplifying medium in the resonator narrows down the apparatus function of the instru-
ment [68]. In the case of a passive spectrometer, the apparatus function, as is well known, is
given by the relation

$$E = E_0 \frac{\sin^2 N\,(\delta/2)}{\sin^2\,(\delta/2)}, \tag{64}$$

where δ is the phase difference ($\delta = 2\pi \Delta S/\lambda$); S is the linear path difference. On the other
hand, the apparatus function of a laser system with the same grating is determined by the dependence of the output generation power P on $\Delta\nu$, whereas E now determines the effective reflection coefficient at the lasing wavelength λ. Thus [69],

$$P = \frac{I_0 \tau}{(1 + E)} \left[\frac{K_0 l}{\ln\,(E\rho_0)^{-1/2}} - 1 \right], \tag{65}$$

where I_0 is the saturation flux, τ is the reflection coefficient of the laser radiation from the
grating in the direction of the zeroth order of the diffraction, K_0 is the gain, l is the length of
the active region, and ρ_0 includes the reflection coefficient of the mirror and the other losses
in the resonator, $\rho_0 \sim 0.9$. The relation (65) is valid in the case of homogeneous broadening
[70], and also if the width of the gain line is much smaller than the width of the apparatus function of the grating. The result of a calculation by means of (64) and (65), i.e., the function
$P(\Delta\nu)$, where $\Delta\nu = \nu - \nu_0$, ν_0 is the frequency to which the grating is tuned, and ν is the frequency corresponding to the maximum of the lasing line, is given in [68]. The calculation shows
that when a grating is used in a laser the width of the apparatus function is greatly decreased,
thus ensuring the observed increase of the effective resolution of the grating. The calculation
also shows that for lines with small gain the resolution is higher than for high gains, as was
also observed in our experiments. The foregoing analysis is only qualitative in character. In
a real situation, the resolution and the form of the spectrum are greatly influenced by the
competition between the different modes and the rotational lines.

On the other hand, the experimentally obtained dependences of the lasing power on $\Delta\nu =
\nu - \nu_0$ at different gains of the system (in the case of sufficiently low pressures, when the width
of the rotational line is much less than the width of the apparatus function of the grating) enable
us to determine the dependence of the effective losses in the resonator on the detuning of the
grating relative to the maximum of the lasing line (the value of $\Delta\nu$). Indeed, the condition

$$K_0 = \frac{1}{2l} \ln\,[E_{\text{eff}}(\Delta\nu)\,\rho_0]^{-1} = K_{\text{thr}}(\Delta\nu), \tag{66}$$

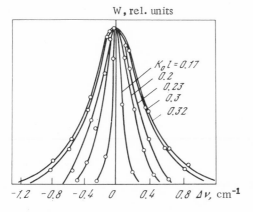

Fig. 47. Dependence of the lasing power for the P(20) line at a pressure 1 atm on $\Delta \nu = \nu_0 - \nu_{P(20)}$ for different values of the gain $K_0(\nu_0) l$. ν_0 is the frequency to which the grating is tuned, and $\nu_{P(20)}$ is the frequency corresponding to the maximum gain for the P(20) line.

which corresponds to termination of the lasing, determines the losses in the resonator at a known gain K_0 if the detuning is by an amount $\Delta \nu$.

The dependence of the lasing power for the P(20) line at a pressure 1 atm at different gains (pump energies) is illustrated in Fig. 47. The gain at the center of the P(20) line is determined from an analysis of the emission spectra, as indicated above. On the basis of the experimental results shown in Fig. 47, we plotted the dependence of the resonator losses [$K_{thr} = (1/2l) \times (E_{eff}\rho_0^{-1})$] on the value of $\Delta \nu$ in accordance with relation (66). This dependence is illustrated in Fig. 48.

At high pressures of the working mixture (several atmospheres), besides a discrete tuning of the lasing frequency, it becomes possible to tune the frequency within strongly broadened vibrational−rotational lines, and if the gain is high enough ($K_0 l \geq 0.5$, P = 6 atm, mixture $CO_2 : N_2 = 1 : 2$) it is possible to tune within the limits of several overlapping lines in the P and R branches without interrupting the lasing between the lines. Figure 49 shows the gain contour (K_{gain}) near the P(20) line, calculated from Eq. (59). The half-width of the vibrational− rotational transition was assumed equal to $\Delta \nu = 0.1$ cm^{-1}-atm^{-1} (the experimental value for the mixture $CO_2 : N_2 = 1 : 2$ and for the pressure 8 atm). The same figure shows the resonator loss curves (see Fig. 48). It is seen from the figure that at a sufficiently strong detuning of the grating relative to the maximum of the gain line ($\nu_0 - \nu_{P(20)} > 0.6$ cm^{-1}, curve a), the lasing should be interrupted, inasmuch as at this grating-rotation angle the losses at all frequencies exceed the gain. This was indeed observed in experiment. The case when the loss curve (K_{thr}) is tangent to the gain contour (curve b) corresponds to a grating position at which the lasing sets in, and the point of tangency determines the generation frequency (ν_g). The quantity $2(\nu_g - \nu_{P(20)})$ determines the range of tuning within the broadened lines and, as seen from Fig. 49, amounts to ~ 0.4 cm^{-1} for the P(20) line. In the case when $\nu_0 - \nu_{P(20)} < 0.6$ cm^{-1}

Fig. 48. Dependence of the threshold gain on the detuning of the diffraction grating from the line center.

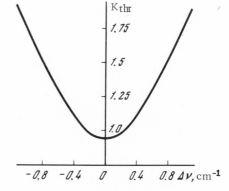

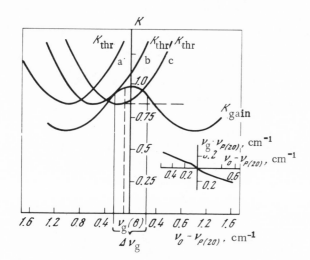

Fig. 49. Dependence of the change in the lasing frequency $\Delta\nu_g$ on the detuning $\nu_0 - \nu_{P(20)}$ of the grating relative to the center of the gain line.

(curve c) lasing takes place at a frequency corresponding to the maximum of the expression

$$K(\nu - \nu_{P(20)}) - K_{thr}(\nu - \nu_0) = f(\nu, \nu_{P(20)}, \nu_0), \tag{67}$$

where $\nu_{P(20)}$ is the frequency corresponding to the maximum of the P(20) line and ν_0 is the frequency to which the grating is tuned. To obtain the dependence of the lasing frequency on $\Delta\nu = \nu_0 - \nu_{P(20)}$, as shown in Fig. 49, we plotted the function $f(\nu, \nu_{P(20)}, \nu_0)$ for different grating positions. Similar calculations were performed for other vibrational lines observed in the emission spectrum of the electroionization CO_2 laser at a pressure of 8 atm (mixture $CO_2 : N_2 = 1 : 2$). The obtained dependence of the generation frequency on $\Delta\nu = \nu_0 - \nu_{P(20)}$ makes it possible to calculate the dependence of the output power on $\Delta\nu_i$, using Eq. (65):

$$P(\nu_0 - \nu_{P(20)}) = \frac{I_0 \tau}{[1 + E_{eff}(\nu_0 - \nu_g)]} \left\{ \frac{K(\nu_g) l}{\ln [E_{eff}(\nu_0 - \nu_g) \rho_0]^{-1/2}} - 1 \right\}, \tag{68}$$

in which the values of the gain and the reflection coefficient of the grating are taken at the frequency ν_g determined by the frequency ν_0 to which the grating is tuned.

Figure 50 shows the calculated and experimental dependences of $P(\nu - \nu_{P(20)})$ for the P(20) line; sufficiently good agreement between calculation and experiment is seen.

Fig. 50. Dependence of the lasing power of the P(20) line in the mixture $CO_2 : N_2 = 1:2$ at a pressure 8 atm on the detuning of the grating from the line center. Solid curve, calculation; points, experiment.

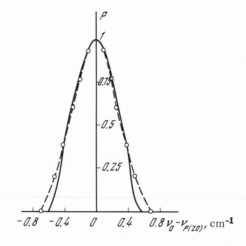

As indicated above, the lasing is interrupted between the rotational lines. On the basis of the results we can estimate the parameters of an electroionization CO_2 laser in which the lasing frequency can be tuned within several overlapping lines of the P and R branches. Using the obtained values of the maximum gain at various pressures, we plotted the gain contours for the mixture $CO_2 : N_2 = 1 : 2$ near the frequency $\nu = \nu_{opt}$ at different pressures (Fig. 51). It is seen from the figure that to obtain a tunable frequency in a range of several lines it is necessary to increase the gain of the active medium per pass (i.e., the quantity Kl). The gain of the active medium of an electroionization CO_2 laser can be increased by increasing the degree of ionization and the duration of the ionizing-radiation pulse, and also by increasing the length of the active region of the laser.

The results enable us to estimate the minimum length of the active region (L_{min}) of an electroionization CO_2 laser, which is sufficient to tune the lasing frequency within the range of several rotational lines, under the condition that the excitation density, which is determined by the parameters of the electron beam (energy $\simeq 600$ keV, current density $j \simeq 10$ A/cm^2, and current pulse duration $\tau \simeq 10^{-8}$ sec) remains the same. The minimum length is determined from the relation

$$l_{min} = 1.5 \frac{\ln (E_0 \rho_0)^{-1/2}}{K \, (\nu_{P(20)} - \nu_{P(18)})/2},$$

(69)

i.e., the threshold gain at the frequency to which the grating was tuned was assumed to be 1.5 times smaller than the gain at the frequency $\nu = (\nu_{P(20)} - \nu_{P(18)})/2$.

The obtained pressure dependence of the minimum length of the active region, necessary to tune the frequency over a range of several pressure-broadened vibrational−rotational lines, is shown in Fig. 52. The minimum length for the given excitation density is $l_{min} \simeq 10$ cm, and the pressure needed to realize continuous tuning at such an active-region length is ~6 atm. With relation (69) satisfied and at a pressure P = 6 atm of the mixture $CO_2 : N_2 = 1 : 2$, tuning is possible in a range of ~25 cm^{-1} in the P branch and in a range ~15 cm^{-1} in the R branch.

One of the methods that make it possible to relax the requirements on the resonator Q needed for a smooth tuning of the lasing frequency is to use compounds such as $C^{12}O^{16}O^{18}$, $C^{13}O_2^{16}$, $C^{13}O^{18}$ etc. as the active media. As is well known, the isotope ^{16}O, the abundance of which is 99.76%, has a nuclear spin I = 0, and in the $C^{12}O_2^{16}$ molecule there exists therefore only symmetrical rotational states with even values of J for the lower level [71]. The frequencies of the vibrational−rotational transitions in the emission spectrum of isometrical CO_2 molecules (for example, $C^{12}O^{16}O^{18}$) are twice as frequent as in the emission spectrum of the symmetrical molecule $C^{12}O_2^{16}$, and correspond to states with even and odd values of the quantum

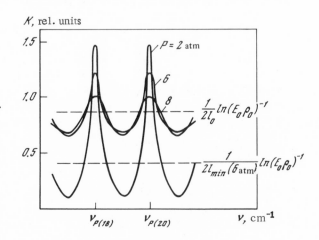

Fig. 51. Dependence of the gain on the frequency for the mixture $CO_2:N_2 = 1:2$ at different pressures. The horizontal dashed lines correspond to the threshold values of the gain.

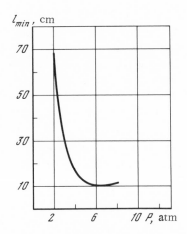

Fig. 52. Dependence of the minimum length
of the active region at which smooth tuning
of the frequency is possible on the pressure,
for the mixture $CO_2:N_2 = 1:2$ and for experimental conditions close to those realized in
the present study.

numbers J (I ≠ 0 for the isotope O^{18}). The emission spectrum of the compounds indicated above
consist of several series of vibrational–rotational lines, corresponding to different isotopes
and shifted from one another by the frequencies $\Delta \nu_{ij}$ [$m^j(O)/m^i(O)$, $m^j(O)/m^i(C)$, $m^j(C)/m^i(C)$],
determined by the mass ratios of the carbon and oxygen isotopes [72]. Therefore the use of
mixtures of these compounds makes it possible to relax further the requirements imposed on
the resonator $Q (Q_{min})$ needed for a smooth tuning of the lasing frequency, and it makes it
possible to obtain at a specified resonator $Q (Q > Q_{min})$ a smooth tuning at pressures lower
than 6 atm.

4. Relaxation of Upper Laser Level at High Pressures

Prior to the advent of CO_2 lasers with electroionization excitation of the gas medium, it
was not clear whether it is possible to obtain lasing in gas lasers at pressures exceeding 1
atm. It was feared that at high pressures there appear new types of quenching processes, due in particular to triple collisions, and that the high electron concentration leads to
an intense population of the lower laser level. This would make it difficult to obtain population
inversion and accordingly lasing at high pressures of the gas mixture. The first report of lasing in mixtures with carbon dioxide at pressures higher than 1 atm (15 atm, to be exact) was
in the journal Kvantovaya Élektronika (Soviet Journal of Quantum Electronics) in 1971 [9].
By now, lasing at higher pressures, up to 100 atm, has been achieved. The results of the
measurements of the magnitude and of the time dependence of the inverted populations of an
electroionization CO_2 laser at high laser-mixture pressures (up to 20 atm), which lead to conclusions on the kinetics of the population and relaxation of the laser levels, are given in Table 3.

The population inversion was measured by measuring the gain of the active medium
($l = 10$ cm) of the electroionization CO_2 laser at various instants of time relative to the pump
pulse with the aid of a testing TEA laser ($\lambda = 10.59$ μm).

Measurements of the distribution of the gain over the cross section of the active medium
have shown that the gain is practically uniform over the entire cross section. The gain decreases in a direction from the anode towards the cathode by approximately 25%, and it decreases in the perpendicular direction, from the center to the edge of the active region, by
20% (Fig. 53). Figure 54 shows the obtained gain averaged over the cross section, as a function
of the mixture pressure, measured at the largest electric field intensity possible for each
given pressure and mixture composition; this intensity was limited by the spark breakdown of
the gas gap. It is seen that an increase of the nitrogen and helium concentration in $CO_2 : N_2 : He$
mixtures leads to an increase of the pressure at which the gain reaches its maximum value.
The gains shown in Fig. 54 were obtained at a distance d = 0.5 cm between electrodes. The
measurements have shown that larger values of the distance between electrodes correspond to

TABLE 3

Mixture He:N$_2$:CO$_2$	E/P, V-cm^{-1}-Torr^{-1}	K, cm^{-1}	W$_H$, J/cm^3	P, atm	Measurement method
0:1:1	26.0	0.04	0.19	1.2	A
	26.0	0.08	0.36	1.0	T
	15.5	0.064	0.79	3.0	T
	13.0	0.05	0.8	6.0	T
0:2:1	27.5	0.026	0.18	1.2	A
	14.0	0.052	1.08	5.0	T
	12.5	0.016	0.55	10.0	A
0:4:1	28.0	0.04	0.36	1.0	T
	22.0	0.43	0.80	2.0	T
	14.0	0.04	1.22	6.0	T
	27.5	0.05	0.35	1.0	T
0:3:1	22.0	0.057	0.78	2.0	T
	14.5	0.042	0.95	5.0	T
	15	0.01	0.06	1.2	T
5:2:1	8.5	0.012	0.24	5.0	A
	7.0	0.015	0.37	10.0	A
	5.5	0.012	0.4	20.0	A
	6.0	0.012	0.24	10.0	A
8:2:1	5.0	0.012	0.27	20.0	T
	4.5	0.012	0.28	35.0	T
	4.5	0.012	0.20	10.0	T
32:2:1	3.5	0.012	0.23	20.0	T
	3.0	0.012	0.25	50.0	T

* The letters A and T designate measurements performed under conditions of weak-signal amplification or on the basis of the lasing thresholds, respectively.

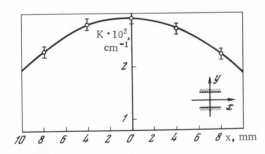

Fig. 53. Spatial distribution of gain in the cross section of the active region ($l = 10$ cm, chamber 3).

Fig. 54. Dependence of the gain of an electro-ionization CO$_2$ laser on the pressure of the working mixture CO$_2$:N$_2$:He ($l = 10$ cm, chamber 3), for different mixture compositions.

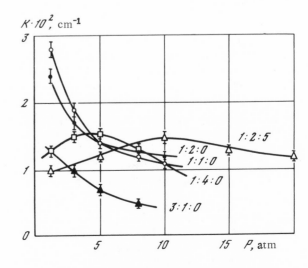

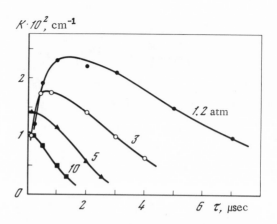

Fig. 55. Dependence of the gain on the time for the mixture $CO_2:N_2 = 1:2$ (chamber 3) at various pressures.

larger values of the specific pump energy and accordingly to larger values of the gain. The maximum gain in the laser chamber ($l = 10$ cm) was obtained for the mixture $CO_2:N_2 = 1:1$ at a pressure P = 1 atm and amounts to $4 \cdot 10^{-2}$ cm^{-1} (d = 1 cm). This is a larger value than obtained in the laser chamber 1 ($l = 4$ cm) (Fig. 44). The fast-electron beam current density, which determines the specific pump energy, was approximately 5 A/cm^2 in laser chamber 3, as compared with 30 A/cm^2 in laser chamber 1. Figures 55 and 56 show plots of the gain against the time reckoned from the pump pulse, for different pressures and mixtures $CO_2:N_2 = 1:2$ and $CO_2:N_2:He = 1:2:5$. It is seen from the figures that up to a certain instant of time the gain reaches a maximum value, after which it decreases. The lifetime of the inversion decreases with increasing pressure. Similar relations were obtained for the remaining mixtures. Figures 57 and 58 show plots of the gain against the pump energy and the lifetime $\tau_{\Delta N}$ of the inversion against the mixture pressure. The value of $\tau_{\Delta N}$ was measured at small values of the pump energy, at which the heating of the working mixture was negligible: $\Delta T = 1/2 \times W_p/C_m < 30°$, where C is the heat capacity of the mixture, m is the mass of the gas, and ΔT is the average change of the gas temperature during the vibrational−translational relaxation of the gas. The lifetime of the inversion, just as in low-pressure CO_2 amplifiers [73] and in TEA amplifiers [74], increases with increasing percentages of the nitrogen and helium in the laser mixture, and amounted in the present experiments to 10–20 μsec-atm. The linear dependence of the gain on W_p/P shows that the population inversion is determined, just as in gas-discharge lasers of low pressure [73], by the pump energy. The experimental values of the population inversion, determined from the measured values of the gain and the line width of the vibra-tional−rotational transition P(20) (mixture $CO_2:N_2 = 1:2$, $2\Delta\nu_j = 0.21 \pm 0.04$ cm^{-1}) in ac-cordance with Eq. [59], are described with good accuracy by Eq. (54), obtained under the assumption that the population of the lower laser level by electron impact can be

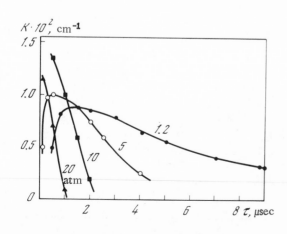

Fig. 56. Dependence of the gain on the time for the mixture $CO_2:N_2:He = 1:2:5$ (chamber 3) at various pressures.

neglected. For the mixture $CO_2 : N_2 = 1 : 2$ at $P = 1.2$ atm and at a pump energy $Q_p = 0.17$ J/cm^3, the experimental value of the population inversion is $\Delta N_{exp} = 10^{18}$ cm^{-3}. Calculations by means of Eq. (54) for this value of the specific pump energy yield for the population inversion, which is determined by the population of the upper laser level, the close value $\Delta N_{theor} \simeq N_{000_1} - N_{100_0} \approx N_{000_1} = 1.3 \cdot 10^{18}$ cm^{-3}. These values agree with each other at the measurement accuracies of the gain ($\pm 10\%$), the vibrational−rotational transition line width ($\pm 25\%$), and the pump energy ($\pm 10\%$). Thus in the active medium of the electroionization CO_2 laser, population by electron impact of the lower laser level is negligible ($N_{100_0} \ll N_{000_1}$) and the measured lifetime of the inversion is determined by the relaxation time of the upper laser level. The time required to obtain maximum inversion in $CO_2 : N_2$ systems, as determined in the present experiments, is $\tau_{N_2 \rightarrow CO_2} = 4 \pm 0.1$ μsec-atm and is determined by the rate of excitation transfer from the nitrogen to the carbon dioxide. The cross section for the transfer of the excitation from N_2 to CO_2, calculated from these experimental data, is $\sigma_{N_2 \rightarrow CO_2} \approx (1.6 \pm 0.4) \cdot 10^{-18}$ cm^2, which is practically the same as the cross section obtained in other studies at low pressures [75]. From Fig. 58, which shows the dependence of the lifetime of the inversion on the laser-mixture pressure, it it seen that the rate of relaxation of the upper laser level increases linearly with increasing mixture pressure. This indicates that there are no new types of quenching processes up to 20 atm pressure. The results of the measurements allow us to conclude that increasing the working-mixture pressure to dozens of atmospheres does not lead to qualitative changes in the excitation or relaxation of the laser levels.

It is seen from Fig. 57, which shows the dependence of the gain of the active medium of an electroionization CO_2 laser on the pump energy, that when the working mixture pressure is increased to 3 atm, the gain remains constant at a fixed value of W_p/P, but increases with further increase of the pressure. At $P \lesssim 3$ atm, the increase of the population inversion is offset by the increase, in direct proportion to the pressure, of the line width of the vibrational−

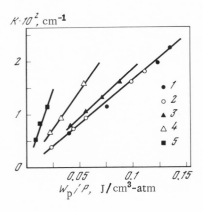

Fig. 57. Dependence of the gain on the pump energy (chamber 3) at different pressures of the mixture $CO_2 : N_2 :$ He. 1) 1.2 atm; 2) 3 atm; 3) 5 atm; 4) 10 atm; 5) 20 atm. 1-4) 1:2:0; 5) 1:2:5.

Fig. 58. Dependence of the lifetime of the inversion in the active medium of an electroionization CO_2 laser on the pressure of the working mixture $CO_2 : N_2 :$ He of varying composition.

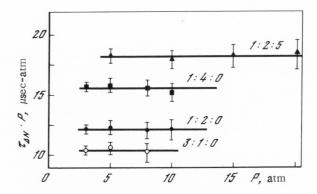

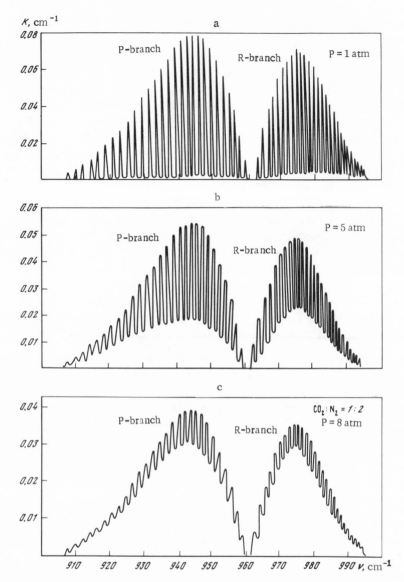

Fig. 59. Dependence of the gain of the active medium of an electroionization CO_2 laser on the frequency for the mixture $CO_2:N_2 = 1:2$.

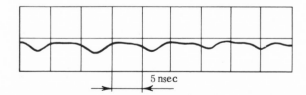

Fig. 60. Oscillogram of radiation spikes of electroionization CO_2 laser in the regime of passive mode locking.

rotational transition (57). At pressures higher than 3 atm, the contribution made to the gain of the vibrational–rotational line by the wings of the neighboring line becomes appreciable, and the gain, at the same values of W_p/P, increases [see Eq. (59)]. From the data of Fig. 57, using Eq. (59), we can calculate the widths of the vibrational–rotational transitions, assuming a Lorentzian line shape, for the line P(20): $2\Delta\nu \simeq 0.21 \pm 0.04$ cm^{-1}-atm^{-1} (mixture $CO_2:N_2 = 1:2$). The obtained values of the gain at the center of the vibrational–rotational P(20) line and of the line width enable us to plot the gain of the active medium of the electroionization CO_2

laser of high pressure against the frequency. The spectral distribution of the gain for pressures 1.5 and 8 atm is shown in Fig. 59 (mixture $CO_2 : N_2 = 1 : 2$). At pressures larger than 10 atm, owing to the overlap of the large number of vibrational—rotational lines, the gain contour becomes continuous, and its width increases to ~50 cm^{-1} ($\lambda = 10.6 \mu$m, P branch), and both a smooth tuning of the lasing frequency and shaping of picosecond pulses become possible. At a pressure of 20 atm, it apparently becomes possible to obtain high-power pulses of 10^{-11} sec duration with an approximate efficiency 10%, which is close to the efficiency in the case of long pulses, and with duration 10^{-12} sec at an efficiency lower by one order of magnitude. The decrease of the efficiency at pulses shorter than 10^{-11} sec is due to the insufficient rate of rotational relaxation [76]. We used the present experimental setup to obtain short electroionization CO_2 laser pulses in a scheme similar to that used in TEA lasers [77], by means of a cell 1-2 m long filled with the working gas. Mode locking was observed at all the investigated pressures and compositions of the $CO_2 : N_2 : He$ mixture. Figure 60 shows a photograph of the radiation spikes of an electroionization CO_2 laser for a mixture $CO_2 : N_2 : He = 1 : 2 : 5$, at a pressure P = 10 atm. The width of the spikes (~3 nsec) is determined by the resolution of the recording apparatus and of the receiver (Ge—Sb, Zn, T = 77°K).

5. Operating Regimes of Electroionization CO_2 Lasers

The operating regimes of electroionization optical CO_2 amplifiers are characterized by the amplified signal pulse duration, while the laser regimes are characterized by the pump pulse durations. We present some numbers that describe qualitatively some of the most interesting lasing and amplification regimes. First of all, all the lasing and amplification regimes can be divided into two groups: pulsed, or quasicontinuous, and continuous. We consider the first group.

Among the pulsed amplification regimes greatest interest attaches to amplification of ultrashort radiation pulses. In this regime, the duration of the amplified signal is much shorter than the duration of either the upper or the lower laser level, and is shorter than the time of excitation transfer from the nitrogen molecules to the CO_2 molecules.

The active medium of the amplifier should obviously be pumped under these conditions by a sufficiently long pulse but for a time not exceeding the relaxation time of the upper laser level. For the most effective utilization of the pump energy, the mixture must contain a large amount of CO_2 or should be purer CO_2. To decrease the probability of population of the lower laser level by electron impact it is necessary to add nitrogen to the mixture, the optimal amount of which is 20-100% of the CO_2 molecule concentration. Addition of He to a mixture operating in the regime of amplification or generation of short pulses is needed only to relax the requirements imposed on the electron-beam source and on the storage battery, which call for principally decreasing to a minimum the density of the electron current (and consequently the ionization density) and the operating speed of the capacitor bank. For a high-power system, it is difficult to shorten the operating time of the capacitor bank below 5-10 μsec, so that a working mixture with a pressure ~1 atm and higher must contain a certain amount of helium. According to [20, 78, 79], the optimal mixture for these conditions is $CO_2 : N_2 : He$ with a component ratio $2 : \frac{1}{4} : 3$. For pressures exceeding atmospheric, the amount of helium must be increased.

It is obvious that at ultrashort radiation pulses, the amplifying system must operate in the saturated-gain region, and the maximum energy of the amplified pulse is determined by the number of particles stored in the upper laser level. Consider the case when the duration of the amplified signal is $\tau_{pul} > \tau_{rot}$, where τ_{rot} is the time of rotational relaxation. The value of τ_{rot} for typical mixtures is ~ $0.2 \cdot 10^{-9}$ sec-atm^{-1} [76], so that at pulse durations $\tau_{pul} > \varkappa \tau_{rot}$ the energy is drawn from the entire spectrum of the rotational components of the 00^01 level of the CO_2 molecule. Here $\varkappa$ is a coefficient larger than unity and characterizes the rate of energy migration over the gain spectrum.

The energy stored on the upper laser level can apparently be effectively drawn if the ultrashort pulses are amplified and the inverse inequality is satisfied, as is the case when ultrashort pulses are amplified in neodymium glass, where there is practically no energy migration over the gain spectrum within the time $\sim \tau_{pul}$.

If the pulse duration τ_{pul} exceeds the time $\tau_{10^00-02^00}$ during which equilibrium population is established between the levels 10^00 and 02^00, the number of stimulated emission quanta in the amplified signal will amount to $\sim 80\%$ of the number of molecules on the 00^01 level of the CO_2 molecule.

This high efficiency of utilization of the energy stored on the 00^01 level is due to the large statistical weight of the group of levels 10^00 and 02^00 in comparison with the weight of the 00^01 level. Taking into account the doubly degenerate level 02^20, the total statistical weight of the group of the lower laser levels is equal to 4. An estimate of the efficiency η and of the specific radiation energy W_{rad} of the active medium of the amplifier, working in the described regime, yields values

$$\eta_{max} \simeq 17\%, \quad W_{rad\,max} \simeq 0.05 \text{ J-cm}^{-3}\text{-atm}^{-1}. \tag{70}$$

In the case of amplification of pulses of shorter duration ($\tau_p < 2\varkappa \cdot 10^{-10}$ sec-atm^{-1}), the efficiency of the amplifier and the energy output will decrease.

An experiment performed in [79] has shown that the delivered energy is $\sim 70\%$ of the energy stored in the upper laser level, in the case of amplification of a signal of duration $\sim 10^{-9}$ sec by an amplifier with a working gas at atmospheric pressure. In a later study by the same group [80], however, the energy efficiency decreased by a factor of 5 when the duration was decreased from 70 to 5 nsec. This fact can be explained either as being due to the very large value of $\varkappa$, or to an experimental inaccuracy due to the presence of a time-dependent fine structure inside the amplified pulse.

Both the efficiency and the energy output increase to the maximum possible values in the case of amplification of a pulse whose duration exceeds the relaxation time of the lower laser level (in essence, larger than the collision relaxation time of the 01^10 level), is larger than the time of excitation transfer from N_2 to CO_2, and is at the same time shorter than the relaxation time of the upper laser level:

$$\tau_{pul} > \tau_{01^10}, \quad \tau_{pul} > \tau_{N_2 \to CO_2}, \quad \tau_{pul} < \tau_{\Delta N}. \tag{71}$$

It is obvious that in this situation the limiting efficiency at a low pump level should equal the quantum efficiency*

$$\eta_{lmax} \simeq \eta_{qu} \simeq 36\%,$$

and the limiting radiation energy can reach

$$W_{rad\,max} \simeq 0.12 \text{ J-cm}^{-3}\text{-atm}^{-1}.$$

To realize this regime, the optimum mixture is $CO_2 : N_2 : He$ with a component ratio $1:1:3$ (for $P \sim 1$ atm).

---

* At a radiation pulse energy $W_{rad\,max} \simeq 0.05$ J-cm^{-3}-atm^{-1}, the efficiency turns out to be somewhat lower than η_{max} and amounts to $\sim 12\%$, and analogously at $\eta = \eta_{max} = 17\%$ the emission energy cannot exceed $W_{rad} \simeq 0.03$ J-cm^{-3}-atm^{-1}.

The duration τ_{pul} of the amplified signal and the duration τ_p of the pump signal can greatly exceed the upper-level relaxation level

$$\tau_p \approx \tau_{pul} \gg \tau_{\Delta N}, \tag{72}$$

if the intensity of the amplified signal is so large that the rate of decay of the upper laser level as a result of the induced transitions $(dN_{00^01}/dt)_{ind} = P_{ind}/\hbar\omega$ (here P_{ind} is the stimulated-radiation power per unit volume of the active medium) exceeds the rate of decay due to the collision-dominated nonradiative processes:

$$P_{ind}/\hbar\omega = (dN_{00^01}/dt)_{ind} > (dN_{00^01}/dt)_{col} \simeq N_{00^01}/\tau_{col}. \tag{73}$$

Here τ_{col} is the collision decay time of the 00^01 level.

A similar situation can be realized also in the lasing regime. It is easy to obtain from (73) the requirement imposed on the minimum laser pump power:

$$P_{ind\,min} \geqslant \eta_{qu} \frac{\hbar\omega\Delta\omega}{A\tau_{col}} \frac{\ln(R_1 R_2)^{-1}}{L}. \tag{74}$$

Here $\eta_{qu} = 0.36$ is the quantum efficiency of the system; R_1 and R_2 are the reflection coefficients of the resonator mirrors; L is the length of the active region; A is the constant in the formula for the gain; $K = A\Delta N/\Delta\omega$ (ΔN is the population inversion and $\Delta\omega$ is the width of the gain line). Knowing the minimum pump power and the maximum attainable temperature rise ΔT of the working mixture ($\Delta T \simeq 200°K$ for CO_2 lasers) we can calculate the maximum duration of the pump pulse $\tau_{p\,max}$ and consequently of the lasing pulse $\tau_{pul\,max}$:

$$\tau_{pul\,max} \simeq \tau_{p\,max} \leqslant \frac{C\rho\Delta T A\tau_{col} L}{\eta_{qu}\hbar\omega\Delta\omega \ln(R_1 R_2)^{-1}}. \tag{75}$$

Here C and ρ are the heat capacity and the density of the working gas. From (75) we can obtain data for the determination of the optimal composition of the working mixture if the duration of the lasing pulse is given. For example, at a lasing pulse duration $\tau = 10^{-4}$ sec at atmospheric pressure of the working mixture, the optimum composition of the mixture $CO_2 : N_2 : He$ is close to $1 : 1 : 8$.

Considerations dictated by the geometry of the active region may make it necessary to vary the gain between certain limits. One of the simplest methods of doing this is to decrease the CO_2 content in the mixture, for when the CO_2 content is decreased the gain decreases.

Estimates for the amplification regime of a long pulse can be obtained in similar fashion.

An interesting feature of an amplifier operating in the long-pulse regime is that the gain decreases in proportion to the pulse duration at a constant radiation energy per unit volume. This effect makes it necessary to increase appreciably the energy of the input pulse W_{in}, which is also proportional to the pulse duration:

$$W_{in} \simeq W_S \frac{\tau_{pul}}{\tau_{col}}. \tag{76}$$

Here W_S is the energy of the input pulse required to realize saturated amplification of a short pulse. However, in spite of the increase of the minimum energy of the input signal, in the regime where long pulses are amplified the intensity of the input signal turns out to be independent of the pulse duration.

From the condition for the maximum permissible heating of the working mixture and the maximum possible pump energy per unit volume of the active medium, we can calculate the rate at which the gas must be pumped through in the cw lasing regime. The characteristic parameters that determine the rate at which the working gas is replaced are, besides the maximum permissible pump energy, the geometrical dimensions of the active region and the maximum possible power flux of the coherent radiation through the active medium, through the output mirror of the laser, or through the input mirror of the amplifier. The latter quantity determines to a considerable degree the design of amplifiers and lasers also in the case of short radiation pulses.

When estimating various operating regimes of electroionization lasers and amplifiers, we did not concern ourselves with problems involved in the excitation of the active medium, particularly the parameters of the electron beam. These quantities can be easily calculated from the given values of the energy, the pump power, and the geometry of the active region. By way of example we present a formula for the current density of the electron beam in the case of excitation of mixtures with large helium contents:

$$j_e = \left(\frac{C \rho \Delta T}{\tau_{\text{pul}} E V_{\text{dr}}} \right)^2 \frac{P_0 b}{\gamma P e}. \tag{77}$$

In addition to the symbols used in the present section, we have introduced also the following: b, the coefficient of electron−ion recombination; γ, the number of electron−ion pairs produced by one fast electron on one centimeter of path in the given mixture at a pressure P_0; P, pressure of the mixture; V_{dr}, drift velocity of the electron; E, electric field.

Generally speaking, there exists a lower limit of the electron-current density j_e, determined by the maximum electron density in the working gas, at which the space charge next to the electrodes still does not come into play. This value of the electron-current density depends strongly on the geometry of the active region, on the energy of the electron beam, and on the voltage of the electric power supply.

CHAPTER III

HIGH-PRESSURE GAS LASERS USING OTHER WORKING MEDIA

The electroionization method of pumping is the most effective for the excitation of high-pressure gas laser, in which the lasing is on vibrational−rotational transitions of molecules. The reason is, first, that this type of pumping requires a weak electric field, in which the energy of the electric current is consumed only in excitation of the oscillations, and the energy lost to excitation of the electronic levels and to ionization of the gas is practically zero. Second, the electroionization approach is characterized by an exceedingly high efficiency of utilization of the free electrons. Each electron produced in the gas by external ionization experiences up to 10^5 collisions with the molecules of the working gas before it recombines. During this time it transfers a tremendous amount of energy from the electric field to the molecule vibrations. This energy reaches $\sim 10^4$ eV and greatly exceeds the energy $W \simeq 35$ eV required to produce a free electron.

The potentialities of the electroionization method, however, are not confined to excitation of vibrational−rotational levels of molecules. It was proposed in [9, 21] to use this method also to excite electronic levels. It is seen from Fig. 1 that if the gas particles have no vibrational degrees of freedom (this is the case for monatomic gases and for gases that form stable molecules only in the excited state), the lowest levels that can be excited by electrons accelerated by an electric field are electronic levels. There exists therefore a region of electric fields weaker than the field required to ignite an autonomous discharge, in which electron levels are effectively excited.

In the pulsed regime, when the time required to switch on the electric field is shorter than the time of development of the spark breakdown, this field region can be greatly expanded. Under these conditions it is possible to excite electronic levels also in stable molecules such as N_2, H_2, CO_2 and others. Pulsed excitation of electronic levels is possible here because the electron energy distribution function is significantly altered in a strong pulsed electric field, and there is an appreciable increase in the number of electrons that pass through the barrier of the excitation of the vibrational level into the region of energies where the cross section for the excitation of the vibrations decreases sharply and the cross section for the excitation of the electronic level increases strongly. A shortcoming of such systems is that the choices of the values of the electric field and of the supply pulse duration are highly critical [33]. In the case of excitation of low-lying electron levels it is difficult to obtain population inversion, if the radiative transitions terminate at the ground state. Therefore greatest interest attaches to molecular and atomic systems in which the lower laser level lies higher than the ground level. Among the atomic gases we can point to the vapors of the metals Cu, Pb, Tl, Mn, Ca, with which electric-discharge lasing was obtained [81]. The use of the electroionization method can lift to a considerable degree the limitations on the volume and pressure of the working gases, imposed by the peculiarities of the autonomous electric discharge.

In [4] it was proposed to use the electroionization method to excite electronic transitions in molecules whose ground state is unstable. These include the inert-gas molecules He, Ne, Ar, Kr, Xe, and also molecules of certain metals Hg Cd, Zn, etc.

The electroionization method of excitation was used to initiate a chemical reaction in a chemical laser working with the mixture $D_2 : CO_2 : F_2 : He$ [82]. Owing to the high efficiency of the homogeneous energy input into a dense molecular gas medium, the electroionization method of initiation of chemical reactions offers a number of significant advantages over such customarily employed methods as photoinitiation and initiation with an electric discharge.

It is shown in [83] that the electroionization method can be used to pump rotational–vibrational transitions in molecules with zero dipole moment. The same paper considers transitions in a hydrogen molecule, which are forbidden in the dipole approximation and have low quadrupole-transition probability. Molecules with zero dipole moments have an exceedingly large lifetime of excited vibrational–rotational levels and relatively short time of rotational–translational relaxation. These molecules are capable of accumulating vibrational energy and can therefore be strongly excited by the electroionization method even at a low rate of energy input into the medium. When a strong electric field is applied to excited hydrogen (or deuterium) molecules, the so-called electric-field-induced transitions, which were predicted in [84], appear; these transitions satisfy the selection rules $\Delta j' \simeq 0 \pm 2$. The case of application of an alternating initiating field can be regarded as a two-photon process. The gain of such a system turns out to depend on the square of the initiating field and in the case of induced radiation by a CO_2 laser pulse with duration $\tau_{pul} \simeq 10^{-11}$ sec and energy flux density ~0.1 J/cm^2 it reaches a value

$$K \, [cm^{-1}] = 0.1 \, Q \, [J\text{-}cm^{-3}],$$

where Q is the stored vibrational energy of the H_2 molecules. This estimate was made for hydrogen pressures 10-25 atm. Thus, it becomes possible to obtain a gain K ~ 0.1 cm^{-1} in compressed hydrogen at a wavelength ~2.2 μm. The quantum efficiency of this system turns out to be ~15% [83]. It was also proposed in [83] to use, in an analogous system, the molecules HD and HT instead of hydrogen, in which, since the symmetry is violated, the hindrance of the transitions with $\Delta j = \pm 1$ [85] is lifted, and consequently it becomes possible to obtain lasing without an inducing external field.

The most interesting among the other trends in the development of high-pressure gas lasers is the investigation of the possibility of directly pumping, by means of an electron beam, compressed single-component inert gases and mixtures containing inert gases. The average energy of the free electrons produced in a gas by an external ionizing radiation is ~10-20 eV, so that pumping with an electron beam is effective when it comes to exciting electronic transitions of molecules.

In this chapter we present the results of experimental investigations aimed at developing high power compressed-gas lasers emitting in various spectral bands, namely CO lasers (near infrared, $\lambda_{rad} \simeq 5$ μm), Xe and Xe : Ar mixture lasers (vacuum ultraviolet band, $\lambda_{rad} \simeq 0.17$ μm), and Ar:N_2 mixture lasers (near ultraviolet band, $\lambda_{rad} \simeq 0.357$ μm).

1. Electroionization CO Laser

Of great interest from the point of view of the development of high-power sources of coherent radiation in the near infrared region of the spectrum (~5 μm) are CO lasers. The reasons are the following: 1) the possibility of attaining a high efficiency exceeding 50% (~50% was obtained in experiment [86]); 2) the existence of "transparency windows" of the atmosphere near 5 μm, the absorption in which is much lower than in the $\lambda = 10.6$ μm band [86]; 3) the possibility of obtaining much higher radiation power at the wavelength 5 μm in comparison with $\lambda = 10.6$ μm, since the limitations imposed by the radiation endurance of the optical materials and of the active medium are less stringent in this band [87, 88]; 4) the possibility of smoothly tuning the frequency of CO lasers at pressures $\gtrsim 10$-15 atm in a very wide range, up to 20-30% away from the central frequency [23].

The maximum of the cross section for the excitation by electron impact of vibrational levels of the CO molecule corresponds to an electron energy ~1.8 eV [89]. This electron temperature is reached in the CO : N_2 or CO : N_2 : He mixture customarily used in CO lasers at very low field intensities, 2-3 kV/cm-atm (Fig. 61 [89]), so that the electroionization method is particularly convenient for the development of high-power CO lasers.

A large energy yield and high efficiency are much easier to obtain with a CO laser if the working gas mixture is cooled to T $\simeq$ 100°K. Great interest attaches, however, also to investigation of lasing at room temperature, since the absence of cryogenic units greatly simplifies the laser construction. Estimates have shown that when an uncooled working mixture is used, the electroionization excitation method makes it possible to produce a CO laser with high output characteristics.

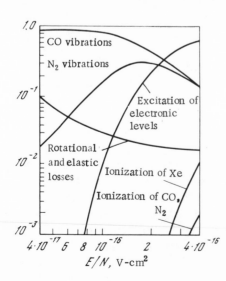

Fig. 61. Pump-power distribution over the degrees of freedom of the molecules following electric excitation of the mixture CO:N_2:Xe:He.

The experiments discussed in the present section were performed for the purpose of studying the possibility of developing an electroionization CO laser operating without cooling the gas mixture.

1. The investigations were performed in chamber 4 (see Chapter I, Sec. 2) with excitation by means of "short" (~1 μsec) and "long" (~30 μsec) pulses. In the former case, the volume of of the excitation region was ~2.7 liters (100 × 10 × 2.7 cm), and in the latter case 10 liters. We investigated the gas mixtures $CO:N_2$; $CO:N_2:He$; $CO:N_2:Xe$ at pressures from 0.5 to 2.5 atm. The mixtures were prepared from gases of commercial purity, the principal impurities in which were, according to the certificates supplied with the flasks, up to 1% air and up to 0.5 g/m³ water. To obtain lasing it was necessary to purify the gas mixture by passing the mixture prior to the entry into the laser chamber through a liquid-nitrogen cooling coil. The gas temperature in the active region was ~20°C under the working conditions.

2. The wave form of the discharge current in both excitation regimes was the same as for the $CO_2:N_2:He$ mixtures (see Fig. 17). The energy output per unit reached 500 J-liter^{-1}-atm^{-1} (Fig. 62). A comparison of the experimental theories of the energy parameters of the discharge with the calculated ones (the calculation method is considered in Chapter I) has shown that the electron−ion recombination in the $CO:N_2$ and $CO:N_2:He$ mixtures at $E/P \simeq$ 2-5 kV/cm-atm and pressures 0.5-2 atm is b $\simeq 10^{-7}$ cm³-sec^{-1} (this agrees well with the data obtained for the active medium of low-pressure CO lasers). When Xe is added to the $CO:N_2$ mixture, the recombination coefficient increases to ~(2-3) · 10^{-7} cm³-sec^{-1} and the energy input decreases (Fig. 62b).

Calculations show that the critical electron density at which the strong-current volume discharge in $CO:N_2:He$ mixture stops under the experimental conditions amounts to 10^{10}-10^{11} cm³, just as in the electroionization CO_2 laser.

3. For laser experiments we used an internal hemispherical resonator made up of a total-reflection gold-coated copper mirror with curvature radius 5-10 m, and a flat output mirror with transmission 2-3%. In the "fast" pumping regime we used an aluminum-coated LiF mirror, and the radiation was extracted through an uncoated window of 2-4 mm diameter at the center of the mirror. In the "long" pulse regime, the radiation was extracted through a multilayer dielectric mirror on NaCl with transmission ~2%.

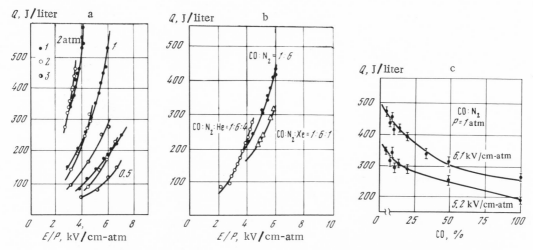

Fig. 62. Dependence of the energy input to the $CO:N_2:He$ mixture on the field intensity in the discharge gap (a, b) and on the CO content in the mixture (c). a: 1) $CO:N_2 = 1:20$; 2) $CO:N_2:He = 1:10:4$; 3) $CO:N_2 = 2:1$; b) P = 1 atm.

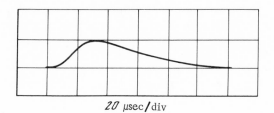

20 μsec/div

Fig. 63. Waveform of electroionization CO laser radiation pulse ($T_{exc} \simeq 3 \cdot 10^{-5}$ sec).

No spectral investigations were made, but a set of optical filters was used to establish that the wavelength of the output radiation was in the interval 4-6 μm. The waveform of the radiation pulse was measured with a cooled PbSnSe photodiode (resolution ~10^{-8} sec). Since the threshold pump energy was high, the lasing started with a large delay relative to the start of the excitation pulse. The total radiation pulse duration was much longer than the pump pulse duration (Fig. 63), this being due to the cascade mechanism of the lasing and to the large lifetime of the vibrational levels of the CO molecule.

The low optical strength of the aluminum mirrors did not make it possible to operate at pump energies much higher than threshold in the "short" excitation-pulse regime (a photograph of the burned-out mirror is shown in Fig. 64). The maximum radiation energy Q_{out} in this regime was therefore 0.5-0.6 J (this corresponds to an energy flux density through the opening in the output mirror ~10 J-cm^{-2}). In the "long" pulse regime, the total output energy exceeds 100 J at an efficiency ~4% (Figs. 55 and 56), while both Q_{out} and the efficiency increase rapidly with the increasing pump.

Just as in an electroionization CO_2 high-pressure laser, the threshold values of the relative field intensity at which lasing begins in the CO:N_2:He mixtures with various compositions decrease with increasing pressure (Fig. 67).

Under the experimental conditions, at a specified field intensity and pressure, the largest output energy was reached by excitation of mixtures with a high nitrogen content ($P_{N_2}/P_{CO_2} \gtrsim 10$). In CO:$N_2$:He mixtures containing less than ~40% nitrogen, no lasing was observed. These facts can be explained in two ways: either the energy transfer from the excited nitrogen to the CO molecules plays an important role in the inversion, or else a more suitable distribution of the CO molecules over the vibrational levels is reached in mixtures with large nitrogen contents in the working range of pump energies and field intensities. The presence of Xe in the CO:N_2 mixture increases the threshold field intensity and lowers the laser efficiency. The

Fig. 64. Photograph of aluminum output mirror damaged by laser radiation.

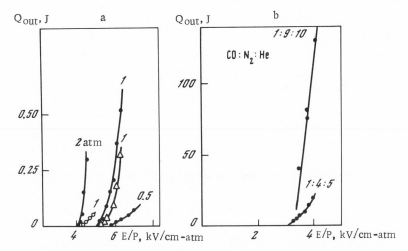

Fig. 65. Dependence of the output energy of an electroionization CO laser on the field intensity in the discharge gap at an excitation pulse duration 10^{-6} sec (a) and $3 \cdot 10^{-5}$ sec (b). a: triangles, $CO:N_2 = 1:20$; circles, $CO:N_2:He = 1:12:4$; points, $CO:N_2 = 1:8$, $j_{beam} \simeq 0.5$ A/cm^2; b) P = 1 atm, $j_{beam} \simeq 0.1$ A/cm^2.

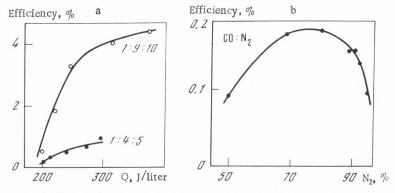

Fig. 66. Dependence of the efficiency of an electroionization CO laser on the excitation energy for the mixtures $CO:N_2:He$ (a) and on the nitrogen content in the mixture $CO:N_2$ (b) under conditions of chamber 4 at P = 1 atm. a) $j_{beam} \simeq 0.1$ A/cm^2, T_{beam} $3 \cdot 10^{-5}$ sec; b) $T_{beam} \simeq 10^{-6}$ sec.

Fig. 67. Pressure dependence of the threshold values of the relative field intensity in the active region of an electroionization CO laser. 1) $CO:N_2:He = 1:10:4$; 2) $CO:N_2 = 1:20$; 3) $CO:N_2:Xe = 1:6:1$.

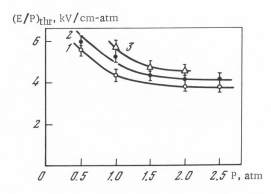

probable reason is the fact that the ionization of the Xe molecules by the plasma electrons, which leads to an increase in the density of the free electrons and to a corresponding increase of the output energy in gas-discharge CO lasers [90], is negligible in fields of the order of $2-6$ kV-cm^{-1}-atm^{-1} (see Fig. 61). Therefore the role of Xe in the electroionization CO laser reduces to the already noted increase of the rate of dissociative electron−ion recombination ($b_{Xe} \gg b_{N_2}$, b_{CO_2}, b_{He} [45]).

It should be noted that at E/P = $(2-6)$ kV-cm^{-1}-atm^{-1} the output energy and the efficiency of the laser were determined by the energy input and not by the field intensity. For example, in both investigated excitation regimes, the lasing begins practically at one and the same value of the energy input, which is reached at different $(E/P)_{thr}$, which depend on the ionization intensity. Moreover, the threshold energy for CO:N_2 mixtures at atmospheric pressure turns out to be the same (~250-300 J/liter) as in an electroionization CO laser with a pump pulse of 300 μsec duration [91], although the threshold field intensity in this laser amounted to only 1.6 kV-cm^{-1}-atm^{-1}. Consequently, the efficiency of excitation of CO molecules in the CO:N_2 mixture depends little on the electron temperature in the investigated 0.7-2 eV interval.

The gain of the active medium of the electroionization CO laser, determined by measuring the threshold characteristics of the resonators with different transmission coefficients of the output mirrors, did not exceed $(3-6) \cdot 10^{-4}$ cm^{-1} under the experimental conditions.

4. Thus, the result of the first experimental investigations of pulsed electroionization CO lasers, carried out in the present study, have shown that the electroionization pumping method makes it possible to develop CO lasers with high output parameters, operating without cooling the working mixture. The obtained values of the radiation energy (130 J) and of the specific energy output (13 J-liter^{-1}-atm^{-1}) are the largest of those published for pulsed CO lasers operating at room temperature.

It should be noted that when the described experiments were performed, reports were published that high-power pulsed electroionization CO lasers were developed in the U.S.A., with the active medium cooled to temperatures ~100°K [86, 92]. These results, as well as recent calculations of the lasing kinetics of atmospheric-pressure CO lasers [93, 94], also point to favorable prospects for electroionization CO lasers.

2. Laser Operating with Compressed Xenon and Ar : Xe Mixture

The use of inert gases as active media for electronically pumped lasers in order to obtain lasing in the vacuum ultraviolet region (VUR) was first proposed in 1966 in [3] and [95]. Experiments on the excitation of condensed xenon by a beam of fast electrons have shown [96] that inert gases are highly efficient active media for lasing in the vacuum ultraviolet region of the spectrum.

Figure 68 shows the scheme of the potential curves of the lowest electronic states of the molecule Xe_2 [97]. The upper laser level is the excited state $^{1,3}\Sigma_u^+$ of the Xe_2 molecule. The excited xenon molecule in the state $^{1,3}\Sigma_u$ is produced principally by three-particle collision of the excited atom Xe($^3P_{1,2}$) with two Xe atoms in the ground state S_0. At an approximate gas pressure of 10 atm, the time of production of the Xe_2 molecule is a fraction of a nanosecond, whereas the lifetime of the excited state of the molecule with respect to spontaneous emission is $\tau_{sp} \sim 30 \pm 10$ nsec [98]. The $^3P_2$ state of the Xe atom is metastable, and all the radiation is dragged along in the case of transitions from $^3P_1$ to the ground state. The transition of the molecule from the excited state to the repulsion ground state $^1\Sigma_g^+$ proceeds in accord with the Franck-Condon principle at fixed internuclear distances. As a result of the transition, a quan-

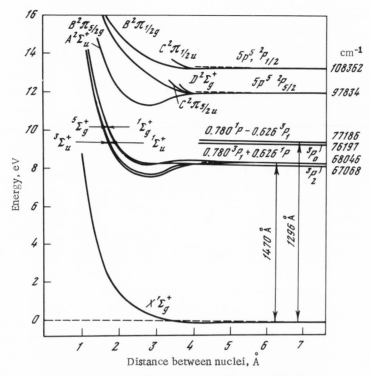

Fig. 68. Potential curves of Xe$_2$ molecule.

tum of light is emitted with an energy lower than the separation of the atomic levels; this quantum cannot be absorbed by the atomic gas. A four-level scheme is realized.

Amplification and lasing in the VUR region of the spectrum were first obtained [4] by exciting liquid xenon with a beam of fast electrons. In the same experiments, stimulated emission was observed following excitation by an electron beam of xenon compressed to 2-3 atm and located over the surface of liquid xenon at T = 165°K. Lasing with gas xenon at room temperature was first reported in [99].

In the present study we have investigated the following characteristics of compressed xenon and xenon-argon mixture excited by a high-density nanosecond beam of electrons: 1) the energy of the spontaneous emission of compressed xenon; 2) the power of spontaneous emission of compressed xenon and xenon-argon mixture; 3) the energy and power of a compressed-xenon laser; 4) the spontaneous emission and lasing spectra of compressed xenon.

In the experiments we used pure xenon with the following impurity content in volume percent: krypton, 0.074; hydrogen, 0.01; oxygen, 0.001; nitrogen, 0.01; hydrocarbons, 0.001; carbon dioxide, 0.001; water vapor, 0.0006. The electron beam from a cold-cathode accelerator [37] ($E_0 \simeq 700$ keV, $j_e \simeq 150$ A/cm^2, $\tau \simeq 10^{-8}$ sec) was introduced into the working chamber through a titanium foil 50 μm thick. The working chamber (Fig. 69) was first evacuated to $\sim 2 \cdot 10^{-3}$ Torr, and then filled with the investigated gas to pressures from 1 to 25 atm. The measurements of the spontaneous-emission energy were performed with a calibrated photo-receiver with quartz entrance window. The radiation power was measured with a single-stage high-speed photomultiplier (ELU-F5, time resolution $\sim 10^{-9}$ sec), which was placed behind the lumogen that converted the VUR radiation into visible radiation, with a time resolution $\sim 2 \cdot 10^{-9}$ sec.

In the experiments on the coherent emission of compressed xenon we used aluminum mirrors with protective coatings of magnesium fluoride. The mirror reflection coefficients were

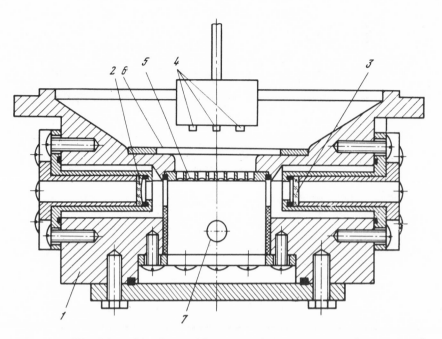

Fig. 69. Working chamber of lasers using compressed xenon and Ar:N_2 mixture. 1) Chamber body; 2 and 3) mirror; 4) accelerator cathode; 5) titanium foil 50 μm thick; 6) accelerator anode; 7) opening for the entry of the gas into the chamber.

$R_1 = 90 \pm 5\%$ for the total-reflection mirrors and $R_2 = 80 \pm 5\%$ (T = 0.5-1%) and $R_2 = 70 \pm 5\%$ (T = 5-10%) for the semitransparent mirrors. The mirrors used in the experiment were 1 cm in diameter. The distance from the titanium foil through which the electron beam entered the working chamber to the optical axis of the resonator was 0.5 cm. The distance between mirrors was 6 cm and the length of the active region was 4 cm. The directivity of the laser radiation was registered by photographing a tube. The spectra of the spontaneous and laser emission were photographed with a vacuum monochromator VM-1 1200 line/mm grating, dispersion 16 Å/mm, on a UF-2T film.

At a xenon pressure 16 atm, the specific energy of the spontaneous emission of the xenon was $Q_{rad} = (0.1 \pm 0.04)$ J/cm^3, and the density of excitation by the electron beam was $Q_p \pm 0.7$ J/cm^3. Thus, the cathode-luminescence efficiency under the conditions of the present experiment was $\eta = (15 \pm 6)\%$. The spontaneous-emission pulse duration at pressures larger than 2 atm was practically independent of the pressure and was $(25 \pm 5) \cdot 10^{-9}$ sec. We have also investigated experimentally the emission of the excited xenon molecules in the Xe : Ar mixture. The measurements have shown high efficiency of energy transfer from the excited argon molecules to the xenon atoms, an efficiency reported also in other papers [4, 100]. Figure 70 shows the dependence of the spontaneous-emission power of xenon in the Xe : Ar mixture on the pressure. It is seen that the efficiency of conversion of the stored electron-beam energy into ultraviolet radiation ($\lambda \simeq 1730$ Å) is approximately 35% less for the Xe : Ar mixture than for pure xenon. This decrease of the efficiency is obviously due to the fact that the ionization potential of argon is 1.3 times larger than the ionization potential of xenon.

The specific energy of the spontaneous emission of compressed xenon determines the population inversion ΔN (the gain) which is produced when xenon is excited by a beam of fast electrons ($\tau_e < \tau_{sp}$)

$$K = \sigma \Delta N \approx \sigma N_2 = \sigma \frac{Q_{rad}}{h\nu} = \sigma \eta Q_p = \sigma \eta \frac{jU\tau_e}{d} \simeq 0.1 \text{ cm}^{-1},$$

(78)

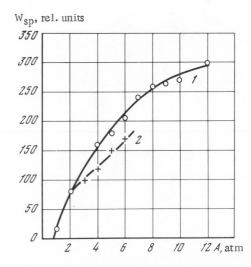

W$_{sp}$, rel. units

Fig. 70. Dependence of spontaneous-emission power of xenon (1) and of the xenon−argon mixture (2) on A = [P$_{Xe}$ + (Z$_{Ar}$/Z$_{Xe}$)P$_{Ar}$]/1.6 atm (Z is the charge of the nucleus).

where d is the depth of penetration of the electrons into the compressed xenon, $\sigma \simeq 10^{-18}$ cm^2 is the cross section of the induced transition, hν is the energy of a radiation quantum with wavelength $\lambda \simeq 1720$ Å, and N$_2$ is the population of the upper laser level.

In experiments aimed at obtaining lasing we use both hemispherical resonators and resonators produced by two spherical mirrors with curvature radii 0.5-5 m. The fact that lasing was attained is indicated by the following: 1) the high directivity and mode structure of the radiation; 2) the presence of a threshold in the dependence of the radiation power on the pressure; 3) the appreciable change in the time dependence of the output radiation before and after the lasing threshold is reached; 4) the narrowing of the spectral width of the radiation from ~140 to 8 Å when a resonator is used.

Microphotographs of the spontaneous and laser emission of xenon compressed to 14 atm and excited with an electron beam are shown in Fig. 71. The laser emission spectrum consists of two lines (this is probably due to transitions from neighboring vibrational levels of the state $^{1,3}\Sigma_u^+$) of width $\Delta\lambda \simeq 3$ Å with distance 5 Å between lines. The output energy of the laser radiation in the present experiments was limited by the optical strength of the mirrors, which burned out near the lasing threshold. A photograph of a mirror burned by laser radiation is shown in Fig. 72. An estimate of the energy of a laser using xenon compressed to 14 atm, based on the amount of evaporated aluminum coating of the mirrors, yields a value 0.2 J, which corresponds to a radiation-energy density ~0.07 J/cm^3 inside the resonator.

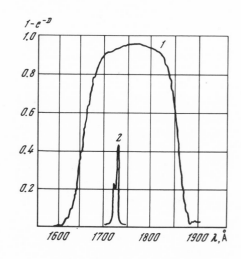

Fig. 71. Emission spectra of xenon compressed to 14 atm. 1) Spontaneous emission spectrum; 2) laser emission spectrum (D is the photographic film density).

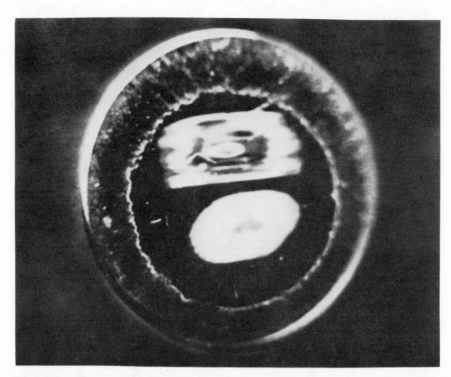

Fig. 72. Photograph of resonator mirror damaged by radiation from a laser with xenon compressed to 14 atm.

As indicated above, the efficiency of pumping compressed xenon by an electron beam was ~15%. An appreciable increase in the efficiency can be expected in the electroionization method of excitation of inert gases, as proposed in [21]. Inasmuch as in such a system the electrons accelerated in the electric field do not encounter a barrier of vibrational levels, the electrons can easily attain, under the influence of the field, levels sufficient for the excitation of the lowest electronic levels ($^{1,3}P_1$) of the inert-gas atoms.

An approximate analytic solution of the kinetic equation for the distribution function of the electrons in xenon was obtained in [101]. In [102] an exact numerical solution was obtained with an M-222 computer by a procedure described in [103]. Calculations show that the efficiency η of conversion of the electric energy into radiation at E/P = 3–6 V-cm^{-1}/mm Hg amounts to 0.65–0.76, which greatly exceeds the efficiency of excitation by only a fast-electron beam. For effective operation of a compressed-xenon electroionization laser it is necessary that the power W_E acquired by the electrons from the electric field exceed the power W_{acc} delivered to the electrons by the accelerator. At a gas pressure P = 10 atm, an electron-beam current density $j_e \simeq 5$ A/cm^2, and E/P = 4 V-cm^{-1} mm Hg, the ratio W_E/W_{acc} is ~5. At these values of the excitation parameters, the gain is $K \simeq 2.5 \cdot 10^{-2}$ cm^{-1}, which is larger than the coefficient of radiation absorption ($\lambda \simeq 1720$ Å) by the impurities in the gas. We note that W_e/W_{acc} decreases with increasing accelerator current in proportion to $j_e^{-1/2}$. The reason is that $W_E \sim j_e^{-1/2}$ and $W_{acc} \sim j$ [see Eqs. (47) and (40)].

An interesting circumstance observed in the present study, in the experiments on the excitation of liquid mixtures of noble gases [4], and in [100], is the exceedingly high efficiency of energy transfer from the Ar atoms to the xenon molecules. This effect can be described by resorting to the following processes:

$$Ar_2^* + Xe \rightarrow Ar + Ar + Xe^*; \tag{79}$$

$$\text{Xe}^* + \text{Xe} + \text{Ar} \xrightarrow{\beta_{\text{Ar}}} \text{Xe}_2^* + \text{Ar}; \tag{80}$$

$$\text{Xe}^* + \text{Xe} + \text{Xe} \xrightarrow{\beta_{\text{Xe}}} \text{Xe}_2^* + \text{Xe}; \tag{81}$$

$$\text{Xe}_2^* \xrightarrow{\tau_{\text{sp}}} \text{Xe} + \text{Xe} + h\nu_{\text{Xe}}. \tag{82}$$

As a result of process (79), since the energy of the molecular state $^{1,3}\Sigma_u^+$ of argon coincides with the energy of the atomic state $^1P_1$ of xenon, excited xenon atoms are produced. The triple-collision processes (80) and (81) lead to the formation of the molecule Xe_2^*. The rate of formation of the Xe_2^* molecules is

$$(dn/dt)_{\text{Xe}_2^*} = \beta_{\text{Ar}} (n)_{\text{Xe}^*} (n)_{\text{Ar}} (n)_{\text{Xe}} + \beta_{\text{Xe}} (n)_{\text{Xe}^*} (n)_{\text{Xe}}^2 - (n)_{\text{Xe}_2^*}/\tau_{\text{sp}}. \tag{83}$$

Here β_{Ar} and β_{Xe} are the rate constants of the collision processes (80) and (81); $(n)_{\text{Xe}^*}$, $(n)_{\text{Xe}}$, $(n)_{\text{Ar}}$, $(n)_{\text{Xe}_2^*}$ are the concentrations of the atoms Xe^*, Xe, Ar and of the molecules Xe_2^*, respectively.

The rate of the disintegration of the excited xenon atoms is determined by the processes

$$\text{Xe}^* + \text{Ar} \xrightarrow{\alpha_{\text{Ar}}} \text{Xe} + \text{Ar}; \tag{84}$$

$$\text{Xe}^* + \text{Xe} \xrightarrow{\alpha_{\text{Xe}}} \text{Xe} + \text{Xe}. \tag{85}$$

Therefore

$$(dn/dt)_{\text{Xe}^*} = -\gamma (n)_{\text{Xe}^*} = -\beta_{\text{Ar}} (n)_{\text{Xe}^*} (n)_{\text{Xe}} (n)_{\text{Ar}} - \alpha_{\text{Ar}} (n)_{\text{Xe}^*} (n)_{\text{Ar}} - $$
$$- \alpha_{\text{Xe}} (n)_{\text{Xe}^*} (n)_{\text{Xe}} - \beta_{\text{Xe}} (n)_{\text{Xe}^*} (n)_{\text{Xe}}^2.$$

The measurements of the time dependence of the spontaneous emission of the xenon molecules at short excitation-pulse durations of pure Xe and of Xe : Ar mixtures make it possible to determine the rates of the considered processes.

An experimental investigation of the time dependence of the spontaneous emission of Xe and of Xe : Ar mixtures was carried out with the aid of a table-top electron gun with a cold cathode, which is a greatly reduced prototype of the accelerator described in [37]. The parameters of the electron beam were the following: $U_e \simeq 150$ keV, $j_e = 50$ A/cm^2, $\tau_e \simeq 2$ nsec. The radiation of the gases was focused with the aid of a LiF lens on the input slit of a DFS-29 vacuum spectrograph, behind the exit slit of which was placed a LiF window covered with a lumogen. The pulses of duration less than 100 nsec were registered with a high-speed photomultiplier (resolution ~1 nsec). Pulses of duration larger than 100 nsec were registered with an FÉU-37 photomultiplier with a cathode follower, which transmitted without distortion pulses from 100 nsec to 40 μsec. The following values were obtained for the rate constants of the processes:

$$\beta_{\text{Xe}} = (3.1 \pm 0.5) \cdot 10^{-32} \text{ cm}^6 \cdot \text{sec}^{-1}; \quad \alpha_{\text{Xe}} = (1.1 \pm 0.2) \cdot 10^{-13} \text{ cm}^3 \cdot \text{sec}^{-1};$$
$$\beta_{\text{Ar}} = (1.8 \pm 0.4) \cdot 10^{-32} \text{ cm}^6 \cdot \text{sec}^{-1}; \quad \alpha_{\text{Ar}} = (2.4 \pm 0.6) \cdot 10^{-15} \text{ cm}^3 \cdot \text{sec}^{-1};$$
$$\tau_{\text{Ca}} = (30 \pm 10) \text{ nsec}.$$

Analogous measurements for the case of a small content of Xe in the Xe : Ar mixture were performed in [104].

The data obtained suggest the possibility of developing a laser using a mixture of compressed argon and a small amount of xenon, pumped by an electron beam or by the electroionization method at a pump duration ~2-3 μsec. The latter circumstance, namely the possibility of pumping for such long times [$\tau \simeq 1/\gamma$, see Eq. (86)], is due to the dragging of the radiation upon decay of the $^1P_1$ state of xenon, which has a short lifetime $\sim 4 \cdot 10^{-9}$ sec.

The considered Ar : Xe laser system is of great practical interest for the following reasons:

(1) The pump pulse duration ~ 2-3 μsec makes it possible to develop installations of very high power.

(2) The low concentration of the active particles leads to relatively small values of the gain, a very important factor for the development of installations of great length.

(3) The high purity of the commercially available argon makes for very small values of nonresonant absorption in active media with large dimensions.

(4) The large width of the Xe_2^* emission line makes it possible to shape very short radiation pulses and to tune the emission frequency.

3. Ultraviolet High-Pressure Laser Using the Mixture Ar : N$_2$

1. At the present time the generation of ultraviolet radiation in the second positive system of nitrogen $C^3\Pi_u \to B^3\Pi_g$ ($\lambda = 3371$ Å) was obtained both by exciting the N_2 with a short discharge-current pulse, and with a beam of fast electrons from an accelerator. The attainable efficiency and the specific lasing power, however, are small (0.6% and 2 kW/cm^3 [105]). To improve the energy characteristics of the laser it was proposed to populate the upper laser level $N_2(C^3\Pi_u)$ by energy transfer from ^3P states of argon excited in a discharge [106]. However, no lasing with the Ar : N_2 mixture was obtained. Recent experiment on the excitation of Xe, Kr, and Ar with an intense electron beam have demonstrated high efficiency of pumping the ^3P levels of inert gases, namely $\sim 20\%$ [107, 108]. We have therefore performed experiments aimed at obtaining lasing in an Ar : N_2 mixture excited by an electron beam.

2. An electron beam from an accelerator with a cold cathode (U $\simeq$ 700 keV, $j_e \simeq$ 100 A/cm^2, $\tau \simeq 10^{-8}$ sec) [37] was introduced into the chamber with the working mixture (see Fig. 68) through a titanium foil 50 μm thick. The length of the active region was 4 cm. The laser resonator was made up of a spherical total-reflection mirror (r = 0.5-3 m) and a flat semitransparent interference mirror (T = 0.5-35%) sputtered on a quartz substrate. The distance between the mirrors was 6 cm. The laser emission energy was measured with a calorimeter and the power with a high-speed vacuum photocell (resolution not worse than 10^{-9} sec). The emission spectra were photographed with the aid of a VM-1 monochromator (dispersion 18 Å/mm) on RF-3 film. The spectral and energy characteristics of the laser and spontaneous emissions were investigated as functions of the composition of the Ar : N_2 mixture at pressures 0-20 atm, and as functions of the Q of the resonator. In the spontaneous emission spectra of the pure nitrogen and of the R : N_2 mixtures, the line of the $C^3\Pi_{uv'=0} \to B^3\Pi_{gv''=1}$ ($\lambda = 3577$ Å) transition has the highest intensity. The spontaneous-emission power increases by approximately 1 order of magnitude when argon is added to the nitrogen. This indicates that direct excitation by electron impact of the $C^3\Pi_u$ level is small in comparison with the population by the argon. The lasing threshold was not reached in mixtures containing more than 50% N_2 at semitransparent mirror reflections less than 90%. Lasing was obtained at the wavelength 3577 Å in a mixture with Ar : N_2 composition from 1 : 1 to 100 : 1. The shape of the laser pulse at pressures larger than 0.7 atm duplicated for all the investigated mixtures the shape of the pump pulse. Figure 73 shows the dependence of the spontaneous emission for the mixture Ar : N_2 = 4 : 1 on the pressure. It is seen that at mixture pressures lower than ~ 5 atm a linear increase of the power of the spontaneous emission with increasing pressure is observed and corresponds to a linear increase of the power pump. At pressures larger than 7 atm, no in-

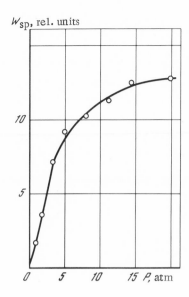

Fig. 73. Dependence of the spontaneous emission power of the mixture Ar:N$_2$ = 4:1 on the pressure.

Fig. 74. Pressure dependence of the specific lasing power (λ = 3577 Å) of an Ar:N$_2$ = 4:1 mixture laser (T = 12.5%).

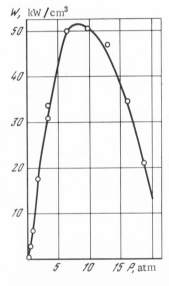

crease of the spontaneous-emission power is observed. In the same pressure region, however, a noticeable decrease in the power and energy of the laser radiation is observed (Figs. 74 and 75). The maximum specific lasing powers and energies reached 70 kW/cm^3 and 0.7 mJ/cm^3 in a mixture of Ar : N$_2$ = 10 : 1 at approximately 8 atm pressure and a mirror transmission coefficient T = 35%. These values greatly exceed those obtained earlier. The laser efficiency reached 3%. Microphotographs of the spontaneous and laser emission spectra at Ar : N$_2$ = 3.5 : 1, mixture pressure 14 atm, are shown in Fig. 76. The width of the lasing line ($\sim$2 Å) is determined by the monochromator resolution. The intensity of the lasing spectrum was decreased in comparison with the intensity of the spontaneous emission spectrum by an approximate factor 500.

3. It can be assumed that the inversion and lasing are produced in the Ar : N$_2$ mixture in accordance with the following scheme (Fig. 77):

$$e + \mathrm{Ar}\,(^1S_0) \xrightarrow{k_1} e + \mathrm{Ar}\,(^3P); \tag{86}$$

$$\mathrm{Ar}\,(^3P) + \mathrm{N}_2\,(X'\,\Sigma^+_{gv=0}) \xrightarrow{k_2} \mathrm{Ar}\,(^1S_0) + \mathrm{N}_2\,(C^3\Pi_{uv'=0}); \tag{87}$$

$$\mathrm{N}_2\,(C^3\Pi_{uv'=0}) \xrightarrow{\tau_{21}} \mathrm{N}_2\,(B^3\Pi_{gv''=1}) + h\nu. \tag{88}$$

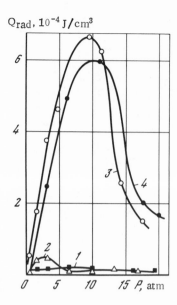

Fig. 75. Dependence of the specific energy of an Ar:N$_2$ mixture laser on the pressure for different mixture compositions (T = 35%). 1) Ar:N$_2$ = 2:1; 2) 20:1; 3) 10:1; 4) 4:1.

Fig. 76. Microphotographs of spontaneous (1) and laser (2) emission spectra (D is the photographic density of the film).

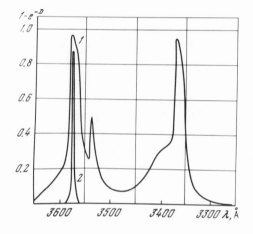

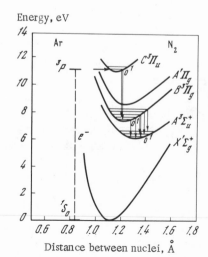

Fig. 77. Level scheme of Ar:N$_2$ mixture laser.

The lifetime of the metastable ^3P levels of argon is ~ 1 sec [108]. As a result of collisions of the second kind, excited molecules $N_2(C^3\Pi_4)$ predominantly in a state with $v' = 0$ are produced. The effective excitation-transfer cross section is $\sim 3.4 \cdot 10^{-16}$ cm^2 [109]. At high pressures, the process competing with (87) is the formation of excited Ar_2^* molecules, which emit in the vacuum ultraviolet region of the spectrum:

$$Ar\,(^3P) + 2Ar\,(^1S_0) \xrightarrow{k_4} Ar_2^* + Ar. \tag{89}$$

The rate of formation of Ar_2^* molecules in triple collisions is $k_4 \sim 10^{-32}$ cm^6-sec^{-1} [110].

The stationary solution of the rate equations describing the reactions (86)-(89) shows that the population of the level $C^3\Pi_{uv'=0}$ at the lasing threshold is equal to

$$n_2 = A\tau_{21}k_2 x\,(1-x)j/[k_2 x + k_4 N\,(1-x)^2], \tag{90}$$

where τ_{21} is the lifetime of the level relative to (88); k_2 is the rate of energy transfer from $Ar\,(^3P)$ to $N_2(X^1\Sigma_g^+)$; j is the accelerator current density; x is the fraction of N_2 in the mixture; N is the number of mixture particles per cubic centimeter; A is a normalization constant. The lower laser level, in accordance with the Franck–Condon principle, is negligibly populated, i.e., $n_1 = 0$. In connection with the collision quenching of the $C^3\Pi_{uv=0}$ level the value of τ_{21} decreases with increasing N_2 pressure [111]

$$1/\tau_{21} = 1/\tau_{21}{}^0 + 3.3 \cdot 10^7 Px, \tag{91}$$

where $\tau_{21}^0 = 4 \cdot 10^{-8}$ sec is the lifetime at $P = 0$ and P is the mixture pressure in atmospheres.

The relations (90) and (91) explain qualitatively the experimental results. The decrease of the gain and of the output power at high pressures is apparently due to collision quenching, to the increased rate of the process (89), as well as to the increased width of the transition line.

In a multistage system with a driving generator, the considered laser medium is promising for the development of high-power subnanosecond pulses with high contrast at an efficiency 2-3%.

The use of the electroionization method is promising for the excitation of lasing in the $Ar : N_2$ mixture. It is possible in this case to decrease appreciably (by one order of magnitude) the accelerator power in comparison with excitation by an electron beam only. An increase in the efficiency is also expected. According to estimates, the efficiency of an electroionization laser with a mixture $Ar : N_2 = 100 : 1$ can reach 7-8%.

CONCLUSION

The development of lasers using condensed and compressed gases uncovers a new type of dense working medium, having a number of significant advantages over other known laser media. Above all, condensed and compressed gases can deliver a large amount of coherent-radiation energy per unit volume at a very high efficiency. Lasing in the vacuum ultraviolet region of the spectrum was obtained with condensed and compressed inert gases.

The unique properties of electroionization CO_2 lasers, the possibility of direct conversion of electric energy into coherent-radiation energy with an efficiency $\sim 30\%$, the possibility of obtaining smooth frequency tuning and the generation of ultrashort pulses, the possibility of developing high-power installations, bring these devices to the forefront when it comes to applications of lasers in science and technology. At the present time, diligent work is being carried out in many laboratories towards the development of ultrahigh power installations for laser thermonuclear fusion [79], laser installations for the material-finishing industry [112], as well as projected installations with an average power 10^9 W [113].

One of the applications of an installation with 10^9 W radiation power is aimed at launching satellites on outer-space orbits with the aid of laser beams. According to estimates in [113], for a mass of 1 ton such a launching on an orbit around the earth will be cheaper than the use of ordinary rocket engines. The use of high-pressure electroionization lasers is also highly promising in chemistry. The generation of high-power radiation that can be tuned smoothly in frequency uncovers a possibility of selective stimulation of chemical reactions, which cannot be effected by ordinary methods in thermodynamic-equilibrium systems [114].

The use of electroionization lasers and of the electroionization method of excitation in scientific research makes it possible to extend considerably the capabilities of laser spectroscopy, makes possible measurements of the constants of many nonequilibrium processes, especially at high pressures, which were investigated so far only by means of expensive experiments carried out in essence under equilibrium conditions in shock tubes or in explosive installations [76]. The electroionization method of excitation of compressed gases is highly promising for the investigation of vibrational–translational, vibrational–vibrational, and rotational relaxation in molecular systems at practically arbitrary pressures.

One of the promising applications of the electroionization method is the investigation of the excitation of electronic levels of molecules [9, 21] and the excitation in molecules of vibrational transitions that are forbidden in the dipole approximation [83].

APPENDIX

THEORY OF CURRENT FLOW THROUGH AN IONIZED GAS

The active medium of an electroionization laser is an ionized gas, and the density of the energy introduced into this medium is determined by its conductivity. The investigation of the conductivity of ionized gases has been the subject of a large number of studies, the first of which were carried out back in 1896 [115]. Dating back to the same time are both the first theoretical papers [116] as well as the experimental investigations, which have demonstrated good agreement between theory and experiment [117]. A sufficiently complete exposition of the ideas concerning the mechanism of the conductivity of ionized gases, which developed during those years, is contained in the books of Thomson and Loeb [27, 28]. It should be stated that these premises served as a basis of modern theory of ionization chambers. The experiments on the conductivity of ionized gases performed to date were aimed at investigating the conductivity of gases under conditions close to those that are optimal for the operation of ionization chambers.

These experiments are characterized by a relatively weak ionization intensity q = $(dn_e/dt)_i \simeq 10^8$-10^{12} cm^{-3}-sec^{-1} and high values of the electric field. The conjunction of weak ionization and a strong electric field determines the electric current flowing through the gas, which turns out to be close to the so-called saturation current [27, 30, 31]. Under the conditions of weak ionization, the current flowing through ionized gas filling the space between metallic electrodes is decisively influenced by the space charges produced near the electrodes. In the case of sufficiently weak ionization the space charge values turn out to be such that the conductivity of the gap between the electrodes is determined by the ion mobility. Inasmuch as the ion mobility is smaller by several orders of magnitude than the electron mobility, the total conductivity of the gap and the energy input σE^2 into a unit gas volume turn out to be very small.

To determine the conditions under which the conductivity of an active medium is determined by the mobility of the electron component, let us consider the approximate theory of charged transport in an ionized gas [9].

Let the working gas be placed between two flat electrodes separated by a distance L. The gas between the electrodes is uniformly ionized by the flux of ionizing particles. Then the equations describing the processes accompanying the charge transport can be written in the form (6) given in Chapter I, Sec. 1.

In the case when the effectiveness of the secondary processes at the negative electrode is high and no positive space charge is produced at this electrode, the conductivity of the medium is determined by the mobility of the electronic component. The conditions under which this case can be regarded will be discussed by us later. We consider first the case of weak secondary processes at the cathode, when

$$[-j_i^+]_{x=0} = [j_e]_{x=L}, \tag{1}$$

In other words

$$[n_e]_{x=L} = [n_i^+]_{x=L} = 0. \tag{2}$$

The system of equations (6) of Chapter I is nonlinear and cannot be solved exactly. We introduce a number of simplifying assumptions. We assume that the voltage across the discharge gap is chosen to be so small that at any point of the gap the first Townsend coefficient is $\alpha = 0$. We assume also that the term responsible for the sticking of the electrons is much smaller than the term describing the recombination:

$$<\sigma^- v_e> N^- \ll b_e n_e n_i^+. \tag{3}$$

This inequality can be realized in the case of very high ionization density in electronegative gases (O_2, CO_2, and others) [45].

Thus, the simplified system of equations takes the following form:

$$\frac{dn_e}{dt} = q - b n_e n_i + \frac{d}{dx} n_e v_e; \tag{4}$$

$$\frac{dn_i}{dt} = q - b n_e n_i + \frac{d}{dx} n_i v_i, \tag{5}$$

$$\frac{dE}{dx} = 4\pi (n_i - n_e) e; \tag{6}$$

$$\int_0^L E dx = V; \tag{7}$$

$$\int_0^L (n_i - n_e) dx = 0; \tag{8}$$

$$[-j_i]_{x=0} = [j_e]_{x=L}. \tag{9}$$

Here q is the number of electron−ion pairs produced by the ionizing agent per unit volume and unit time.

Let k_e and k_i be the mobilities of the electrons and ions, determined respectively by

$$v_i = k_i E, \quad v_e = k_e E. \tag{10}$$

The density of the current flowing through the gap is given by

$$j = n_i e k_i E + n_e e k_e E. \tag{11}$$

Equation (11) presupposes that the current is due only to the influence of the electric field E. From (5) and (11) we obtain

$$n_i e = \frac{1}{k_i + k_e} \left\{ \frac{j}{E} + \frac{k_e}{4\pi} \frac{dE}{dx} \right\};$$ (12)

$$n_e e = \frac{1}{k_i + k_e} \left\{ \frac{j}{E} - \frac{k_i}{4\pi} \frac{dE}{dx} \right\}.$$ (13)

Under equilibrium conditions $dn_e/dt = dn_i/dt = 0$ and the production of electron–ion pairs is balanced by the annihilation of the pairs as a result of recombination and of drift out of the gap due to the motion under the influence of the electric field. The diffusion of the electrons and ions corresponds to rates of losses of electrons and ions per unit volume

$$- D_i \frac{d^2 n_i}{dx^2}, \qquad - D_e \frac{d^2 n_e}{dx^2},$$

where D_i and D_e are the ion and electron diffusion coefficients, respectively. For an electric field that is not too weak, the motion of the ions under the influence of the diffusion is negligible in comparison with the motion under the influence of the field. Thus, neglecting diffusion, Eqs. (4) and (5) can be rewritten in the case of stationary motion in the form

$$\frac{d}{dx} (n_i k_i E) = q - b n_i n_e;$$ (14)

$$- \frac{d}{dx} (n_e k_e E) = q - b n_i n_e.$$ (15)

We assume that k_i and k_e are constant at any point of the gap. This assumption is valid for fields satisfying the inequality $E/P < 10$ V–cm^{-1}–Torr^{-1} [45]. We then have from (6), (14), and (15)

$$d^2 E^2/dx^2 = 8\pi e (q - b n_i n_e) (1/k_i + 1/k_e).$$ (16)

It is seen from (16) that in the case when the ionization rate exceeds the recombination rate, i.e., $(q - b n_i n_e) > 0$, we have

$$d^2 E^2/dx^2 > 0,$$

and the plot of $[E(x)]^2$ is convex; in the case when $(q - b n_i n_e) < 0$ this curve is concave relative to the x axis. Substituting in (16) the quantities n_i and n_e from (12) and (13), we obtain

$$\frac{d^2 E^2}{dx^2} = 8\pi e \left(\frac{1}{k_i} + \frac{1}{k_e} \right) \left\{ q - \frac{b}{e^2 E^2 (k_i + k_e)^2} \left(j + \frac{k_e}{8\pi} \frac{dE^2}{dx} \right) \left(j - \frac{k_i}{8\pi} \frac{dE^2}{dx} \right) \right\}.$$ (17)

Equation (17) has a general solution only if q is constant and $k_i = k_e = k$. This case is of no interest to us, since $k_i = k_e$ corresponds to small values of the negative carrier mobility, in which case the energy input to the gas is very small at reasonable values of the electric field.

The system of equations (6), (11), (14), (15) has a particular solution for the case of stationary motion of the electrons, which is determined by the relations

$$n_i = n_e = \left(\frac{q}{b} \right)^{1/2};$$ (18)

$$k_i n_i E_e = \frac{k_i}{k_i + k_e}\, j;$$

(19)

$$k_e n_e E_e = \frac{k_e}{k_i + k_e}\, j;$$

(20)

$$E = \left(\frac{b}{q}\right)^{1/2} \frac{j}{e\,(k_i + k_e)}\, .$$

(21)

This solution corresponds to a constant value of the electric field between the plates, and indicates that the ratio of the currents transported by the positive and negative charges is equal to the ratio of the drift velocities of the carriers of these charges. While this solution can be valid for the central part of the gap, it cannot be used for the regions near the electrodes. Let x* be a point in the gap between the electrodes, at which relations (18)-(20) are valid. Then a unit surface parallel to the electrodes and passing through the point x* will be crossed in a unit time by $jk_i/(k_i + k_e)_e$ positive carriers. These carriers should come from the region between x* and the positive electrode. If the distance between the point x* and the positive electrode is L_i, then not more than qL_i positive carriers can leave this region in a unit time, and furthermore only if there is no recombination in this region. Consequently, the solution in the form (18)-(21) cannot be correct at distances from the positive or negative electrode smaller than

$$L_i = \frac{k_i}{k_i + k_e} \frac{j}{qe}\, ,$$

(22)

$$L_e = \frac{k_e}{k_i + k_e} \frac{j}{qe}\, ,$$

(23)

respectively.

We assume that the solution (18)-(21) is valid at distances larger than L_i and L_e from the electrodes. We assume also that in the layers next to the electrodes, where relations (18)-(21) do not hold, there is no recombination of the ions. The assumption that there is no recombination is justified by the fact that the concentrations of carriers of opposite sign in the region of these layers are not equal to each other and their product is smaller than $n_e n_i$ in the central part of the gap. Since $n_e = n_i = 0$ on the positive and negative electrodes, in the immediate vicinity of the electrodes the recombination is exactly equal to zero.

Consider now the layer between the positive electrode (x = 0) and the plane $x = L_i$. If there is no recombination in the region (0, L_i), then Eqs. (6) and (14), (15) can be rewritten in the form

$$\frac{dE}{dx} = 4\pi e\,(n_i - n_e);$$

(24)

$$\frac{d}{dx}\,(k_i n_i E) = q;$$

(25)

$$\frac{d}{dx}\,(k_e n_e E) = -q.$$

(26)

If q is constant, we have

$$k_i n_i E = qx;$$

(27)

$$k_e n_e E = j/e - qx.$$

(28)

The integration constant was chosen here such that

$$[n_i]_{x=0} = 0.$$

(29)

Substituting the values of n_i and n_e from (27) and (28) in (24), we obtain

$$E \frac{dE}{dx} = 4\pi \left\{ qx \left(\frac{1}{k_i} + \frac{1}{k_e} \right) - \frac{j}{ek_e} \right\}. \tag{30}$$

Integrating (30) we get

$$E^2 = 8\pi e \left[\frac{1}{2} qx^2 \left(\frac{1}{k_i} + \frac{1}{k_e} \right) - \frac{jx}{ek_e} \right] + C. \tag{31}$$

We determine the constant C from the condition that at $x = L_i$ [in accord with (21)] we have

$$E^2 = \frac{b}{q} \frac{j^2}{e^2 (k_i + k_e)^2}. $$

From this we get

$$C = \frac{b}{q} \frac{j^2}{e^2 (k_i + k_e)^2} \left[1 + \frac{4\pi e}{b} \frac{k_i}{k_e} (k_i + k_e) \right]. \tag{32}$$

At $x = 0$, i.e., at the positive electrode, we have $E^2 = C$. Let $[E]_{x=0} = E_i$, $[E]_{x=L} = E_e$, and $E = E_0$ in the region between the layers. Then

$$E_i = E_0 \left[1 + \frac{4\pi e}{b} \frac{k_i}{k_e} (k_i + k_e) \right]^{1/2}. \tag{33}$$

We obtain similarly an expression for E_e:

$$E_e = E_0 \left[1 + \frac{4\pi e}{b} \frac{k_e}{k_i} (k_i + k_e) \right]^{1/2}. \tag{34}$$

From (33) and (34) it is seen that we always have

$$E_i > E_0 \quad \text{and} \quad E_e > E_0. \tag{35}$$

The potential drop across the space-charge layers near the electrodes is equal to

$$V_i = \int_0^{L_i} E dx; \tag{36}$$

$$V_e = \int_{L-L_e}^{L} E dx. \tag{37}$$

We put

$$\beta_i = \frac{4\pi e}{b} \frac{k_i}{k_e} (k_i + k_e), \quad \beta_e = \frac{4\pi e}{b} \frac{k_e}{k_i} (k_i + k_e).$$

Substituting in (36) and (37) the value of E obtained from (31), we get after integration

$$V_i = \frac{1}{2} E_i L_i + \frac{1}{2} \frac{E_0 L_i}{\sqrt{\beta_i}} \ln \left(\sqrt{\beta_i} + \sqrt{1 + \beta_i} \right)$$
$$= \frac{1}{2} E_0 L_i \left[(1 + \beta_i)^{1/2} + \frac{1}{\sqrt{\beta_i}} \ln \left(\sqrt{\beta_i} + \sqrt{1 + \beta_i} \right) \right]. \tag{38}$$

Using expression (21), which determines the field E_0 at the center of the gap, as well as Eq. (22), we obtain

$$V_i = \frac{1}{2} \frac{b^{1/2}}{q^{3/2}} \frac{j^2 k_i}{(k_i + k_e)^2} \left[(1 + \beta_i)^{1/2} + \frac{1}{\sqrt{\beta_i}} \ln(\sqrt{\beta_i} + \sqrt{1 + \beta_i}) \right]. \tag{39}$$

We obtain similarly an expression for the cathode potential drop

$$V_e = \frac{1}{2} \frac{b^{1/2}}{q^{3/2}} \frac{j^2 k_e}{(k_e + k_i)^2} \left[(1 + \beta_e)^{1/2} + \frac{1}{\sqrt{\beta_e}} \ln(\sqrt{\beta_e} + \sqrt{\beta_e + 1}) \right]. \tag{40}$$

It is seen from (39) and (40) that

$$V_i \sim j^2 q^{-3/2} \frac{k_i}{(k_i + k_e)}; \tag{41}$$

$$V_e \sim j^2 q^{-3/2} \frac{k_e}{(k_i + k_e)}. \tag{42}$$

Thus, the values of the cathode and the anode potential drops can be very small if the current through the gap is small and the ionization is strong enough. It is also seen from (41) and (42) that the potential drop is larger at the electrode having the same sign as the carriers with the high mobility.

Expressions (39) and (40) can be simplified somewhat by recognizing that under real conditions we have

$$\beta_i \gg 1 \quad \text{and} \quad \beta_e \gg 1. \tag{43}$$

Using (38) we obtain

$$V_i = \frac{1}{2} \frac{b^{1/2}}{q^{3/2}} \frac{j^2}{e^2} \frac{k_i}{(k_i + k_e)^2} \beta_i^{1/2} = \frac{\sqrt{\pi} j^2}{k_e^2} \left(\frac{k_i k_e}{q_e (k_i + k_e)} \right)^{3/2}; \tag{44}$$

$$V_e = \frac{1}{2} \frac{b^{1/2}}{q^{3/2}} \frac{j^2}{e^2} \frac{k_e}{(k_i + k_e)^2} \beta_e^{1/2} = \frac{\sqrt{\pi} j^2}{k_i^2} \left(\frac{k_i k_e}{q_e (k_i + k_e)} \right)^{3/2}. \tag{45}$$

Hence

$$V_i / V_e = k_i^2 / k_e^2. \tag{46}$$

Since the carrier mobility is inversely proportional to the pressure:

$$k_i \sim 1/P, \quad k_e \sim 1/P, \tag{47}$$

and the ionization rate is

$$q \sim P, \tag{48}$$

we obtain from (44) and (45)

$$V_i \sim 1/P \quad \text{and} \quad V_e \sim 1/P. \tag{49}$$

If we add L_i and L_e, we obtain

$$L_i + L_e = j/qe.$$

From (7) we have

$$V = V_i + V_e + \int_{L_i}^{L-L_e} E dx.$$

(50)

According to (21), E is independent of x in the interval $[L_i, L - L_e]$, so that the integral in (50) can be written in the form

$$\int_{L_i}^{L-L_e} E dx = E_0 \left(L - \frac{j}{qe} \right) = \left(\frac{b}{q} \right)^{1/2} \frac{j}{e(k_i + k_e)} \left(L - \frac{j}{qe} \right).$$

(51)

Substituting (44), (45), and (51) in (50), we obtain the dependence of the potential difference V between the electrodes on the current density j:

$$V = \frac{1}{2} \frac{b^{1/2}}{q^{3/2}} \frac{j^2}{e^2 (k_i + k_e)^2} \left[k_i (1 + \beta_i)^{1/2} + \frac{k_i}{\sqrt{\beta_i}} \ln (\sqrt{\beta_i}) + \sqrt{1 + \beta_i} + k_e (1 + \beta_e)^{1/2} + \right.$$
$$\left. + \frac{k_e}{\sqrt{\beta_e}} \ln (\sqrt{\beta_e} + \sqrt{1 + \beta_e}) \right] + \frac{b^{1/2}}{q^{1/2}} \frac{j}{e (k_i + k_e)} \left(L - \frac{j}{qe} \right).$$

(52)

Expression (52) can be represented in the form

$$V = A j^2 + B j.$$

(53)

The approximations that have led to (32) are justified only so long as

$$L_i + L_e \ll L$$

(54)

or

$$j \ll qeL.$$

(55)

Expression (55) determines the saturation current, or the maximum current through the ionized gas under conditions when Townsend multiplication processes do not set in:

$$j_s = qeL.$$

(56)

Substituting (56) in (52) we obtain the potential V_s at which the saturation current is reached:

$$V_s \sim L^2 q^{1/2}.$$

(57)

Experiments performed by Seemann [117] have shown splendid agreement between the theory developed above and experiment. These experiments corresponded to the case of weak ionization. However, inasmuch as no assumptions regarding the value of q were made in the derivation of (52), there are grounds for assuming that (52) is valid also for the case of strong ionization.

We change over from the general expressions derived above to an actual case of high electron mobility, when

$$v_e \gg v_i, \qquad k_e \gg k_i.$$

(58)

We then get from (21)

$$E_0 = \sqrt{\frac{b}{q}} \frac{j}{ek_e}.$$

(59)

Taking (58) and (43) into account in expressions (33), (34), (44), (45), we obtain

$$E_i = \left(\frac{\pi e k_i}{k_e^2 q}\right)^{1/2} \frac{j}{e} = \left(\frac{k_i}{k_e^2} \cdot \frac{\pi e}{q}\right)^{1/2} \frac{j}{e};$$

(60)

$$E_e = 2\left(\frac{\pi e}{qk_i}\right)^{1/2} \frac{j}{e} = 2\left(\frac{1}{k_i} \frac{\pi e}{q}\right)^{1/2} \frac{j}{e};$$

(61)

$$V_i = \frac{\sqrt{\pi} j^2}{k_e^2}\left(\frac{k_i}{qe}\right)^{3/2} = \left(\frac{k_i}{k_e}\right)^{3/2}\left(\frac{\pi e}{q^3 k_e}\right)^{1/2} \frac{j^2}{e^2};$$

(62)

$$V_e = \frac{\sqrt{\pi} j^2}{k_i^2}\left(\frac{k_i}{qe}\right)^{3/2} = \left(\frac{\pi e}{q^3 k_i}\right)^{1/2} \frac{j^2}{e^2}.$$

(63)

From (22) and (23) we obtain expressions for the thicknesses of the anode and cathode potential drops

$$L_i = \frac{k_i}{k_e} \frac{j}{qe};$$

(64)

$$L_e = \frac{j}{qe}.$$

(65)

When the conditions (58) are satisfied

$$E_i \ll E_e, \quad V_i \ll V_e, \quad L_i \ll L_e.$$

(66)

It is seen from (66) that in practice it is necessary to take into account only the space charge at the negative electrode. From (52) we can obtain the total potential drop between the electrodes:

$$V = \left(\frac{\pi e}{q^3 k_i}\right)^{1/2} \frac{j^2}{e^2} + \left(L - \frac{j}{qe}\right)\left(\frac{b}{qk_e^2}\right)^{1/2} \frac{j}{e}.$$

(67)

From (67) follows an expression analogous to (53):

$$V = Aj^2 + BLj.$$

(68)

A simple analysis shows that under conditions of weak ionization, i.e., at

$$q \leqslant \frac{1}{L}\left(\frac{\pi k_e^2}{ebk_i}\right)^{1/2} j,$$

(69)

expression (67) can be rewritten in the form

$$V = \left(\frac{\pi e}{q^3 k_i}\right)^{1/2}\left(\frac{j}{e}\right)^2.$$

(70)

In (70) we have retained only one term proportional to $(j/e)^2$. The second term, also proportional to $(j/e)^2$ and dependent on the recombination coefficient, can be neglected in the pressure region P > 1 atm of interest to us. The values of the constants used in this estimate will be given below.

If the inequality sign in (69) is reversed, the dependence of the current density on the voltage between the plates becomes linear:

$$V = L \left(\frac{b}{q} \right)^{1/2} \frac{1}{k_e e} j. \tag{71}$$

Expression (71), rewritten in the form

$$j = \sigma E, \tag{72}$$

shows that Ohm's law is satisfied, and the conductivity of the medium is determined by the mobility of the electronic component.

The conductivity of the medium is here

$$\sigma = n_e e k_e, \tag{73}$$

where $n_e = (q/b)^{1/2}$.

The electric field in (72) is

$$E = V/L. \tag{74}$$

Let us estimate numerically the ionization intensity, the potential between the plates, and the distance between the plates at which Ohm's law is satisfied. To this end we write down the solution of the quadratic equation (67):

$$j = \left(\frac{q}{b} \right)^{1/2} e k_e \frac{V}{L} - \left(\frac{\pi k_e^6 e^3}{k_i b^3 L^2} \right)^{1/2} \left(\frac{V}{L} \right)^2, \tag{75}$$

which is valid if the inequality

$$\frac{q^{1/2}}{V} > \frac{\pi^{1/2} e^{1/2} k_e^2}{k_i^{1/2} L^2 b} \quad \text{or} \quad \frac{q}{V^2} > \frac{\pi e k_e^4}{k_i L^4 b^2} \tag{76}$$

is satisfied. The exact solution of (67) is written in the form

$$j = \frac{q L b^{1/2} e^{1/2} k_i^{1/2}}{2 \pi^{1/2} k_e} \left(\sqrt{1 + \frac{4 \pi^{1/2} k_e^2 e^2}{q^{1/2} k_i^{1/2} L^2 b} V} - 1 \right). \tag{77}$$

At real values of the constants, say, for nitrogen:

$$P = 10 \text{ atm}, \quad k_i = 3 \cdot 10^{-1} \text{ cm}^2 \text{-V}^{-1}\text{-sec}^{-1},$$
$$k_e = 5.9 \cdot 10 \text{ cm}^2\text{-V}^{-1}\text{-sec}^{-1}, \quad V = 3 \cdot 10^4 \text{ volts}, \tag{78}$$
$$L = 1 \text{ cm}, \quad b = 2 \cdot 10^{-7} \text{ cm}^{-3}\text{-sec}^{-1},$$

we have

$$j = 15 \left(\sqrt{1 + 5.1 \cdot 10^{-4} V} - 1 \right) \text{ [A-cm}^{-2}\text{]}. \tag{79}$$

The condition (76) corresponds to the value

$$q > 4.2 \cdot 10^{23} \text{ cm}^{-3}\text{-sec}^{-1}. \tag{80}$$

Since

$$q = j_e \gamma P/P_0 e, \tag{81}$$

where $\gamma = 50$ for N_2 at an electron energy 500 keV and a pressure P = 1 atm, the inequality (80) can be rewritten in the form

$$j_e > 1.3 \cdot 10^2 \text{ A-cm}^{-2}. \tag{82}$$

Satisfaction of the inequality (76) means at the same time that the term linear in V in the expansion of j in powers of V in [75] makes the main contribution to the electric current, and Ohm's law is satisfied in this case. If we assume that $k_i \sim 1/P$, $k_e \sim 1/P$ and that b is constant (as a rule this holds true in experiments [45]), then we can obtain the approximate dependence of the critical value q^*, defined by (76), on the pressure, voltage, and distance between the electrodes:

$$q^* \approx \frac{V^2}{P^3 L^4} \frac{1}{b^2}. \tag{83}$$

Thus, at a high rate of ionization of the gas between the electrodes, the value of which exceeds by several orders of magnitude the typical value of the ionization rate for high-pressure ionization chambers, Ohm's law is satisfied, and the conductivity of the gas is determined by the mobility of the electronic component. The foregoing theory does not take into account secondary processes on electrodes, which lead to the appearance of electron emission from the cathode, nor does it take into account the formation of the "hot cathode layer" with electron multiplication, produced at the cathode at high values of the voltage. The last two effects can greatly raise the voltages at which Ohm's law is satisfied. Let us examine these effects briefly.

Many specialists in the field of gas discharge [28] assume that the main process that leads to electron emission from the cathode is the "potential" knock-out of the electrons by the ions. This process ensures much larger current from the cathode than photoemission, field

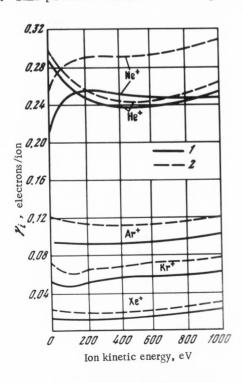

Fig. 78. Electron yield γ_i when atomically pure tungsten (1) and molybdenum (2) is bombarded with inert-gas ions.

emission under the influence of positive space charge, and the kinetic knock-out of electrons by high-energy ions. We shall not dwell on the analysis of these mechanisms, since they are described in many monographs [28, 45].

The potential knock-out from metals is the result of electron interaction between the incident ion and the conduction electrons of the metal, when the ion is at a distance of the order of the dimensions of the atom from the metal surface. The potential knock-out is possible only when the ionization energy is more than double the work function of the material of the surface [118]. The potential knock-out mechanisms are also called direct Auger neutralization and two-step Auger process. Numerous experimental data offer evidence that up to an incident-ion energy of the order of several dozen kiloelectron volts the electron yield depends little on the ion energy and is determined by the ion charge and by its potential. Since the maximum potential is possessed by helium ions, the maximum electron yield, $\gamma_i \sim 0.25$, is observed when helium ions collide with a surface of a metal. Figures 78 and 79 show plots of the number of electrons detached from a metal surface that collides with ions, against the ion energy. It is seen from these figures that the coefficient γ_i can reach large values and that the secondary-electron current can amount to an appreciable fraction of the current through the discharge gap.

At low electron emission from the cathode, when the efficiency of the secondary processes at the cathode is low, an increase of the gap voltage in excess of the value determined by Eqs. (69) and (70) leads to a deviation from Ohm's law; the current density depends in this case on the voltage in the following manner:

$$j \sim \sqrt{V}. \tag{84}$$

If the discharge-gap voltage exceeds a certain critical value at which the field in the cathode layer satisfies the inequality

$$E > E_{cr}, \tag{85}$$

where E_{cr} is the electric field at which a noticeable multiplication of the electrons sets in, then the equations describing the conductivity of the gas in the cathode layer can be written in the form

$$\frac{d}{dx}(n_e v_e) = n_e \alpha(x) v_e. \tag{86}$$

Equation (86) corresponds to a situation wherein the rate of production of the electron–ion pairs as a result of Townsend multiplication of the electrons greatly exceeds the rate q due to the ionizing agent. Then the processes in the cathode layer recall in many respects the processes in the cathode layer of a low-pressure glow discharge. The role of the positive

Fig. 79. Electron yield γ_i when mixtures of hydrogen (1), nitrogen (2), and oxygen (3) ions are used to bombard platinum coated with molecules of the same type as the bombarding ions, as a function of the energy of the incident ion.

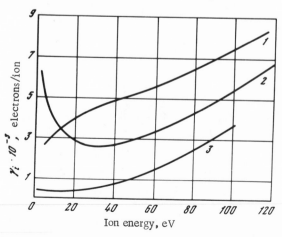

column in the investigated discharge is played by the quasineutral plasma produced by the ionizing radiation.

In accordance with the foregoing, we shall assume that the main process that determines the electron current from the cathode is ion bombardment of the cathode:

$$j_e(x=0) = -\gamma_i j_i \qquad (x=0). \tag{87}$$

From this we obtain

$$n_e(0) = \gamma_i n_i(0) \frac{v_i(0)}{v_e(0)}. \tag{88}$$

The condition for maintaining this type of discharge is the equality of the electron current at the end of the cathode layer to the electron current in the central part of the gap

$$j_e(L_c) = e n_e k_e E \tag{89}$$

and equality of the sum of the electron and ion currents to the cathode and the electron current in the main part of the gap

$$\begin{aligned} j_e(0) + j_i(0) &= j_c, \\ (1 + \gamma_i) j_i(0) &= j_c. \end{aligned} \tag{90}$$

When a discharge with a "hot" cathode layer is maintained, the region in which Ohm's law is satisfied can greatly extend in the direction of higher voltages, and is limited only by the voltage across the gap at which the electron multiplication in the central region of the gap becomes noticeable:

$$V/L > E_{\mathrm{cr}}. \tag{91}$$

As already mentioned, the experimental data obtained in the present study have shown that the region of applicability of Ohm's law is wider than is determined by Thomson's theory given at the beginning of this section. In [9, 33] are presented calculations of the conditions for the existence of the considered type of discharge, performed by Belenkov and Suchkov. We present a brief description of these calculations.

To solve Eq. (86) it is customary to make a number of simplifying assumptions, since the parameters α, v_e, v_i that enter in (86) and in the boundary conditions (89) and (90) depend on the composition of the gas and the analytic expressions that reflect their field dependences turn out to be different for different gases. In accordance with the assumptions already made by us, we assume that k_i and $k_e \sim 1/P$, and that k_i and k_e do not depend on the field E.

Thus

$$v_e(x) = k_e E(x); \tag{92}$$

$$v_i(x) = k_i E(x). \tag{93}$$

Integrating (86) with (92) and (93) taken into account, we obtain the electron density $n_e^k(x)$ and the electronic component of the current $j_e^k(x)$ in the cathode layer

$$n_e^k(x) = n_e^k(0) \frac{E_0}{E(x)} e^{\int_0^x \alpha\, dx'}; \tag{94}$$

$$j_e^k(x) = e k_e n_e^k(0) E(0) e^{\int_0^x \alpha\, dx'}. \tag{95}$$

Just as in the case of a glow discharge [28], we shall assume that the field intensity in the region of the potential cathode drop decreases linearly:

$$E(x) = E(0)(1 - x/L_e), \quad 0 < x < L_e. \tag{96}$$

The assumption that the field intensity E(x) decreases linearly in the cathode layer is confirmed experimentally with good accuracy [36]. We use the empirical relation for α/P, taken from Brown's book [36]:

$$\alpha/P = A(E/P - B)^2. \tag{97}$$

At A = $1.17 \cdot 10^{-4}$ cm-Torr-V^{-2} and B = 32.2 V-cm^{-1}-Torr^{-1}, Eq. (97) for the nitrogen describes the dependence of α/P on the field accurate of several percent in the E/P range

$$40 \text{ V/cm-Torr} < E/P < 180 \text{ V/cm-Torr}. \tag{98}$$

In the same range of the parameter E/P, the drift velocity of the ions can be regarded as constant [36]. Substituting (92), (96), and (97) in (95) and integrating (95) in the range $0 < x' < x_k$, where

$$x_k = L_e - BPL_e/E(0) \tag{99}$$

is determined from the condition $\alpha(x_k) = 0$, we obtain

$$j_e(x_k) = en_e(0)k_e(0)E(0)\exp\left[\frac{AL_eP^2}{3E(0)}\left[\frac{E(0)}{P} - B\right]^3\right]. \tag{100}$$

Using the boundary conditions (89) and (90), we obtain from (100) the equation

$$\frac{AL_eP^2}{3E(0)}\left[\frac{E(0)}{P} - B\right]^3 = \ln\frac{1 + \gamma_i}{\gamma_i}. \tag{101}$$

The values of L_e and E(0) are determined by the space charge. In view of the fact that the mobility of the ions is low in comparison with the mobility of the electrons, the ions make the main contribution to the space charge and, in accordance with the assumed linear field drop, are uniformly distributed in the cathode layer. The Poisson equation (6) is rewritten in this case in the form

$$E(0) = 4\pi en_i L_e. \tag{102}$$

In the practical system of units, which will be used in the final formulas, the Poisson equation (102) is written with an additional factor

$$E(0) = 4\pi en_i L_i c^2 \cdot 10^{-9}\left[\frac{\sec^2 \cdot V}{C\text{-cm}}\right]. \tag{103}$$

In formula (103), E(0) has the dimensionality of V/cm, e is in Coulombs, L_e is in centimeters, and $c^2 = (3 \cdot 10^{10} \text{ cm} \cdot \sec^{-1})^2$ is the square of the speed of light in vacuum. Using (88), (99), and (103) we obtain

$$\frac{E(0)}{P} = \frac{4\pi eL_e}{(1 + \gamma_i)}\frac{n_e k_e(E_0/P)}{v_i(0)}9 \cdot 10^{11}. \tag{104}$$

In (104) the parameters E_0/P, k_e, n_e were taken for the main volume of the ionized gas far from the cathode layer.

Substituting (104) in (101), we obtain an expression for the parameter E(0)/P at the cathode

$$\frac{E(0)}{P} = \left[\ln\left(\frac{1+\gamma_i}{\gamma_i}\right) \frac{108\pi \cdot 10^{11}e}{A v_i(0)(1+\gamma_i)} \right]^{1/3} \left[\frac{k_e n_e}{P}\left(\frac{E_0}{P}\right) \right]^{1/3} + B. \tag{105}$$

From (104) and (105) we can obtain the thickness of the cathode layer:

$$L_e = \frac{(1+\gamma_i)\,v_i(0)}{108\pi \cdot 10^{11} e n_e k_e(E_0/P)} \left[\frac{E(0)}{P}\right] \tag{106}$$

and the cathode potential drop V_e:

$$V_e = \frac{E(0)\,L_e}{2} = \frac{(1+\gamma_i)\,v_i(0)}{216\pi \cdot 10^{11} e} \left[\frac{P}{n_e k_e(E_0/P)}\right]\left[\frac{E(0)}{P}\right]^2. \tag{107}$$

At large values of the electron density n_e in the main volume of the gas (this corresponds, for example, to the case $n_e \sim 10^{15}$ cm^{-3} at $P \sim 10^4$ Torr of N_2 and $(E_0/P) \sim 10$ V-cm^{-1}-Torr^{-1}), the field intensity in the cathode layer can exceed the limit of the region where the approximation (97) is valid. In this case we can use another approximation of the dependence of the parameter α on E/P. For the E/P interval

$$200 \text{ V-cm}^{-1}\text{-Torr}^{-1} < E/P < 800 \text{ V-cm}^{-1}\text{-Torr}^{-1} \tag{108}$$

we can express, accurate to α/P 20%, the dependence of α/P on E/P in the form [36]:

$$\alpha/P = A'(E/P - B'). \tag{109}$$

Using the approximation (109), we obtain the following expressions for E(0)/P, L_e and V_e:

$$\frac{E(0)}{P} = \left[\ln\left(\frac{1+\gamma_i}{\gamma_i}\right) \frac{72\pi \cdot 10^{11}e}{A' v_i(0)(1+\gamma_i)} \right]^{1/2} \left[\frac{k_e n_e}{P}\left(\frac{E_0}{P}\right) \right]^{1/2} + B; \tag{110}$$

$$L_e = \frac{(1+\gamma_i)\,v_i(0)}{108\pi \cdot 10^{11}e} \frac{P}{n_e k_e(E_0/P)} \left[\frac{E(0)}{P}\right]; \tag{111}$$

$$V_e = \frac{(1+\gamma_i)\,v_i(0)}{216\pi \cdot 10^{11}e} \frac{P}{n_e k_e(E_0/P)} \left[\frac{E(0)}{P}\right]^2. \tag{112}$$

In formula (109), A' = 1.25 · 10^{-2} V^{-1} and B' = 50 V-cm^{-1}-Torr^{-1}.

The results of the calculations of L_e and V_e for nitrogen at a pressure of 13 atm and at different values of the electron density n_e are given in Table 4.

TABLE 4

E_0/P, V/cm-Torr	$n_e = 2 \cdot 10^{12}$ cm^{-3}				$n_e = 5 \cdot 10^{14}$ cm^{-3}			
	5	10	15	20	5	10	15	20
$E(0)/P$, V/cm-Torr	79	92	101	107	450	615	740	850
L_e, cm	$8.8 \cdot 10^{-3}$	$5.1 \cdot 10^{-3}$	$3.9 \cdot 10^{-3}$	$3 \cdot 10^{-3}$	$2 \cdot 10^{-4}$	$2.36 \cdot 10^{-4}$	$1.1 \cdot 10^{-4}$	$0.94 \cdot 10^{-4}$
V_e, volts	3500	2330	2870	1610	450	428	410	400

It is seen that for an interelectrode distance $L = 1$ cm the cathode potential drop is not more than several percent of the applied voltage. With increasing density n_e, as follows from (106), (107), (111), (112), the cathode potential drop V_e and the thickness of the cathode layer decrease while the field intensity at the cathode increases. When the electron density is decreased to a value $n_e \simeq 10^{10}$ cm^{-3}, the drop V_e increases to 25 kV and can constitute an appreciable fraction of the applied voltage.

LITERATURE CITED

1. N. G. Basov, Advances in Quantum Electronics, Columbia University Press, New York (1961), p. 506.
2. Yu. M. Popov, Trudy FIAN, 31:3 (1965).
3. O. V. Bogdankevich, Doctoral Dissertation, Physics Institute, Academy of Sciencies of the USSR, (1966).
4. N. G. Basov, E. M. Balashov, O. V. Bogdankevich, V. A. Danilychev, G. N. Kashnikov, N. P. Lantzov, and D. D. Khodkevich, Intern. Conf. on Luminescence, Newark, Delaware, August 20-28 (1969), p. J-7; J. Luminescence, 1:834 (1970); N. G. Basov, V. A. Danilychev, Yu. M. Popov, and D. D. Khodkevich, Pis'ma Zh. Éksp. Teor. Fiz., 12:473 (1970); N. G. Basov, V. A. Danilychev, A. G. Molchanov, Yu. M. Popov, and D. D. Khodkevich. Izv. Akad. Nauk SSSR, Ser. Fiz., 37:494 (1973).
5. H. A. Koehler, L. J. Ferderber, D. L. Redhead, and P. J. Ebert, Appl. Phys. Lett., 21: 198 (1972).
6. M. L. Bhaumik, W. B. Lacina, and M. M. Mann, IEEE J. Quant. Electron., QE-8:150 (1972).
7. A. V. Eletskii and B. M. Smirnov, Dokl. Akad. Nauk SSSR, 190:809 (1970).
8. Yu. V. Afanas'ev, É. M. Belenov, O. V. Bogdankevich, V. A. Danilychev, S. G. Darznek, and A. F. Suchkov, Kratk. Soobshch. Fiz., No. 11:23 (1970).
9. N. G. Basov, É. M. Belenov, V. A. Danilychev, and A. F. Suchkov, Kvant. Élektron., No. 3:121 (1971); N. G. Basov, É. M. Belenov, V. A. Danilychev, O. M. Kerimov, I. B. Kovsh, and A. F. Suchkov, Pis'ma Zh. Éksp. Teor. Fiz., 14:421 (1971); N. G. Basov, É. M. Belenov, V. A. Danilychev, and A. F. Suchkov, Vestn. Akad. Nauk SSSR, No. 3:12 (1972).
10. Yu. A. Tkach, Ya. B. Fainberg, L. I. Bolotin, Ya. Ya. Bessarab, N. P. Gadetskii, Yu. N. Chernen'kii, and A. K. Berezin, Pis'ma Zh. Éksp. Teor. Fiz., 6:956 (1967).
11. V. M. Andriyakhin, E. P. Velikhov, S. A. Golubev, S. S. Krasil'nikov, A. M. Prokhorov, V. D. Pis'mennyi, and A. T. Rakhimov, Pis'ma Zh. Éksp. Teor. Fiz., 8:346 (1968); S. A. Golubev, V. D. Pis'mennyi, T. V. Rakhimov, and A. T. Rakhimov, Zh. Éksp. Teor. Fiz., 62:458 (1972).
12. G. G. Dolgov-Savel'ev, V. V. Kuznetsov, Yu. L. Koz'minykh, and A. M. Orishich, Zh. Prikl. Spektrosk., 12:737 (1970).
13. R. Dumanchin and J. Rocca-Serra, C. R. Acad. Sci. Paris, 269:916 (1969).
14. J. A. Beaulieu, Appl. Phys. Lett., 16:504 (1970).
15. K. B. Persson, J. Appl. Phys., 36:3086 (1965).
16. C. A. Fenstermacher, M. J. Nutter, J. P. Rink, and K. Boyer, Bull. Am. Phys. Soc., 16:42 (1971).
17. K. Boyer, Presentation at Japan − U. S. Seminar on Laser Interaction with Matter, Kyoto, Japan, September 24-29 (1972).
18. J. D. Daugherty, E. R. Pugh, and D. H. Douglas-Hamilton, Bull. Am. Phys. Soc., 17:399 (1972); Phys. Today, 25(1):18 (1972).
19. N. G. Aulstrom, G. Inglesakis, J. F. Holzrichter, T. Kau, J. Jensen, and A. Kolb, Appl. Phys. Lett., 21:492 (1972).
20. N. G. Basov, V. A. Danilychev, A. A. Ionin, I. V. Kovsh, and V. N. Sobolev, Zh. Tekh. Fiz., 43:2357 (1973).

21. N. G. Basov, É. M. Belenov, V. A. Danilychev, O. M. Kerimov, I. B. Kovsh, A. S. Podsosonnyi, and A. F. Suchkov, "Electroionization Lasers," Preprint No. 56, FIAN (1972); N. G. Basov, Seventh Intern. Quantum Electron. Conf., Montreal, May 8-11 (1972); N. G. Basov, Laser Focus, No. 9:45, September (1972).

22. E. V. Locke, E. D. Hoag, and R. A. Hella, IEEE J. Quant. Electron., QE-8:132 (1972).

23. N. G. Basov, É. M. Belenov, V. A. Danilychev, and A. F. Suchkov, Pis'ma Zh. Éksp. Teor. Fiz., 14:545 (1971).

23a. V. N. Bagratashvili, I. M. Knyazev, and V. S. Letokhov, Opt. Commun., 4:154 (1971).

24. N. G. Basov, É. M. Belenov, V. A. Danilychev, O. M. Kerimov, A. S. Podsosonnyi, and A. F. Suchkov, "Emission spectra of CO_2 electroionization lasers," Preprint No. 58, FIAN (1972); N. G. Basov, É. M. Belenov, V. A. Danilychev, O. M. Kerimov, A. S. Podsosonnyi, and A. F. Suchkov, Kratk. Soobshch. Fiz. FIAN, No. 5:44 (1972).

25. N. G. Basov, É. M. Belenov, E. P. Markin, A. N. Oraevskii, and A. V. Pankratov, Zh. Éksp. Teor. Fiz., 64:485 (1973).

26. W. L. Nighan, Phys. Rev., A2:1989 (1970).

27. J. J. Thomson and G. P. Thomson, Conduction of Electricity through Gases, Vol. 1, Cambridge University Press, London (1928).

28. L. Loeb, Basic Processes of Electrical Discharges in Gases [in Russian], GITTL (1950).

29. V. L. Granovskii, Electrical Currents in Gases, Steady-State Current [in Russian], Nauka, Moscow (1971).

30. V. Veksler, L. Groshev, and B. Isaev, Ionization Methods of Investigating Radiation [in Russian], GITTL (1949).

31. B. Rossi and H. Staub, Ionization Chambers and Counters, Experimental Techniques, McGraw-Hill (1949).

32. E. P. Velikhov, I. V. Novobrantsev, V. D. Pis'mennyi, A. T. Rakhimov, and A. N. Starostin, Dokl. Akad. Nauk SSSR, 205:1328 (1972).

33. N. G. Basov, É. M. Belenov, V. A. Danilychev, O. M. Kerimov, I. B. Kovsh, and A. F. Suchkov, Zh. Tekh. Fiz., 62:2540 (1972).

34. C. B. Mills, Digest of Technical Papers, Seventh Intern. Quantum Electron. Conf., Montreal, May 8-11 (1972), p. 78.

35. G. A. Mesyats, Yu. I. Bychkov, B. B. Kremnev, D. D. Korolev, Yu. A. Kurbatov, and V. V. Savin, Preprint No. 3, IOA SO AN SSSR, Tomsk (1972).

36. S. Brown, Basic Data of Plasma Physics, MIT Press (1966).

37. V. A. Danilychev and D. D. Khodkevich, Prib. Tekh. Éksp., No. 3:153 (1971).

38. V. V. Andreev, V. D. Pavlova, E. P. Medvedeva, I. V. Leonov, L. V. Makarova, V. K. Eroshev, and I. B. Kovsh, Supplement to the Report "Investigation of electroionization lasers" [in Russian], FIAN (1973).

39. N. G. Basov, É. M. Belenov, V. A. Danilychev, A. A. Ionin, O. M. Kerimov, I. B. Kovsh, A. A. Lobanov, V. A. Sobolev, A. F. Suchkov, and B. M. Urin, "Investigation of electroionization lasers," Report [in Russian], Chapters 1-4, FIAN (1973).

40. N. G. Basov, V. A. Danilychev, A. A. Ionin, I. B. Kovsh, V. A. Sobolev, A. F. Suchkov, and B. M. Urin, Kvant. Élektron., 2(11):2000 (1975).

41. V. A. Danilychev, O. M. Kerimov, and I. B. Kovsh, Prib. Tekh. Éksp., No. 1:184 (1973).

42. A. Glass and A. Guenther, Appl. Opt., 12:637 (1973).

43. E. M. Vorontsova, B. I. Grechushnikova, G. I. Dastler, and I. P. Petrov, Optical Materials for Infrared Engineering [in Russian], Nauka, Moscow (1966).

44. N. G. Basov, É. M. Belenov, V. A. Danilychev, O. M. Kerimov, I. B. Kovsh, and A. S. Podsosonnyi, "Electroionization Lasers," Report [in Russian], FIAN (1971).

45. E. W. McDaniel, Collision Phenomena in Ionized Gases, Wiley (1964).

46. R. L. Platzman, Radiation Biology and Medicine, Massachusetts (1958).

47. S. V. Starodubtsev and A. M. Romanov, Penetration of Charged Particles through Matter [in Russian], Izd. Akad. Nauk Uzb. SSR, Tashkent (1962).

48. A. G. Molchanov, Usp Fiz. Nauk, 106:165 (1972).
49. R. C. Smith, Appl. Phys. Lett., 21:352 (1972).
50. D. H. Douglas-Hamilton, J. Chem. Phys., 58:4820 (1973).
51. R. T. G. Flynn, Proc. Phys. Soc. London, B-69:748 (1956).
52. G. A. Mesyats and D. I. Proskurovskii, Pis'ma Zh. Éksp. Teor. Fiz., 13:7 (1971).
53. J. J. Lowke, A. V. Phelps, and B. W. Irwin, J. Appl. Phys., 44:4664 (1973).
54. G. Roeter, Electronic Avalanches and Breakdowns in Gases [Russian translation], Mir, Moscow (1968).
55. C. K. N. Patel, Phys. Rev. Lett., 12:588 (1964).
56. N. N. Sobolev and V. V. Sokovikov, Usp. Fiz. Nauk, 91:425 (1967).
57. V. P. Tychinskii, Usp. Fiz. Nauk, 91: 389 (1967).
58. K. C. N. Patel, Sci. Am., 219(2):24 (1968).
59. R. L. Taylor and S. Bitterman, Rev. Modern Phys., 41:26 (1969).
60. B. F. Gordiets, N. N. Sobolev, and L. A. Shelepin, Zh. Éksp. Teor. Fiz., 53:1822 (1967).
61. P. K. Cheo and R. L. Abramo, Appl. Phys. Lett., 14:47 (1969); 15:177 (1969); T. K. McCubbin, Jr., R. Darone, and J. Sorvell, Appl. Phys. Lett., 8:116 (1966); R. R. Patty, E. E. Mauring, and J. A. Gardner, Appl. Opt., 7:224 (1968).
62. R. Dumanchin and J. Rocca-Serra, C. R. Acad. Sci. Paris, 269B:916 (1969).
63. K. A. Laurie and M. M. Hale, IEEE J. Quant. Electron., QE-6:530 (1970).
64. V. V. Ragul'skii and F. S. Faizullov, Opt. Spektrosk., 27:707 (1969).
65. A. E. Siegman, Prof. IEEE, 53:277 (1965); A. N. Chester, Appl. Opt., 12:2139 (1973).
66. M. Born and E. Wolf, Principles of Optics, 5th ed., Pergamon Press (1975).
67. Yu. A. Anan'ev, Usp. Fiz. Nauk, 103:705 (1971).
68. V. P. Avtonomov, E. T. Antropov, N. N. Sobolev, and Yu. V. Troitskii, Preprint No. 20, FIAN (1973).
69. E. F. Ishchenko and Yu. M. Klimkov, Lasers [in Russian], Vysshaya Shkola, Moscow (1969).
70. W. Witterman, IEEE J. Quant. Electron., QE-5:92 (1969).
71. M. A. El'yashevich, Atomic and Molecular Spectroscopy [in Russian], Fizmatgiz, Moscow (1962).
72. N. G. Basov, É. M. Belenov, V. A. Danilychev, O. M. Kerimov, I. B. Kovsh, A. S. Podsosonnyi, A. F. Suchkov, and S. I. Sagitov, "Investigations of electroionization lasers," Report [in Russian], FIAN (1972).
73. R. C. Crafter, A. F. Gibson, M. J. Kent, and M. Kimmett, Brit. J. Appl. Phys., D2:183 (1969).
74. A. M. Robinson, Can. J. Phys., 48:1996 (1970); Brit. J. Appl. Phys., D4:882 (1971).
75. R. L. Taylor, M. Camae, and R. M. Feinberg, "Measurements of vibration–vibration coupling in gas mixtures," Proc. Eleventh Sympos. on Combustion, August 14-20, 1966, Combustion Institute, Pittsburg (1967), p. 49; C. B. Moore, R. E. Wood, B. L. Hu, and J. T. Yardley, J. Chem. Phys., 46:4222 (1967).
76. Ya. B. Zel'dovich and Yu. P. Raizer, Elements of Gas Dynamics and the Classical Theory of Shock Waves, Academic Press, New York (1968).
77. A. F. Gibson, M. F. Kimmitt, and C. A. Posito, Appl. Phys. Lett., 15:546 (1971).
78. H. G. Aulstrom, G. Ingesakis, J. F. Holarichter, T. Kau, J. Jenaon, and A. Kolb, Appl. Phys. Lett., 21:492 (1972).
79. K. Boyer, Presentation at Japan–U. S. Seminar on Laser Interaction with Matter, Kyoto, September 24-29 (1972).
80. J. F. Figueira, W. H. Reicheli, G. T. Schappert, J. F. Stratton, and C. A. Fenstermacher, Appl. Phys. Lett., 22:216 (1973).
81. G. G. Petrash, Usp. Fiz. Nauk, 105:645 (1971).
82. N. G. Basov, V. A. Danilychev, K. Mal'tsev, E. P. Markin, O. M. Kerimov, A. N. Oraevskii, and A. S. Podsosonnyi, Kratk. Soobshch. Fiz. FIAN, No. 7:25 (1973).

83. N. G. Basov, A. N. Oraevskii, and A. F. Suchkov, Pis'ma Zh. Éksp. Teor. Fiz., 16:301 (1972).
84. E. V. Condon, Phys. Rev., 41:579 (1932).
85. R. A. Durie and G. Herzberg, Can. J. Phys., 38:806 (1960).
86. Laser Focus, 9:32 (1973).
87. S. D. Rockwood, G. U. Canavan, and W. A. Proctor, IEEE J. Quant. Electron., QE-9:154, 782 (1973).
88. A. Glass and A. Guenther, Appl. Opt., 12:637 (1973).
89. W. L. Bighan, Appl. Phys. Lett., 20:96 (1972).
90. T. F. Kacheva, V. I. Ozhin, and N. N. Sobolev, Kvant. Élektron., No. 6(18):58 (1973).
91. W. B. Lacina and M. M. Mann, Appl. Phys. Lett., 21:224 (1972).
92. R. E. Center, Digest of Tech. Papers, Seventh Intern. Quantum Electron. Conf., Montreal (1973), p. 70.
93. W. B. Lacina, M. M. Mann, and C. L. McAllister, IEEE J. Quant. Electron., QE-9:588 (1973).
94. S. D. Rockwood, J. E. Brau, W. A. Proctor, and G. H. Canavan, IEEE J. Quant. Electron., QE-9:120 (1973).
95. N. G. Basov, IEEE J. Quant. Electron., 2:354 (1966).
96. N. G. Basov, V. A. Danilychev, O. V. Bogdankevich, A. G. Devyatkov, G. N. Kashnikov, N. P. Lantsov, Pis'ma Zh. Éksp. Teor. Fiz., 7:404 (1968).
97. R. S. Milliken, J. Chem. Phys., 52:5170 (1970).
98. A. V. Phelps, "Tunable gas lasers utilizing ground state dissociation," JILA Report 110, Univ. Colorado, September 15 (1972).
99. H. A. Kochler et al., Appl. Phys. Lett., 21:1 (1972).
100. A. Wayne Johnson and J. B. Gerardo, Preprint 01345, Sandia Laboratories, Albuquerque.
101. A. G. Molchanov and Yu. M. Popov, Kvant. Élektron., 1:134 (1974).
102. N. G. Basov, V. A. Danilychev, V. A. Dolgikh, O. M. Kerimov, A. N. Lobanov, A. S. Podsosonnyi, and A. F. Suchkov, Kvant. Élektron., 2:28 (1975).
103. A. N. Lobanov and A. F. Suchkov, Kvant. Élektron., 1:1527 (1974).
104. D. C. Lorents, R. M. Hill, and D. I. Eckstron, "Molecular metal laser," Preprint 50021, Standford Research Institute.
105. D. A. Leonard, Appl. Phys. Lett., 7:4 (1965); R. W. Dreyfus and R. J. Hodson, Appl. Phys. Lett., 20:195 (1972).
106. A. V. Eletskii and B. M. Smirnov, Gas Lasers [in Russian], Atomizdat, Moscow (1971).
107. J. B. Gerardo and A. W. Johnson, IEEE J. Quant. Electron., QE-9:748 (1973).
108. B. Vandich, Jr., C. E. Johnson, and H. A. Shugart, Phys. Rev., A5:991 (1972).
109. O. P. Bychkova, N. V. Chernysheva, and Yu. A. Tolmachev, Opt. Spektrosk., 36:36 (1974).
110. E. Ellis and N. D. Twiddy, J. Phys., B2:1366 (1969).
111. J. M. Calo and R. C. Axtmann, J. Chem. Phys., 54:1332 (1971).
112. E. V. Locke, E. D. Hoag, and R. A. Hella, IEEE J. Quant. Electron., QE-8(2):132 (1972).
113. A. Kantrowitz, Astronautics and Aeronautics, May, 74 (1972).
114. N. G. Basov, É. M. Belenov, E. P. Markin, A. N. Oraevskii, and A. V. Pankratov, Zh. Éksp. Teor. Fiz., No. 1 (1973).
115. J. J. Thomson, Nature, April, 13 (1896).
116. G. Mie, Ann. Phys., 13:857 (1904); R. Seeliger, Ann. Phys., 33:319 (1970); D. Welker, Philos. Mag., Nov. (1964); K. Robb, Phil. Mag., Aug. and Dec. (1905).
117. A. Seemann, Ann. Phys., 38:781 (1912).
118. H. D. Hagstrom, Phys. Rev., 96:325, 336 (1954).

EXPERIMENTAL INVESTIGATION OF THE REFLECTION AND ABSORPTION OF HIGH-POWER RADIATION IN A LASER PLASMA

O. N. Krokhin, G. V. Sklizkov, and A. S. Shikanov

Results are presented of experimental investigations of the processes of interaction of high-power radiation of a neodymium laser with bulky and thin targets in the flux density interval 10^{10}-10^{15} W/cm^2. A number of observed effects point to an anomalous character of the interaction of the radiation with the plasma, including generation by the plasma of harmonics of frequency ω_0 of the incident radiation, of the type $m\omega_0/2$ (m is an integer), an oscillatory character of the reflected radiation, and anisotropy of the continuous x radiation relative to the electric-field vector of the light wave. A comparison is made of the experimental data with the results of the theory of parametric phenomena in a plasma. Procedures and results of experiments on the absorption of radiation in thin targets are described.

The suggestion to use laser radiation of high intensity to heat plasma to thermonuclear temperatures [1-3] has brought into being a new trend in thermonuclear physics, which is now rapidly developing, namely laser-controlled thermonuclear fusion (LCTF). The progress made in this direction, such as observation of the production of an appreciable number of neutrons [4, 5] in a laser plasma, an indication of the existence of strong compression at the center of an isolated spherical target [6] ($N_e \sim 10^{24}$-10^{25} cm^{-3}), and a lowering to 10^4-10^5 J of the theoretical value of the laser energy needed for a physically profitable thermonuclear reaction [7], etc., have raised the interest of the researchers in the problem of the interaction of intense laser radiation with a plasma.

This article is aimed at clarification of one of the main problems of laser-controlled thermonuclear fusion, namely the absorption and reflection of intense laser radiation in a plasma corona. The importance of research of this type is due, first of all, to the need for creating optimal conditions in the target laser plasma, ensuring both a high coefficient of laser-radiation absorption by the plasma and an effective transformation of the absorbed optical energy into thermal motion of the plasma ions.

CHAPTER I

REFLECTION OF LASER RADIATION FROM A PLASMA (SURVEY OF THE LITERATURE)

One of the most important questions in LCTF is that of the absorption and reflection of optical radiation in the plasma "corona" of a thermonuclear target.

It must be stated that the reflection of light from a plasma is a parasitic phenomenon, since it decreases the optical energy input to the plasma, leading to an increase of the minimum laser energy required for a useful thermonuclear reaction. In addition, the reflected laser radiation returns to the laser system, is amplified there, and contributes to the damage of the active elements, while introduction of various decoupling devices leads to an appreciable decrease of the laser energy and lowers its efficiency.

On the other hand, a study of the radiation reflected and scattered by a plasma makes it possible to determine many parameters of the expanding plasma, to draw some conclusions concerning the mechanisms whereby laser radiation is absorbed, since knowledge of these mechanisms is essential for the optimization of plasma heating, i.e., to obtain a plasma with the required parameters at a minimum laser energy.

In this chapter we survey the experimental studies devoted to one degree or another to reflection and scattering of laser radiation by a dense plasma. The results of the theoretical investigations are presented only when needed for the interpretation of the experimental results.

1. Experimental Conditions Realized in Research on Laser-Plasma Parameters

In this section, without dwelling in detail on the methods used to obtain high-intensity light pulses, we consider briefly the parameters of the light beam and of the focusing systems which are of greatest importance from the point of view of the comparison of the experimental data (a review of modern laser devices for high-temperature plasma heating is contained, for example, in [8]).

1. Focusing System. Dimension of Focal Spot. Length of Caustic.
The overwhelming majority of the heretofore reported experimental studies of the interaction of intense laser radiation with a plasma were performed for the case of sharp focusing, when the laser radiation is "gathered" by a focusing system consisting of one or several lenses and objectives, onto the surface of a flat target placed in a vacuum chamber. The minimum transverse dimension of the light beam in the focal plane is bounded both by the divergence of the laser beam ($d = \alpha f$) and by the spherical aberration of the focusing optical system.

When the divergence of the laser beam is small, the principal role in the formation of the focal spot is played by spherical aberrations, the influence of which is decreased either by aspherical lenses [9] or by focusing systems consisting of two or several lenses [6], which make a focal spot of diameter 20–30 μm possible. Another important parameter of a focusing system is the length of the caustic along which the transverse dimension of the light beam, and consequently the flux density, can be regarded as constant. Simplifying the situation somewhat, we define the caustic length as $L = d \cot \varphi$, where d is the diameter of the focal spot and φ is the angle of incidence of the radiation on the target (Fig. 1).

2. Energy. Flux Density. Radiation Contrast.
From the point of view of concentrating energy in small volumes of matter, the laser method of plasma heating greatly

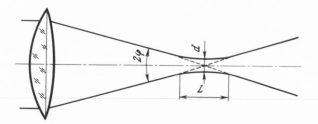

Fig. 1. Definition of the length of the caustic of the lens.

exceeds other known energy sources. The presently existing laser installations with energy 10^2 and even ~10^3 J [10-15] can deliver energy at a rate as high as 10^{17} W/cm^3 and more.

Although the energy of modern lasers is still much lower than that needed for a thermo-nuclear reaction (which ranges from 10^4 to 10^8 J in various theoretical papers, depending on the target structure and the waveform of the laser pulse [16-22]), the rates of energy release and the radiation flux density are in agreement even now with the values that will be realized in a laser thermonuclear installation. For example, for a thermonuclear target of 1 mm diameter and a laser pulse of energy 10^5 J and duration 10^{-9} sec, energy will be delivered at a rate 10^{17} W/cm^3 at a flux density $q \simeq 3 \cdot 10^{15}$ W/cm^2 on the target surface. Thus, the processes that occur in the corona of such a target can be partially simulated and investigated at present or in the nearest future by using sharp focusing.

In experiments on laser-plasma parameters, stringent requirements must be imposed on the value of the radiation contrast, that is, on the ratio of the energy in the heating pulse to the energy of the preceding background. At low values of the contrast, the target becomes damaged before the heating pulse arrives, a cloud of relatively cold plasma is produced, and the picture of the interaction between the radiation and the plasma becomes highly complicated. By now, a contrast of the order of 10^5-10^6 has been reached (e.g., [15]), but to increase the energy of laser installations further, this parameter must be improved in proportion to the laser energy.

2. Energy Composition of the Reflected Radiation; Anomalous Character of the Interaction of Laser Radiation with a Plasma in a Wide Range of Flux Densities

The question of the efficiency at which laser radiation of high power is absorbed when focused on a surface of a condensed target was first raised in [23, 24], devoted to the reflection and scattering of radiation of a neodymium laser (pulse duration 15 nsec, power up to 200 mW) from targets having different atomic numbers in the flux-density range $3 \cdot 10^7$-$3 \cdot 10^{10}$ W/cm^2. (Investigations at low fluxes are reported in [25].) The measurements, performed with a light-measuring sphere that made it possible to register the energy reflected and scattered by a plasma into a solid angle 4π, have demonstrated the following. For all the substances investigated in the experiments, a distinction can be made between three characteristic regions of flux density:

(1) The region of flux-density values at which the reflection coefficient is close to the usual constant of the substance, the target exhibiting no noticeable traces of damage

(2) The region where the reflection coefficient decreases quite rapidly with increasing flux density, and the damage to the target surface is noticeable

(3) The regions of high flux densities ($q \simeq 10^{10}$ W/cm^2), where the reflection coefficient amounts to several percent and a crater is produced on the target surface

Following this study, interest in the problem of the reflection of laser radiation from a plasma arose anew only several years after methods were developed for shortening laser pulses and made it possible to reduce the pulses of Q-switched lasers to several nanoseconds, and installations with radiation energy 10^1-10^2 J were developed.

Of particular interest was the interaction with targets containing deuterium atoms. A group of French scientists [9] investigated the reflection of laser radiation interacting with targets of hydrogen and deuterium ice. In this experiment, the laser energy reached 85 J at a pulse duration 3.5 nsec.

A large-aperture aspherical focusing lens produced a focal spot measuring ~30 μm, but owing to the small beam divergence ($\alpha \sim 2$-$4 \cdot 10^{-3}$) this value was not realized. It was observed that for the solid-hydrogen target the reflection coefficient integrated over the pulse increases monotonically with laser energy as $R \simeq 0.18 \, E_{las}^{1/6}$ and becomes larger than 40% at flux densities ~10^{14} W/cm^2. Here, just as in the experiments that will be discussed below, the reflection coefficient is taken to mean the ratio of the radiation energy reflected into the solid angle of the focusing system to the optical energy incident on the target, i.e., $R = E_{ref}/E_{inc}$, since it followed from the experimental results that the predominant fraction of the radiation scattered by the plasma is reflected "backwards." In particular, the cited experiment has demonstrated that reflection at 45° to the laser beam is smaller by a factor of more than 100 than the strictly "backward" reflection.

It was also observed that at a constant laser energy, both the reflection coefficient R and the neutron yield (~$5 \cdot 10^4$ neutrons) correlate well with each other, and become maximal when the target is exactly in the focal plane of the lens, decreasing strongly even at small defocusing (~100 μm). Analogous results were obtained in [26], where the reflection coefficient was decreased by defocusing, as did also the electron temperature T_e. (In later studies by the same group, values $R \simeq 50\%$ [27] were obtained, and even 60% [28] at flux densities $q \lesssim 10^{15}$ W/cm^2.)

Computer calculations showed very good agreement with the experimental results. However, no nonlinear processes were taken into account in the calculations (the authors had believed that the nonlinear absorption mechanisms should not come into play at $q < 10^{14}$ W/cm^2), and the only mechanism responsible for the absorption of the radiation in the plasma was assumed to be inverse bremsstrahlung, the reflection from the point of the plasma profile with the critical density ($N_e \simeq 10^{21}$ cm^{-3}) being taken equal to unity. Under these assumptions, the monotonic growth of the reflection coefficient can be explained also semiqualitatively, as was done in [29]. Assume that the laser radiation is incident on a plasma with a density profile that falls off from the density ~10^{23} cm^{-3} of the solid to zero (Fig. 2).

On this profile there is a point a at which the plasma frequency $\omega_p = (4\pi e^2 N/m)^{1/2}$ is equal to the frequency of the heating radiation ω_0 ($\omega_p \gg \nu_{ei}$). For a neodymium-laser frequency $\omega_0 = 1.8 \cdot 10^{15}$ sec^{-1}, the dielectric constant ε becomes equal to zero at a density $N_e \simeq 10^{21}$ cm^{-3}, and an effective reflection of the incident radiation takes place in the vicinity of this point. The absorption of the radiation from zero to the point a is determined by the inverse bremsstrahlung coefficient for the plasma, which is given by the expression [30] $K = \omega^2/c\omega_0\tau \, (\omega_0^2 - \omega^2)^{1/2}$, where $\tau \simeq 2.4 \cdot 10^4 T_e^{3/2}/N_e Z$ is the time of electron–ion collisions, Z is the charge of the nucleus, and T_e is the electron temperature in electron volts.

The integrated absorption coefficient is given by

$$K_{int} = 2 \int_{-\infty}^{a} \frac{\omega_p^2(x)\,dx}{c\omega\tau(x)\,[\omega_0^2 - \omega_p^2]^{1/2}}, \tag{1}$$

where the factor in front of the integral sign takes into account the double passage of the radiation through the layer. In a laser plasma at densities $3 \cdot 10^{20}$-10^{21} cm^{-3}, i.e., in the small re-

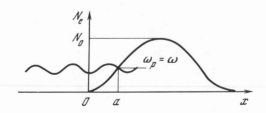

Fig. 2. Schematic density profile of laser plasma.

gion of the profile where the effective absorption takes place, the density profile can be re-garded as linear. Substituting then in the integral (1)

$$\omega_p^2(x) = \frac{4\pi N_e(x) e^2}{m} = \frac{x}{a}\,\omega_0^2, \qquad \tau(x) = \tau(a)\frac{N_e(a)}{N_e(x)}$$

and integrating, we obtain for the reflection coefficient

$$\ln R = -\frac{32}{15}\frac{a}{c\tau(a)}. \tag{2}$$

As seen from the last expression, the reflection coefficient in the inverse bremsstrahlung mechanism increases with increasing plasma temperature (i.e., with the laser energy) in ac-cord with the results of [9].

Some of the results were obtained also in the experimental study [26]. The laser energy was varied between 0.2 and 20 J, but the flux density at the focal plane even exceeded that reached in [9], because of the better divergence of the laser radiation. The dependence of the reflec-tion coefficient on the laser energy turned out to be absolutely the same $E_{ref} \sim E_{inc}^{1.28}$ (the reflection coefficient changed from 8 to 30%). This enabled the authors to draw the rather pessimistic conclusion that when the laser energy is increased to ~1 kJ the reflection coefficient will ap-proach unity and the laser method of plasma heating will become impossible.

Some additional experiments of this study, in which the directivity of the reflected radia-tion was investigated, have yielded, however, more information on the mechanism whereby radiation is reflected from a dense plasma, and led to the conclusion that the appearance of the reflected radiation can be due not only to "classical" reflection from the point with a criti-cal density, but possibly also to stimulated Mandel'shtam−Brillouin scattering (SMBS).

It was observed that the cone of the angle in which the scattering is reflected is not broader, but may even be possibly narrower than the cone in which the radiation is incident on the target (see also [31]), and that each light ray incident on the target is reflected back from it in the same direction; i.e., the target does not reflect like a flat mirror, but like a spherical mirror whose center of curvature must coincide with the focal plane of the lens. This behavior of the reflected radiation was observed not only for solid deuterium but also for carbon. On the basis of all this, the authors have assumed that "classical" reflection is unlikely and pointed to the possibility of SMBS.

Great attention to the problem of reflection of radiation from a plasma was paid also in the experiments of the Japanese scientists at Osaka and Nagoya Universities, who investigated interactions both with hydrogen and deuterium targets and with substances having larger Z [32-38]. In [32] they investigated the reflection of radiation from a $(CH_2)_n$ target. It was observed that the reflection coefficient remains constant during the first 1-1.5 nsec of the laser pulse, and then decreases rapidly within 2-3 nsec from ~10% to ~2%. Simultaneous measurement of the electron temperature with the aid of the absorber [39] has shown that the electron tempera-ture does not change when the laser pulse duration is changed (at a constant flux density q). Starting with flux densities ~$2 \cdot 10^{12}$ W/cm^2, the reflected signal becomes amplitude modulated with frequency ~10^{-10} rad/sec (the waveforms of the incident and reflected signals were regis-tered with coaxial photocells having a time resolution ~10^{-9} sec). The appearance of oscilla-tions of this type were attributed by the authors to excitation in the plasma of ion-acoustic waves following development of parametric instability [34], the threshold for which was esti-mated at $q_{thr} \simeq 10^{13}$ W/cm^2 at $N_e = 10^{21}$ cm^{-3} and $T_e = 200$ eV. Subsequently [36], however, the authors started to ascribe effects of this type to "macroscopic fluctuations of the plasma sur-face."

Similar results are cited in [34]. Oscillations of frequency $1.4 \cdot 10^9$ Hz appear, starting with a flux density $q \simeq 6 \cdot 10^{11}$ W/cm^2 (for $(CH_2)_n$). The reflection coefficient decreases from 2% to 0.5% for Be and from 0.14% to 0.05% for C. For a solid-hydrogen target, the reflection coefficient increases with flux density as $R \sim E_{las}$ [35, 38] and reaches ~10-15% at $q \simeq 10^{14}$ W/cm^2, while in [33, 36, 37] the value of R changed rapidly from 4 to 20% at $q \simeq 10^{13}$ W/cm^2. At the same flux density, neutron production sets in [35-38], the dependence of the electron temperature on the flux takes the form $T_e \sim q^{1/2}$ [35-38], and fast ions with energy ~10 keV appear and their number increases with increasing plasma temperature. The authors believe therefore that the flux density $q \simeq 10^{13}$ W/cm^2 is the critical value starting with which the inverse bremsstrahlung mechanism ceases to play any role. This is motivated by the fact that the absorption length in a plasma near the critical density $l = 5 \cdot 10^{35} \cdot T_e^{3/2} n^{-2}$ cm becomes larger than the characteristic dimension of the plasma at $T_e \simeq 1$ keV, and the laser radiation must be reflected from a point with critical density.

At these temperatures, an important role in the absorption of their radiation can be played by two parametric instabilities that develop at $\omega_0 \simeq \omega_p$ and lead to absorption of the optical energy [40]. The first of them, oscillating two-stream instability (called in the Soviet literature usually aperiodic instability), occurs at $\omega_{ek} > \omega_0$, where $\omega_{ek} = \omega_p \{1 + {}^3/_2 K^2 r_{De}\}$ is the Bohm−Gross frequency of the electron Langmuir oscillations with wave number K.

The threshold of this instability q_{thr} is given by

$$q_{thr} = 2\gamma_e \left(\omega_{ek} \nu_{ek} / \omega_p^2\right) n_0 ck T_e, \tag{3}$$

where ν_{ek} is the damping decrement of the plasma waves including electronic Landau damping, and c is the speed of light.

The other instability develops at $\omega_0 > \omega_{ek}$ (decay of the pump wave into ion-acoustic and plasma waves, $t \to l + s$). The threshold for this instability is

$$q_{thr} = \begin{cases} \dfrac{2\sqrt{3}}{9} \gamma_e \dfrac{\nu_{ek}}{\omega_p} \dfrac{\nu_{ik}}{\Omega_k^2} n_0 ck T_e \ (\text{W}/\text{cm}^2) & (\nu_{ek} > \Omega_k), \\[3mm] \dfrac{\gamma_{ek} \nu_{ek} \nu_{ik}}{\omega_p \Omega_k} n_0 ck T_e \ (\text{W}/\text{cm}^2) & (\Omega_k > \nu_{ek}), \end{cases} \tag{4}$$

where ν_{ik} is the damping decrement of the ion-acoustic waves, including the ion Landau damping, and Ω_k is the frequency of the ion-acoustic wave.

At a temperature $T_e = 400$ eV, $N_e = 10^{21}$ cm^{-3}, and $\omega_0 = 1.8 \cdot 10^{15}$ rad/sec the threshold for this instability is $q_{thr} = 10^{13}$ W/cm^2 (T_e amounted to 400 eV in the experiment of [36] at $q = 3 \cdot 10^{13}$ W/cm^2), and the threshold value for the oscillating two-stream instability turned out to be several times higher at these temperatures, i.e., both instabilities were perfectly feasible in the experiments of [35-38], where the flux density reached $q = 10^{14}$ W/cm^2.

The rapid growth of the reflection coefficient from 4 to 20% is attributed by the authors to the fact that the developing parametric instability produces a sharp dense-plasma boundary, from which strong reflection takes place. This is in all probability an incorrect conclusion, since the aperiodic and decay instabilities lead to a strong absorption of the energy in the plasma.

In addition to [34], low values of the reflection coefficient for laser pulses of nanosecond durations were obtained in experiments carried out in Sandia [41] and the Lawrence Livermore Laboratory [48].

In [41] they investigated the interaction with a $(CD_2)_n$ target. For targets 0.25 and 3 mm thick, the reflection coefficient into the solid angle of the focusing system was the same $R \simeq 0.2\%$ ($E_{las} = 33$ J, $\tau = 3.5$ nsec, $q \simeq 2 \cdot 10^{14}$ W/cm^2).

It was noted that scattering at other emission angles at the wavelength 1.06 μm is small, but integration of the scattered signal over the solid angle 4π sr yielded a value $\sim 10\%$ of the incident energy.

In [42] they investigated the reflected, x-ray, and neutron radiation from $(CD_2)_n$- and C_2H_3Cl (polyvinylchloride) plasma. The reflected signal has a waveform different from the incident-radiation pulse, and in particular, there were sharp spikes of reflected radiation (reaching 3-4% of the incident power) towards the end of the laser pulse, that is, results opposite to those of [32, 34]. The reflection coefficient integrated over the pulse was in the range $R \sim 0.3$-1% at $q \simeq 3 \cdot 10^{14}$ W/cm^2. In addition, the reflected signal was intensity-modulated with a period ~ 2-2.5 nsec, something to which the authors paid no attention. Another characteristic feature of this experiment is the qualitative correlation between the appearance of the reflected-radiation spike, the neutron pulses from the $(CD_2)_n$ (the number of neutrons was 10^3-10^4), and the intensity of the hard x rays with $h\nu \simeq 100$ keV. Similar correlations between the energy of the reflected light, the x radiation, and the appearance of fast ions and neutrons was reported also in [9, 36, 43].

To interpret the observed effects, the authors consider the possibility of exciting in a plasma parametric instabilities, primarily the one connected with the decay of the incident wave into a plasma wave and an ion-acoustic wave. The threshold for this instability is estimated in fact from Eq. (4) for a homogeneous plasma and amounts to $q_{thr} \simeq 6 \cdot 10^{12}$ W/cm^2 at $N_e = $ 0.9 at N_{cr} and $T_e = 0.5$ keV $= T_i$. Excitation of the instability ($t \rightarrow l + s$) in the plasma can explain (see the results of the numerical experiment [44]) the appearance of an appreciable high-energy "tail" in the electron-velocity distribution function. The tail consists of electrons that are drawn out from the thermal distribution and are accelerated to the phase velocity of some of the plasma waves. This is how the authors explain the appearance of high-energy electrons in the experiment, while the behavior of the reflected signal can be explained within the framework of SMBS. It should be noted that the behavior of the reflected signal in [42] differed significantly from the results of the experiments discussed above, where the reflection coefficient was either maximal at the start of the laser pulse and decreased towards its end, or was practically constant during the laser pulse.

At the end of 1973, however, results were published [45] in which the behavior of the reflection coefficient during the laser pulse agreed with [42]. In these experiments the frequency of a neodymium laser was doubled with a KDP harmonic which in turn was used to heat the plasma. It was observed that the reflection coefficient was $\sim 10\%$ and decreased when the radiation was defocused from the surface of the target (in analogy with [9]). The waveform of the reflected radiation duplicated the waveform of the incident pulse at small values of the defocusing, but when the target was exactly in the focal plane the maximum reflection coefficient occurred at the end of the laser pulse. In the same experiments, about 500 neutrons per pulse were registered, apparently for the first time in the case of the second harmonic of a neodymium laser.

The results of the experiments on the reflection coefficient from a plasma are summarized in the survey plots (Fig. 3). In addition to the experiments described above, Fig. 3 shows also the results of [46-50], in which the behavior of the reflection coefficient was not investigated in detail. The figure shows also the results of [51-53], which will be discussed in detail in subsequent chapters.

Despite the large scatter in the presently available results (the differences in the values of the reflection coefficient R reach two to three orders of magnitude at the same flux density

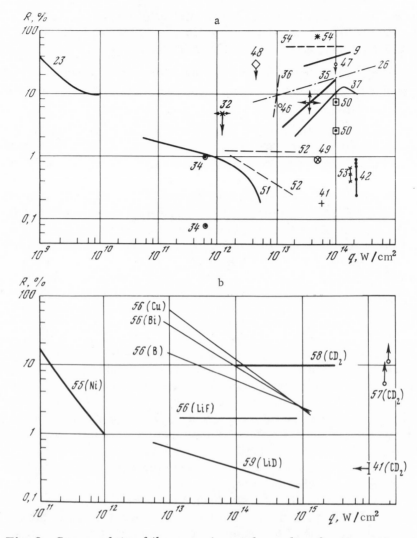

Fig. 3. Survey plots of the experimental results of an investigation of the coefficient of reflection of neodymium-laser radiation from a plasma at nanosecond (a) and picosecond (b) durations into the solid angle of the focusing system. The numbers on the figures are the literature references.

q), it can be concluded that the reflection coefficient exceeds 10% only for targets of solid hydrogen and deuterium, and for these targets R increases monotonically in the flux density interval $q \simeq 10^{13}$-10^{15} W/cm². For energies with $Z > 1$, on the other hand, the reflection coefficient is less than 10%, in some cases less than 1%, and in the experiments of [51, 52] the reflection coefficient for an aluminum target even decreases with increasing flux density q. It can thus be concluded that at $Z > 1$ the overwhelming part of the laser energy is absorbed in the plasma, at any rate up to fluxes $q \lesssim 10$ W/cm².

It should be noted that this conclusion is contradicted in [54], where the reflection coefficient reached 80% for a $(CH_2)_n$ target. The radiation pulse of a mode-locked laser consisted of individual spikes each $\sim 10^{-9}$ sec long with a total duration $(30-35) \cdot 10^{-9}$ sec. The plasma was heated both at the fundamental frequency and at the second harmonic. At the fundamental frequency, the reflection coefficient was $\sim 50\%$ and was independent of either the target material (LiD; $(CH_2)_n$; $\cdot$ C; Al), or the laser energy. In the case of heating at the second harmonic, the reflection coefficient decreases appreciably to ~5% (for LiD), 6% for $(CH_2)_n$) and 7% (for all). The authors attribute this to the fact that the inverse bremsstrahlung coefficient is proportional to ω_0^2 and the absorption should increase strongly with increasing frequency of the heating radiation.

An investigation of the directivity of the reflected radiation was carried out at the second-harmonic frequency by placing diaphragms in the path of the light beam (the radiation flux in this case was $q \simeq 10^{13}$ W/cm^2). As was observed when a round diaphragm was moved along the beam closer to the source, the reflected beam moved farther from the source, and vice versa, in contradiction to the results of [26].

A large number of experimental data on the reflection of optical radiation from a plasma was obtained also with lasers operating in the mode-locking regime. A mode-locked laser generates a train of light pulses with spikes each 10^{-10}-10^{-12} sec long. Electrooptical shutters are usually used to cut out from the sequence of such pulses one pulse or several pulses, which are then amplified and used to produce the plasma.

A characteristic feature of laser installations of this type is the low contrast of the radiation, and manifests itself in the fact that the principal heating pulse is preceded by a precursor with a duration corresponding to the operating time of the electrooptical shutter (usually more then 10^{-9} sec), and this circumstance greatly distorts the picture of the interaction of the ultrashort pulse with the target, since actually the ultrashort light pulse interacts not with the solid target but with the plasma produced by the precursor. The results of an investigation of the reflection coefficient, for neodymium lasers of picosecond duration, is shown in Fig. 3b. From this plot we can draw the following conclusions: In the flux-density interval 10^{14}-10^{16} W/cm^2 the reflection coefficient does not exceed 10% (it must be taken into account, however, that the overwhelming majority of the results were obtained for targets with Z > 1), and in practically all the experiments R is either constant or may even decrease with increasing flux density.

We shall not treat in detail all the experimental results on the reflection of the radiation for light pulses of picosecond duration [55-62], and will dwell only on those studies [57, 61, 62] which were performed at thermonuclear flux densities $q \sim 10^{16}$ W/cm^2, which are of great interest.

In the experiments of [61, 62], a study was made of the reflection of one or several light pulses of 120 picosecond duration from spherical targets of LiD and $C_{36}D_{74}$ (interval between pulses 10 nsec, prepulse energy 0.5%, its duration 6 nsec).

It was observed that for a single pulse the reflection coefficient R increases monotonically from 1 to 4% in the flux-density interval $q \sim 2 \cdot 10^{15}$-10^{16} W/cm^2. For the second pulse, the reflection coefficient increases from 4 to 10% at $q \simeq 5 \cdot 10^{15}$ W/cm^2 and remains constant up to $q \simeq 2 \cdot 10^{16}$ W/cm^2.

The mechanisms that contribute to the reflection, in the authors' opinion, can be, besides reflection from the critical density, also stimulated Raman scattering, stimulated Mandel'shtam–Brillouin scattering (SMBS), and stimulated Compton scattering. The threshold for SMBS for an inhomogeneous plasma was taken from [63]: $P_B > (m^2c^4\omega_0/\pi e^2 a)\,(v_e/c)^2$ ergs/cm^{2}sec, where a is the characteristic dimension of the inhomogeneity and is connected with the density profile $N_e(x) = N_{cr}(1 + x/L)$, and v_e is the thermal velocity of the electrons. Measurement of the electron temperature yielded a value ~1 keV. The characteristic dimension a was taken to be $3 \cdot 10^{-3}$ cm from an estimate of the density profile produced by prepulse. The threshold values of the stimulated Mandel'shtam–Brillouin scattering is in this case $3 \cdot 10^{14}$ W/cm^2. The threshold for the Raman scattering should be c/v times larger and amounts under the same conditions to $3 \cdot 10^{16}$ W/cm^2. This exceeds the flux density reached in the discussed paper.

An estimate of the effective stimulated Compton scattering can be made by following the analysis of [64]. (Notice should be taken here also of the experimental study [65], in which amplification of laser radiation was observed upon interaction of opposing laser beams in a plasma, a fact that can be explained also within the framework of stimulated Compton scattering.)

The reflection coefficient due to the Compton scattering can be expressed, following [64, 65], in the form $R = R_0 \exp\{\beta q a (1 - R)\}$, where R_0 is the reflection from the critical density in the absence of Compton scattering and

$$\beta = \frac{c r_0}{\pi \omega_0^3} \frac{v_0}{k T_e} N_e \left(\frac{m}{2\pi k T_e}\right)^{1/2} \exp\left(-\frac{m v_0^2}{2 k T_e}\right),$$

where r_0 is the classical radius of the electron; m is the electron mass; k is Boltzmann's constant; v_0 is the plasma velocity. For the experimental conditions of [61, 62], namely $v_0 = 2 \cdot 10^7$ cm/sec, $T_e = 1$ keV, $N_e \simeq 4 \cdot 10^{20}$ cm^{-3}, $a \simeq 3 \cdot 10^{-3}$ cm, and $\beta \simeq 1.4 \cdot 10^{-4}$ cm/W, we obtain

$$R \simeq R_0 \exp\{(4.7 \cdot 10^{-17}) q (1 - R)\}.$$

For $R_0 = 0.01$, the value of R is 0.013 at $q = 5 \cdot 10^{15}$ W/cm^2 and $R = 0.03$ at $2 \cdot 10^{16}$ W/cm^2. Estimates of this type indicate that stimulated Compton scattering is also capable of contributing to the observed backward reflection at these flux densities.

A low value of the reflection coefficient from a pulse with $\tau \sim 100$ psec and $q \simeq 10^{16}$ W/cm^2 was obtained also in experiments [57] for a $(CD_2)_n$ target. The value of R was ~10% at flux densities $q \sim 10^{16}$ W/cm^2. It should be noted, however, that the surface of the flat target was mounted in this case at an angle 13° to the laser-beam axis, and the reflection was measured only in the solid angle of the focusing system. Then, in the case of specular reflection of radiation from the target (it will be shown in subsequent chapters that this situation is in fact realized, albeit at $q \sim 10^{14}$ W/cm^2), the greater part of the reflected radiation does not strike the lens and the reflection coefficient can be larger than ~10% (double, in the authors' opinion).

3. Spectral Composition of Reflected and Scattered

Radiation

Less attention has been paid until recently to another very important aspect of the problem of the reflection of radiation from a laser plasma, namely, to the investigation of the spectral composition of the reflected energy. At the same time, investigations of this type can lead to conclusions concerning the processes that occur in the plasma "corona," and concerning the parameters of the produced plasma.

The change of the spectrum of the heating light wave upon reflection from the plasma, in the case of nanosecond laser pulses, was investigated in [26, 38, 48, 49, 54, 66].

It was observed [26, 38, 48, 49] that the spectrum of the incident light wave broadens upon reflection and shifts to the long-wave region by several angstrom units, this shift being weakly dependent on the target material and on the flux density q. It seemed natural at first to identify this shift with the Doppler shift [48, 49] due to the motion of the reflecting layer (where the density has a critical value N_{cr}) in the direction of propagation of the laser beam with a velocity of several units of 10^6 cm/sec. This explanation is based on the fact that the pressure of the expanding hot plasma [67] produces a shock wave that propagates into the interior of the solid (see, e.g., [68]). Then [49], using the Hugoniot relations, we can obtain the velocity v_b

of the boundary between the plasma and the compressed solid (which, in the authors' opinion, is the velocity of the reflecting layer):

$$v_{\mathrm{b}} \simeq 0.7 \left(N_{\mathrm{cr}} \frac{M}{Z} \right)^{1/6} q^{1/3} \rho^{1/2}.$$

For example, for a lithium target [49] (Z = 3, q = 10^{13} W/cm^2) we have $v_{\mathrm{b}} = 3 \cdot 10^6$ cm/sec, which is in sufficiently good agreement with the value obtained from the observed shift.

However, in the already cited paper [26] the authors, starting from spectral measurements and from a study of the directivity of the reflected radiation, point to the possibility of explaining the shift of the wavelength within the framework of SMBS. For example, for a homogeneous plasma the long-wave shift of the wavelength in SMBS is defined by $\Delta\lambda = 2\lambda\, n v_{\mathrm{s}}/c$. Assuming a speed of sound $v_{\mathrm{s}} = 2 \cdot 10^7$ cm/sec (at $T_{\mathrm{e}} = 0.5$ keV $\gg T_{\mathrm{i}}$), we find that the experimentally observed shift $\Delta\lambda = 2.5$ Å corresponds to a plasma with a refractive index n = 0.18 for the heating wave. It must be noted, of course, that for the case of a laser plasma a quantitative description is much more complicated because of the presence of a strong gradient of the electron density (n varies in the range 0 < n < 1 between the point with critical density and vacuum) and to the motion of the plasma into the vacuum.

Entirely different experimental results are cited in [54], where the spectral composition of the reflected radiation was investigated, and the heating was produced with a neodymium laser operating at both the fundamental and the second harmonics. The targets used were LiD, $(CH_2)_n$, $(CD_2)_n$, D_2O ice, and aluminum. The reflected-radiation spectra of all the targets revealed a large number of equidistant lines (whereas the spectrum of the incident light shows only one line), the reflected-radiation lines being located in both the long-wave and the short-wave parts of the spectrum relative to the heating-radiation line. This "multiplication" of the lines was observed starting with a flux density 10^{12}–10^{13} W/cm^2. The spectral interval between the lines was the same ($\Delta\lambda_1 = 0.23$ Å for $\lambda_1 = 1.06$ μm and $\Delta\lambda_2 = 0.46$ Å for $\lambda_2 = 0.53$ μm) and was independent of either the energy or the material of the target. It was also observed that the fundamental line has weak spectral satellites, the distance between which corresponds to the spectral interval between the reflected-radiation lines. The appearance of satellites is connected with the selection of the axial modes inside the laser resonator, into which two Fabry–Perot interferometers were introduced to narrow down the line.

A linear structure of the reflected radiation was observed also in [66], but in a later paper [49] the authors have indicated that the linear character of the reflected spectrum is due to free lasing in the amplification stages of the laser installation (the radiation contrast was very low, only 2-3) and becomes smoothed out when the contrast is improved.

An important role in the understanding of the interaction of the laser radiation with the plasma was played by a study [69] that revealed the presence of lines of wavelength $\lambda = 0.53\,\mu$m, corresponding to the second harmonic $2\omega_0$ of the neodymium-laser radiation incident on hydrogen ice, in the spectrum of the radiation reflected from the plasma. This result was the first direct evidence that the bremsstrahlung mechanism is not the only one that determines the absorption of the radiation in the plasma corona, and furthermore even at not very high flux densities (q $\simeq 10^{12}$ W/cm^2).

It must also be noted that in contradiction to such facts as, for example, the behavior of the reflection coefficient, the deviation of the electron velocity distribution function from Maxwellian, the appearance of fast ions, etc., which may be interpreted ambiguously, the presence in the reflected spectrum of certain harmonics of the heating radiation can identify exactly the actual mechanism whereby the radiation is absorbed. For example, the only explanation available at present for the appearance of the $^3/_2\omega_0$ harmonic in the spectrum of the reflected radiation is that parametric instability, due to decay of the incident light wave into two plasmons $(t \rightarrow l + l)$, is present in the reflected radiation.

In this connection, investigation of the generation of harmonics of the heating radiation in the plasma is of great interest, and several recent papers deal with the spectral composition of the reflected light not only near the frequency ω_0, but also in other regions of the spectrum.

Mention should be made here of an investigation [70] of the polarization of the $2\omega_0$ line. Without dwelling in detail on the results of this study, we note that the second harmonic was observed not only in the backward direction and perpendicular to the target, but also in radiation passing through a 2-mm target of deuterium ice. The latter circumstance has not been explained theoretically to date (see, e.g., [71]).

In studies discussed in detail in later chapters [51, 52, 72], the second harmonic was observed at flux densities $q < 10^{12}$ W/cm^2 not only for targets of solid hydrogen, but also for targets with large Z (Be, Al, $(CH_2)_n$, Cu, Zn, Ta, Pb). The $2\omega_0$ energy depended little on the target material and increased rapidly in the flux-density interval $q \sim 5 \cdot 10^{11}$-$5 \cdot 10^{13}$ W/cm^2 with increasing laser energy ($E_{las}^{1.5 \pm 0.1} \sim E_{2\omega}$). The dependence of the second-harmonic energy was investigated also in [38, 73] for a solid-hydrogen target at $q \sim 10^{13}$-$2 \cdot 10^{14}$ W/cm^2, and turned out to be of the form $E_{2\omega} \sim E_{las}^{2.0}$. The authors therefore drew the justified conclusion that second-harmonic generation is due only to the development of parametric instability at the point with critical density. In the same paper, the reflected-radiation second-harmonic line shapes were investigated for different incident-wave line shapes. To this end, the active material used for the laser was not only neodymium glass but also yttrium aluminum garnet (YAG), the emission line width of which was 2-3 Å. It was observed that the spectrum of the reflected radiation at the fundamental frequency broadens to 100 Å for a neodymium-glass laser, and to 10 Å for YAG, and that the maximum shifts towards larger wavelengths by an amount of the order of 5 Å. The authors present a very interesting result: it turns out that the reflection coefficient R is smaller for a broad heating-radiation line (R $\simeq$ 18-20%) than for a narrow one (R $\simeq$ 20-25%). Unfortunately, the authors do not state the measurement accuracy, so that it is possible for the difference between the values of the reflection coefficient R to lie within the error limits. The maximum wavelength of the second-harmonic lines was practically at half the wavelength of the heating-radiation line, the line was broadened in the longer-wavelength region, and the broadening was larger for an H_2 target than for D_2. This fact is called by the authors the isotopic effect.

A similar behavior of the $2\omega_0$ line for a solid-hydrogen target was observed also in [74] at flux densities $q \simeq 10^{13}$ W/cm^2 and at a heating-radiation line width ~ 30 Å. In addition to the second harmonic, lines were observed corresponding to $\frac{3}{2}\omega_0$ (just as in [50]) and $\frac{1}{2}\omega_0$. The $\frac{3}{2}\omega_0$ line was broadened in analogy with $2\omega_0$ into the long-wave region, and the broadenings were practically equal in the scale of wavelengths. The $\frac{1}{2}\omega_0$ line was symmetrical, but in view of the absence of photographic material and sensitive image converters for this spectral range ($\lambda > 2$ μm), its shape was determined with the aid of a monochromator and a photomultiplier in a larger number of flashes, and this can apparently lead to an error. All three harmonics were investigated in this experiment at an angle 45° to the incident laser beam. It was noted that the intensity of the $\frac{3}{2}\omega_0$ line was larger than that of the $2\omega_0$ lines, but since the results were not recorded directly on photographic film but with the aid of an electron optical converter (the maximum gain was $\sim 10^3$), it can be concluded that the absolute intensity of these lines was much less than, for example, in [52], where the registration was effected directly on photographic film, and there was a sufficient reserve of light. The investigation of the temporal behavior of the lines $2\omega_0$, $\frac{3}{2}\omega_0$, and $\frac{1}{2}\omega_0$ has shown that they appear on the front of the laser pulse, their maximum intensities occur at the laser maximum intensity, and they vanish more rapidly than the trailing edge of the laser pulse.

The next results of the same group [75] were obtained with a somewhat different experimental scheme. First, introduction of two Fabry−Perot interferometers into the laser resonator decreased the spectral width of the emission line to 0.5 Å; second, the time behavior of the shapes of the lines $2\omega_0$ and $^3/_2\omega_0$ was investigated with the aid of a photorecorder with large amplification. Just as in [74], the lines $2\omega_0$ and $^3/_2\omega_0$ have maxima, but these maxima were less sharp than before and were broadened toward the long-wave region. For $2\omega_0$, this broadening was quite critical to the focusing and energy of the laser, although the short-wave part of the line was always present. The lack of a relation between the intensity of this line and its broadened part (the authors call it the line of the ion-acoustic spectrum) possibly indicates that two different effects are mixed, and the variation of the intensity of the broadened line from flash to flash is attributed to the threshold character of the parametric excitation.

Generation of the $2\omega_0$ and $^3/_2\omega_0$ lines was investigated also in [27] as a function of the defocusing of the radiation on the target surface at flux densities $q \sim 10^{14}$ W/cm². At the same time, the neutron yield and the soft and hard [28] x rays were investigated. It turned out that the reflection coefficient R and the intensity of the $^3/_2\omega_0$ line have a maximum when the target (D_2 ice) is shifted by 100-150 μm behind the focus, and the neutron yield and the intensity of the $2\omega_0$ line have two maxima. The smaller one occurs when the target is exactly at the focus, and the other occurs when the target is placed 300-400 μm away from the focus. It must be stated here that this experimental procedure is highly questionable. The point is that the variation, for example, of the reflection coefficient R as a function of the focusing condition is mainly a geometrical effect. This can be explained with the aid of Fig. 4.

If the target is in front of the focus (Fig. 4a), the reflected radiation diverges past the focusing lens, and since the detectors are usually placed several meters away from the vacuum chamber (to avoid parasitic reflection from the focusing system), another increasing fraction of the reflected radiation fails to strike the detector when the focusing is increased. If the target is placed behind the focus (Fig. 4b), the decrease of the reflection coefficient as a result of defocusing can be explained, for example, in the following manner. Absorption of the light energy by the plasma takes place only in a ray-convergence cone whose apex lies in the focal plane. The plasma electron temperature T_e, despite the high thermal conductivity, is maximal inside this cone and decreases in a direction perpendicular to the laser beam. Since the inverse bremsstrahlung coefficient is $K \sim 1/T_e^{3/2}$, it increases in the plasma situated outside the light-beam cone. Thus, the radiation reflected from the target surface is observed in this case as if through a pinpoint camera with small aperture. The foregoing causes of the change of the reflection coefficient R and of the intensities and of the various harmonics as a result of defocusing are apparently not the only ones. It must also be borne in mind that the caustic

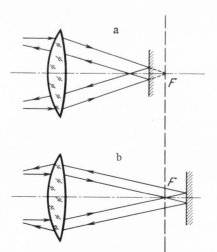

Fig. 4. Change of experimental geometry in the course of defocusing.

of the light beam of a real laser setup differs from that drawn on the paper, and there can exist along it several intensity maxima. The reason is that the laser is an extended source, and the laser beam has astigmatism. The described picture is very strongly complicated by the fact that the reflecting layer does not stand in place but undergoes a complicated evolution along the caustic during the time of the laser pulse. This motion depends strongly on the focusing condition, on the target material and thickness, on the laser radiation parameters, etc. All this leads to the conclusion that experiments similar to those of [27] are useless (at any rate in investigations of the reflection coefficient R and of harmonic generation) because such results cannot be interpreted.

Let us continue the discussion of the results of [27]. It was observed that the maximum intensity of the $2\omega_0$ and $^3/_2\omega_0$ lines is approximately 1% of the maximum reflected energy ($R_{max} = 50\%$), and that the polarizations of the lines ω_0, $2\omega_0$, $^3/_2\omega_0$ are equal within 10%. The hard x-ray quanta ($h\nu \leq 100$ keV) appeared only when the intensity of the $^3/_2\omega_0$ line was maximal, and were absent when the neutron yield was maximal. Also investigated was the spectrum of the ions whose maximum energy (< 20 keV) and whose number correspond to thermal equilibrium. Ultimately, the authors draw two conclusions. The first is that the parametric instability (its development is evidenced by the $^3/_2\omega_0$ line) gives rise to a high reflection coefficient. This conclusion is most likely incorrect, since parametric instability of the type ($t \rightarrow l + l$), which explains the appearance of $^3/_2\omega_0$, leads to an increase of the absorbed energy. The second conclusion is that no anomalous acceleration of the ions takes place in the laser plasma, and the maximum neutron yield is observed in the absence of parametric instability (it is borne in mind that the intensity of the $^3/_2\omega_0$ line is in this case close to zero). This conclusion would be very important for the problem of laser-controlled thermonuclear fusion were it not for the serious doubts concerning the possibility of interpreting the results of [27].

For laser installations in the mode-locking regime the spectral composition of the resultant radiation was investigated in [57, 61, 62, 76]. In [61] the width of the lines of the radiation on the target (($(CD_2)_n$, LiD) was ≤ 2 Å. In almost all cases the maximum of the reflected-radiation line was shifted into the red region by 2-6 Å and the line broadened to 2-15 Å. For the $2\omega_0$ line, the shift increased to 10-15 Å, and the broadening remained the same. The intensity of $2\omega_0$ was at most 10^{-4} of the laser energy ($E_{las} = 30$ J, $\tau = 120$ nsec; $q = 10^{16}$ W/cm^2, the prepulse was 5% of the laser energy), while the $^3/_2\omega_0$ intensity was half as large. The presence of a prepulse in laser radiation of picosecond duration is a very important circumstance and calls for great caution in the performance of the spectral measurements and in their interpretation. The point is that in a prepulse of duration up to several nanoseconds the energy can be large enough, and the flux density q can reach values of the order of 10^{13} W/cm^2 [59]. At these flux densities and even lower ones harmonics are generated in the plasma, for example, $2\omega_0$ [49, 70] and $^3/_2\omega_0$ [52]. In spectral measurements integrated over time it is therefore impossible to exclude the influence of such a prepulse on the shape of the investigated line.

In the experiment of [57] the laser had likewise a narrow emission line (> 1-2 Å). The spectrum of the reflected radiation at the frequency ω_0 ($R \sim 5$-15%), just as in [61, 62], was strongly broadened, and the line maximum shifted towards longer wavelengths; the reflected-radiation spectrum which is relatively smooth at a flux density $q \geq 10^{16}$ W/cm^2 acquires at lower flux densities a clearly pronounced spiked character. A shift in the red direction and a spike structure were traced also for the $2\omega_0$ line. The second harmonic contained 1% of the energy of the total reflected radiation.

It must be stated that more and more attention is being paid of late to the investigation of the spectral composition of the reflected and scattered radiations, and this leads to observation of a large number of experimental facts. Thus, in [77], besides the already well-known $2\omega_0$ and $^3/_2\omega_0$ harmonics, the spectrum of radiation reflected by a $(CD_2)_n$ plasma revealed two new harmonics, $^5/_2\omega_0$ at an angle 90° to the heating radiation and $3\omega_0$ emitted into the solid angle of the focusing system. There is no doubt that an interpretation of these facts will complicate even more the picture of the processes occurring in a dense laser plasma.

In concluding this section, let us mention an interesting but not yet confirmed result [78]. It was observed that when ruby-laser radiation ($q \le 2 \cdot 10^{12}$ W/cm^2) interacts with solid hydrogen and deuterium targets, the lines 9750 Å (H$_2$) and 8750 Å (D$_2$) appear in the spectra of the reflected radiation and of the radiation transmitted through the ice. They were identified as the Raman line of hydrogen and deuterium, due to the transition of the molecules from the ground state to the first excited molecular-vibrational level. An investigation of their temporal variation has allowed us to conclude that they are emitted prior to the start of the giant pulse of the ruby laser, when the flux density is low, $q \sim 10^9$ W/cm^2.

CHAPTER II

INVESTIGATION OF THE ABSORPTION OF LASER RADIATION IN THIN TARGETS

In this chapter we present the results of investigations of the interaction of laser radiation with thin targets.

It is quite clear that a planar irradiation geometry is not best from the point of view of attaining thermonuclear conditions. At the same time, the process of plasma formation by an intense laser beam has been investigated theoretically by various authors for one-dimensional planar geometry [79-84], and when flat foils were used (especially when the foil thickness was much less than the diameter of the focal spot), the conditions were sufficiently close to those of the theoretically considered problem. Experiments of this type (e.g., [85-90]) help refine the picture of the interaction. Finally, the importance of an experimental investigation of interaction with thin targets is due to the fact that hollow spherical targets, say of (CD$_2$)$_n$ [91], are quite promising when it comes to attaining high compressions and plasma temperatures, the walls of these targets being several microns thick.

In this chapter, by measuring the transmission of a laser pulse through foils of various thicknesses, we investigate the mass of the evaporated matter (M) and the mass flux ($\dot{M}$) during the time of the laser pulse. To investigate the shock waves produced in the gas surrounding the target, we used both multiframe shadow photography and a specially developed procedure for multiframe high-speed Schlieren photography in a ruby-laser beam.

1. Experimental Setup

The experimental setup and the diagnostic apparatus are illustrated in Fig. 5. The beam of a Q-switched neodymium laser system consisting of generators (3) and two amplifiers (4, 5), with energy up to 30 J, duration ~25 nsec at half-power level, and rise time ~15-20 nsec was focused with a lens of focal length $f = 50$ mm on the investigated foil (12), which was placed in a vacuum chamber. The radiation contrast in this case was ~10^4. The diameter of the focal spot was determined by the laser divergence and did not exceed 150 μm. As a result, the flux density q averaged over the time and over the focal spot was as high as $5 \cdot 10^{12}$ W/cm^2 in this series of experiments. The pressure in the vacuum chamber was varied in accordance with the experimental conditions in the range 10^{-5}-10 Torr.

The parameters of the incident, reflected, and transmitted (through the target) radiation were registered with calorimeters (11) and calibrated coaxial photocells (6). The signals from the latter were fed to a six-beam oscilloscope 6LOR-02 (9).

The energy of the radiation incident on the foil was varied in a wide range with the aid of calibrated neutral light filters.

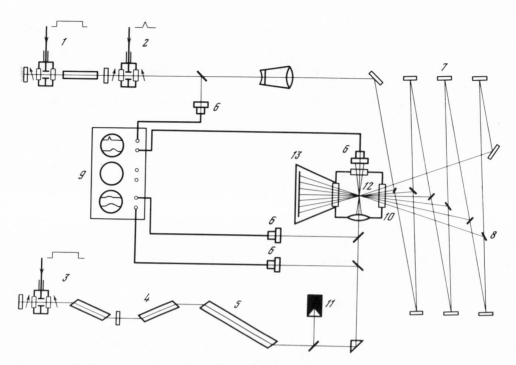

Fig. 5. Experimental setup for the investigation of the absorption of radiation in thin targets. 1) Ruby laser; 2) shaping shutter; 3) neodymium laser; 4, 5) amplifiers; 6) coaxial photocells; 7, 8) mirrors and optical-delay wedges; 9) six-beam oscilloscope 6LOR-02; 10) vacuum chamber; 11) calorimeter; 12) investigated foil; 13) photographic film.

For the multiframe shadow photographs [92–95] we used a Q-switched ruby laser. The Q-switching, just as for the neodymium-glass laser, was with the aid of a Kerr cell. The cells were controlled by a single modulator of high-voltage nanosecond pulses, thus ensuring reliable synchronization of the neodymium and ruby lasers.

The ruby-laser radiation pulse, the duration of which was ~20 nsec, was then shortened by means of a high-speed Kerr shuttter [92]. The shortened pulse was then guided through a telescopic system to an optical delay, with the aid of which [92, 93] seven frames, separated in space and in time, were formed. In this experiment [95], the maximum optical time delay (in the case when flat mirrors were used for the delay) was $5 \cdot 10^{-7}$ sec and the exposure time of each frame was determined by the duration of the ruby-laser radiation pulse and was equal to ~0.5 nsec. To use shadow photography, wherein a parallel light beam is passed through the inhomogeneity and its "shadow" image is formed on a screen placed at a certain distance from the object, there is a certain inaccuracy in the determination of the position of the shock-wave front (of the order of 0.5 mm in some cases). It is due to the strong refraction of the probing radiation in regions with large electron-density gradient on the shock-wave front. In a number of problems, this spatial resolution is insufficient, for example when it comes to determining the fraction of the laser radiation absorbed by the plasma from the velocity of the shock waves. The point is that in some cases, for example, in spherical irradiation of an isolated target, it is practically impossible to use calorimetric and photometric methods for the measurement of the reflected energy, and hence of the energy absorbed by the plasma. This is due to the strong refraction of the heating radiation in the plasma corona, which leads to an uneven scattering of the radiation into a solid angle 4π, the measurement of which is extremely difficult because of the large number of diagnostic devices in the vacuum chamber and the many elements required in the complicated focusing system.

In [6, 15, 96] there was proposed and used a different method of measuring the absorbed light energy. The experimental dependence of R(t) of the shock wave during the later stage, when the mass of the enclosed gas exceeds the mass of the target material, is compared with a theoretical family of curves that are self-similar solutions of the problem of a spherical explosion from a point in a homogeneous atmosphere [97, 98].

It must be stated, however, that the strong dependence $E \sim r^5$ calls for high accuracy in the measurement of the shock-wave radius r. For example, a 10% error in the determination of r leads to a 50% error in the absorbed energy E. We have developed and used for the investigation of shock waves a multiframe Schlieren-photography procedure [99] that made it possible to obtain the necessary spatial resolution.

2. Multiframe Schlieren Photography in Ruby-Laser Light; Spatial Resolution

The Schlieren method was first proposed by the French astronomer Leon Foucault [100] in 1853 to control the quality of large astonomical objectives of high resolution. In that discipline it is known to this day as the "Foucault knife-edge method" [101]. The German physicist August Toepler used this method in 1864 to investigate inhomogeneities in gases [102]. As a result of the increasing role of gas dynamics in the general development of science, this method is frequently called the Toepler method. The principle of this method is described in detail in [100-105]. We shall dwell in detail on the diagram shown in Fig. 6. The ruby-laser beam duration was shortened by an additional Kerr shutter controlled by a laser-ignited discharge gap [106-108]. As usual, the formation of seven spatially separated frames was with the aid of optical delays, and the total optical-delay time was approximately 300 nsec.

The seven transmitted beams intersected in the target region at a small angle ($\alpha \simeq 5°$) to one another. A separate Schlieren system was placed in each beam and consisted of a Schlieren lens of focal length $f = 20$ cm and a visualizing diaphragm placed in the focal plane of the Schlieren lens. A cassette with the photographic film was placed in the image plane.

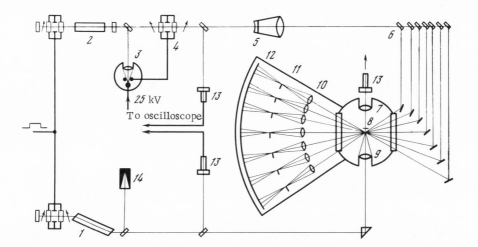

Fig. 6. Experimental setup for multiframe Schlieren photography in ruby-laser light. 1) Q-switched neodymium laser; 2) Q-switched ruby laser; 3) laser-ignited discharge gap; 4) Kerr shutter; 5) collimator; 6) optical-delay mirrors; 7) vacuum chamber; 8) target; 9) focusing lens; 10) Schlieren lenses; 11) visualizing diaphragms; 12) photographic film; 13) coaxial photocells; 14) calorimeter.

The visualizing diaphragm was either a Foucault knife or a "hairline in the focus," consisting of a blackened metallic wire measuring from 100 to 200 μm. The construction makes it possible to move the visualizing diaphragm in the focal plane of the Schlieren lens in three directions with accuracy to within 5 μm. The installation of the knife in the focal plane of the Schlieren lens, and also the focusing of the image in the plane of the photographic film, were effected with the aid of a microscope. The adjustment of the system was in the light of a helium-neon laser ($\lambda = 0.63$ μm).

It should be noted that the chosen experimental setup has enabled us, using the same number of frames, to increase the number of experimental points in the investigation of the shock-wave dynamics, owing to the use of stroboscopic photography. To this end, the system consisting of the Kerr shutter and the laser-ignited discharge gap were adjusted in such a way that two short electric pulses were separated on the Kerr-cell plates, with a variable interval between them. The duration of the ruby-laser pulse was artificially stretched, the Kerr shutter was opened twice during the pulse, and the optical delay system received two light pulses. Thus, by suitable choice of the discharge system it was possible to ensure any required number of light pulses.

The spatial resolution in the image plane is determined in Schlieren photography mainly by the speed of motion of the object and by diffraction, i.e., the resolution is the smaller of the quantities $N_1 = 1/d_1$ and $N_2 = 1/v\tau$, where d_1 is the smallest inhomogeneity dimension that can be registered in the object's plane. In the diffraction approximation [109, 110] the minimal resolvable element is given by the expression $d_1 = 2\alpha_3 L/6$, where $2\alpha_3 = 2\alpha_1 + 2\alpha_2$, $2\alpha_1$ is the divergence angle of the transmitted beam, and $2\alpha_2$ is the angle of diffraction of the beam by an element of dimension d_1 and length L.

If the blurring of the investigated inhomogeneity as a result of the diffraction is of the order of its dimension, then $2\alpha_2 = (L/\lambda)^{1/2}$. The divergence of the transmitted beam in this experiment was $2\alpha_1 \sim 10^{-4}$; at $L \leq 1$ cm the divergence of the transmitted beam can be neglected and $2\alpha_3 = 2\alpha_2$. For the minimum resolvable element we obtain $d_1 = (\lambda L/6)^{1/2}$.

When the front of a spherical shock wave is registered we have in this experiment $L = 2.82(r\Delta r)^{1/2}$ (if $r \gg \Delta r$), where r is the radius of the shock wave and Δr is the width of the front. Then $d_1 = 0.28\lambda^{1/2}R^{1/4}\Delta R^{1/4}$. If $r = 1$ cm, $\Delta r = 10^{-2}$ cm, and $\lambda = 7 \cdot 10^{-5}$ cm, then $d_1 = 7.8 \cdot 10^{-4}$ cm, and $N_1 = 1350$ lines/cm. At a shock-wave velocity $v = 10^7$ cm/sec and an exposure time $\tau = 5 \cdot 10^{-10}$ sec, we have $N_2 = 200$ lines/cm and consequently the spatial resolution in our experiment is determined by the velocity of the object (we neglect the operations of the Schlieren lenses and the resolution of the photographic film, which in our case, at $N_2 = 200$ lines/cm, do not play any role).

3. Determination of the Time of Bleaching

of a Thin Target

We investigated the interaction of laser radiation with polyethylene foils 15 to 110 μm thick and aluminum foil 5 and 10 μm thick. The measurements have shown [95] that in the case of polyethylene, at a flux density $q \simeq 10^{10}$ W/cm^2, there is practically no absorption in the target, the target shows no traces of damage, and the entire laser radiation strikes the coaxial photocell behind the target. We note that a similar result was obtained in [86], but for a ruby laser.

Starting with a flux density $q \sim 10^{10}$ W/cm^2, the foil begins to absorb effectively, and the coaxial photocell that registers the light passing through the target produces no signal, i.e., the incident radiation is fully absorbed in the target. With further increase of the laser energy, the foil begins to become transparent during the time of the laser pulse. The instant

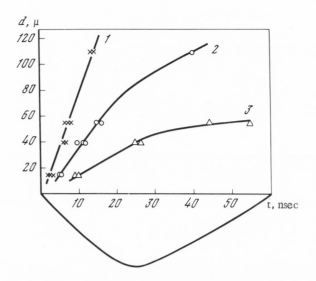

Fig. 7. Experimental dependence of the time of "bleaching" of a polyethylene foil on the thickness for various values of the laser-radiation flux density. 1) $q_0 = 2 \cdot 10^{12}$ W/cm^2; 2) $q_0 = 8 \cdot 10^{11}$ W/cm^2; 3) $q_0 = 4 \cdot 10^{11}$ W/cm^2.

of time t_1 up to which absorption takes place (or starting with which the foil becomes transparent) increases with increasing foil thickness and decreases with decreasing energy.

Figure 7 shows the dependence of the time of "bleaching" of a polyethylene foil on the thickness at various values of the laser-radiation flux density (at different laser energies). The rate of "bleaching" of the foil turns out to be approximately constant on the front of the laser pulse, just as in [86, 88] at lower flux densities, and decreases towards the end of the pulse.

If the diameter of the focal spot, the foil thickness, and the time of "bleaching" are known, it is possible to determine the mass flux $\dot{M}$ through the heated surface of the target. Figure 8 shows the dependence of $\dot{M}$ on the front of the laser pulse, for polyethylene and aluminum, on the average laser-radiation flux density q_0. For comparison, the same figure shows the results of a theoretical paper [79] dealing with the one-dimensional gas dynamic problem of heating a medium with high-power laser radiation. The rate of evaporation of the material mass is given by the expression $\dot{M} = 0.28 a^{1/4} q^{1/2} t^{1/4}$ (the curve in the figure corresponds to 10 nsec). The same figure shows the results of an experimental paper [67], in which the slit image of the interference pattern of a carbon "flare" (for a solid target) was scanned

Fig. 8. Experimental dependence of the mass flux $\dot{M}$ for $(CH_2)_n$ (1) and Al (2) on the front of a laser pulse on the average flux density q_0. The point 4 corresponds to the result of [67]. The line 3 shows the result of [79] at 10 nsec.

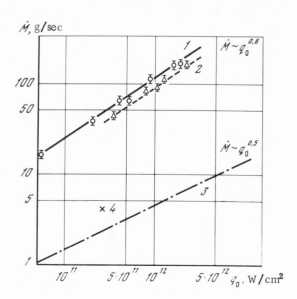

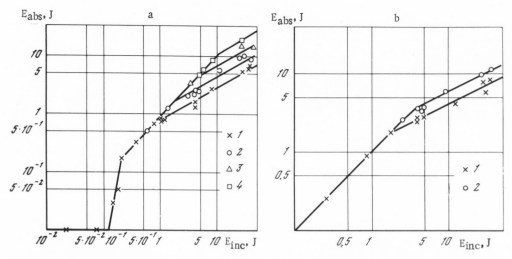

Fig. 9. Dependence of the energy absorbed by polyethylene (a) and aluminum (b) foils on the laser energy at different target thicknesses. a: 1) 15 μm; 2) 40 μm; 3) 55 μm; 4) 110 μm; b: 1) 5 μm; 2) 10 μm.

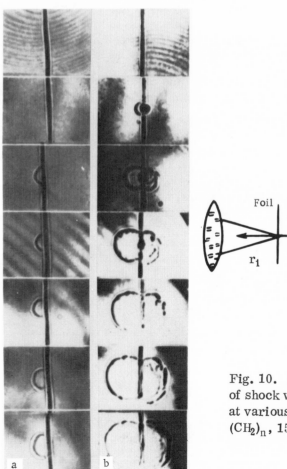

Fig. 10. Characteristic seven-frame shadow photographs of shock waves produced in the gas surrounding the target, at various laser energies. a) Al, 10 μm; $E_{las} = 0.2$ J; b) $(CH_2)_n$, 15 μm, $E_{las} = 5$ J.

in time by the photorecorder. The reduction of the interference pattern has made it possible to determine the density profile of the expanding plasma at different instants of time, and consequently the mass flux $\dot{M}$. Thus, Fig. 8 shows a significant difference between the values of $\dot{M}$ in the present study and in the theoretical [79] and experimental [67] papers for solid targets. We note that the values obtained for $\dot{M}$ in the cited paper [86], devoted to the absorption of ruby-laser radiation by polyethylene foils, are almost double the values obtained in our experiment. This difference can apparently be attributed to the different absorption coefficients for the wavelengths 6943 and 10,600 Å, and also to the different structures of the focal spots. We emphasize, however, that the mass flux $\dot{M}$ for solid targets and the flux determined for thin foils from the "bleaching" time differ by approximately one order of magnitude.

Figure 9a shows the dependence of the absorbed energy on the laser energy for polyethylene foils of various thicknesses. It turns out that starting with the instant of bleaching we have $E_{abs} \sim E_{las}^{0.5}$, so that the absorbed energy per particle is proportional to $q^{0.5}$, in agreement with [111]. The corresponding dependence for aluminum foils is similar in form (Fig. 9b).

4. Investigation of the Dynamics of Motion of Shock Waves in the Gas Surrounding the Target; Absorbed Energy

When a laser plasma expands into the atmosphere of the residual gas surrounding the target, a shock wave is produced [15, 93, 94, 112–115] and propagates with velocities 10^6–10^7 cm/sec and even $\sim 10^8$ cm/sec [6], which are equal to or larger than the maximum shock-wave velocities attained under laboratory conditions by other methods [116]. The shock waves obtained in this manner are therefore in themselves an object of investigation, the interest in which is due to the fact that a laser can be used to produce shock waves of practically any configuration, spherical, cylindrical, etc., and to investigate the cumulation of shock waves [117–120]. On the other hand, a study of the laws of motion of shock waves makes it possible to determine certain important parameters of a heated plasma, for example, the ion temperature [6], the energy absorbed by the plasma, etc. Figure 10 shows seven-frame shadow photographs produced when laser radiation is focused on surfaces of foils. At low values of the energy, the shock propagates only in a direction opposite to that of the laser beam. With increasing energy, a second shock wave is produced, propagating in the opposite direction.

Figure 11 shows an (r−t) diagram of motion of a shock wave at $E_{las} = 30$ J. The slope of the curve in this figure determines the relation $r \sim t^{\gamma}$, which can be written in the range from 50 to 250 nsec in the form $r \sim t^{0.6}$. Subsequently, γ decreases and at 400 nsec, it becomes equal to $\gamma \simeq 0.4$.

Figure 12 shows the experimental dependence of the radius of a shock wave propagating counter to the laser radiation (at 100 nsec) on the energy absorbed by an aluminum foil of 10 μm thickness. This dependence can be expressed in the form $r \sim E_{abs}^{0.4}$. Thus, the law of motion of the shock wave produced when laser radiation is focused on the surface of the foil can

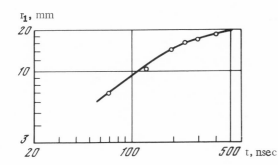

Fig. 11. r−t diagram of motion of a shock wave in a direction opposite to the laser beam for $E_{las} = 30$ J. The target is an aluminum foil 10 μm thick.

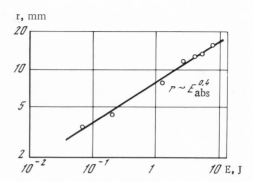

Fig. 12. Dependence of the radius of the shock wave propagating in opposition to the laser radiation on the energy absorbed by an aluminum foil 10 μm thick (at 100 nsec).

be approximated, during the first 250 nsec after the end of the heating pulse, by the expression $r \sim E_{abs}^{0.4} \cdot t^{0.6}$.

Figure 13 shows the dependence of the ratio r_1/r_2 of the radii of shock waves propagating in opposite directions on the incident energy at 100 nsec. It is seen that this ratio tends to a constant value with increasing laser energy.

Investigations of the dependence of the velocity of shock waves in air (at a pressure 1 Torr) on the atomic number of the target were carried out with the aid of multiframe Schlieren photography. The energy of the neodymium laser was $E_{las} \sim 1$ J. The targets were beryllium ($Z = 4$), aluminum ($Z = 13$), zinc ($Z = 30$), tin ($Z = 50$), tantalum ($Z = 73$), lead ($Z = 82$), and Teflon.

Figure 14 shows a typical seven-frame Schlieren photograph of a shock wave in air for an aluminum target. The Foucault knife placed parallel to the target surface covered one half of a focal spot with dimension of the order of 50 μm. In frames III and IV, the energy of the second stroboscopic pulse was sufficient to register the shock waves.

It turned out that the velocity of the shock waves produced in expansion of the plasma with small Z is somewhat larger than for $Z > 30-40$. Estimates show that for a light target approximately 80-90% of the laser emission energy goes over into the shock-wave energy, as against, for example, 40-50% for lead. This indicates that the method used to determine the absorbed energy from the velocity of the shock wave, mentioned in the first section of this chapter, gives only the lower value of the energy input to the plasma and does not take into account, for example, the losses to recombination radiation, which increase strongly with increasing Z ($E_{rec} \sim Z^4$).

The investigations of the dynamics of shock-wave motion by the method of multiframe shadow photography were carried out with an optical delay system consisting of flat mirrors. It must be noted, however, that the possibility of obtaining large times τ in optical delays with flat mirrors are limited, first, by the energy loss due to the vignetting of the beam by the

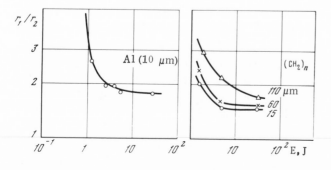

Fig. 13. Ratio r_1/r_2 of the shock wave propagating towards the laser and away from the laser vs. the laser energy at 100 nsec, for aluminum and polyethylene.

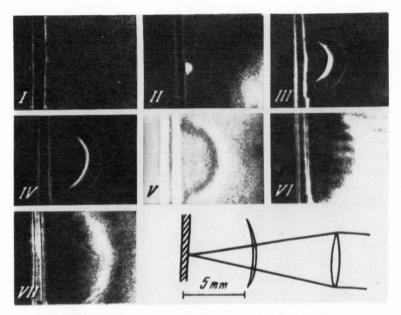

Fig. 14. Seven-frame Schlieren photograph of shock wave. Laser energy 0.3 J, pressure in vacuum chamber 1 Torr, target aluminum. The instants of exposure of the frames after the maximum of the heating pulse: I) t = 33 and 7 nsec; II) 12 and 52 nsec; III) 56 and 96 nsec; IV) 101 and 141 nsec; V) 147 nsec; VI) 192 nsec; VII) 236 nsec.

mirrors, in view of the finite divergence of the ruby-laser radiation, and, second, by the complexity of the system. Usually the maximum delay time in such a system does not exceed 500–600 nsec [15]. To investigate the dynamics of the motion of the shock waves at much later times, use was made of a small-size optical delay system especially developed in the Optical Design Division of the Physics Institute of the Academy of Sciences of the USSR, using spherical mirrors [121] (see also [122, 123]), and the use of this system has made it possible to raise the maximum optical delay time to 3 μsec.

An investigation of the dynamics of motion of shock waves during the later stages has shown that starting with approximately 300 nsec after the end of the laser pulse the shock wave velocity decreases strongly and the dependence of the shock-wave radius on the time in the interval 600–3000 nsec takes the form $r \sim t^{1/5}$.

5. Discussion of Results

As already mentioned, the values obtained for M and $\dot{M}$ in the present study exceed by approximately one order of magnitude the data of [67, 79] for solid targets. It appears that an important role in the understanding of this result is played by the following experimentally observed fact: when laser radiation is focused on a foil, there exists an energy interval in which shock waves propagate in both directions away from the target, although bleaching has not yet set in. For example, for an aluminum foil 10 μm thick, the shock wave in air on the opposite side begins to be produced at an energy larger than 0.5 J, even though the foil begins to be bleached only at an energy $E_{las} \simeq 3$ J. This circumstance allows us to conclude that even though the radiation does not penetrate into the deep layers of the foil, a hot plasma is produced on its rear side, expands, and forms in the air a shock wave that propagates in the opposite direction. The mechanisms responsible for the energy transport may be electron thermoconductivity and a shock wave in the solid.

At flux densities q $\simeq 10^{12}$ W/cm^2, the produced plasma has an electron temperature $T_e \sim$ 100-150 eV [124]. We can estimate approximately the time during which the thermal wave produced by the electron thermal conductivity passes through the foil. At a thermo-conductivity coefficient $\varkappa = 2 \, (kT_e)^{5/2}/m_e^{1/2}e^4 \ln \Lambda$ [125] (where $\ln \Lambda$ is the Coulomb logarithm), at a temperature $T_e \sim$ 100 eV, the time to heat a foil 10 μm thick will exceed 10^{-8} sec.

The "back pressure" of the plasma on the target was determined in [67] at heating radiation flux densities 10^{12} W/cm^2 from the results of the measurements of the density profiles and the velocities of individual plasma regions, and was found to be approximately $2.5 \cdot 10^6$ atm. The action of such a pressure on the target should produce in the solid a shock wave propagating into the interior of the foil. At these pressures, according to the measurements of [126], the compression for aluminum amounts to $\beta = \rho_{comp}/\rho_0 \simeq 1.5$. The expression for the shock-wave velocity in the solid is [97]

$$D = \sqrt{\frac{P}{\rho_{comp}(1 - 1/\beta)}}.$$

The shock-wave velocity is therefore D = $1.6 \cdot 10^6$ cm/sec and the shock wave reaches the rear wall of an aluminum foil 10 μm thick within 0.5 nsec.

Notice should also be taken of a theoretical paper [84], which gives the results of a numerical calculation of the expansion parameters of a plasma produced by the action of laser radiation on a foil of solid hydrogen, with allowance for the thermoconductivity and compression of the material. At a flux q $\simeq 10^{12}$ W/cm^2, the shock wave produced in such a foil 50 μm thick will reach the opposite side of the target in 2 nsec. The compression in this case is equal to $\beta = 4$. The foregoing estimates explain the large values obtained for $\dot{M}$ from the results of bleaching the thin targets.

Thus, it can be concluded that at flux densities q $\sim 10^{11}$-10^{12} W/cm^2 the main mechanisms that determine the time of bleaching of a thin target is the shock wave produced in the solid and causing the destruction of the foil. It must be emphasized, however, that at higher temperatures, i.e., at higher flux densities, the electron thermal conductivity will play a major role, in view of the strong dependence of the thermal conductivity coefficient on the temperature ($\varkappa \sim T^{5/2}$).

CHAPTER III

REFLECTION OF LASER RADIATION FROM A DENSE PLASMA

1. Experimental Setup

In this chapter we describe experiments [51-53] on the reflection of the radiation from a plasma at the heating-radiation frequency in the case of sharp focusing.

A diagram of the experimental installation and of the diagnostic apparatus is shown in Fig. 15. A system was used consisting of a Q-switched driving laser, a light-pulse sharpener, and four amplification stages. The radiation pulse of approximate duration 20 nsec was fed from the output of the driving laser to the light-pulse sharpener consisting of a laser-ignited discharge gap and a Kerr shutter. The latter was controlled by the discharge gap and was opened, when the maximum intensity of the light pulse from the laser was reached, for a time 3-4 nsec determined by the parameters of the electric circuit of the discharge−cell system. To increase the contrast, the radiation flux passed through the Kerr shutter twice. It is obvious that such a pulse-shortening system is equivalent to a system with two ideally synchronized shutters and offers great simplicity and high reliability. From the sharpener the

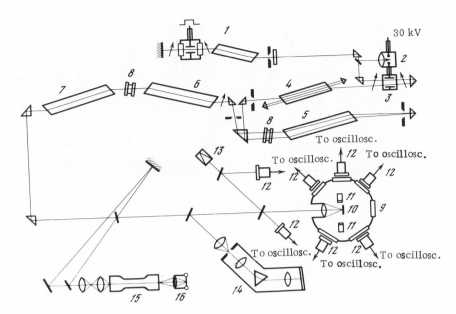

Fig. 15. Experimental setup. 1) Neodymium laser Q-switched with a Kerr cell; 2) laser-ignited discharge gap; 3) Kerr shutter; 4-7) amplification stage; 8) bleaching filter; 9) vacuum chamber; 10) target; 11) x-ray detectors; 12) coaxial photocells; 13) calorimeter; 14) prism spectrograph; 15) FÉR-2 photoelectronic recorder; 16) photographic camera.

light pulse was fed to a system of four amplification stages. To increase the amplification efficiency, the radiation passed three times through the rod in the first stage and twice through the rod in the second. A negative lens was placed past the first two amplifiers and the amplification in the last stage was in a diverging beam. As a result, in the presence of a target at the focus of the lens, the laser made it possible to obtain light pulses of duration $\tau \simeq 3\text{-}4$ nsec at half-intensity level, with an energy up to 30 J, with radiation contrast not worse than 10^5, and with a divergence up to $2 \cdot 10^{-4}$ rad. The energy and the shape of the pulses incident and reflected from the target were registered with colorimeters and coaxial photocells, the signals from which were fed to six LOR-02 and I2-7 oscilloscopes. The high temporal resolution of the system was the result of the use of a photoelectronic recorder FÉR-2, the image on the screen of which was photographed. To investigate the spectral composition of the reflected radiation in the employed band, an ISP-51 prism spectrograph was used, with a dispersion in the vicinity of the second harmonic of the neodymium laser (5300 Å) amounting to 20 Å/mm.

2. Behavior of the Coefficient of Reflection of Laser Radiation from a Plasma in the Flux-Density Interval $10^{10}\text{-}10^{14}$ W/cm^2

An experimental investigation of the reflection coefficient was carried out as a function of the laser-radiation energy (or of the flux density q) for solid targets of $(CH_2)_n$, $(CD_2)_n$, Be, Al, Cu, Zn, Ta, Pb, and also aluminum and polyethylene foils with thickness from 5 to 200 μm. By the reflection coefficient is meant here the ratio R = E_{ref}/E_{las}, where E_{ref} is the energy reflected from the plasma into the aperture of the focusing lens. Typical oscillograms of the incident and reflected pulses are shown in Fig. 16. Figure 17 shows plots of the reflection coefficient R against the flux density q incident on an aluminum target for two series of ex-

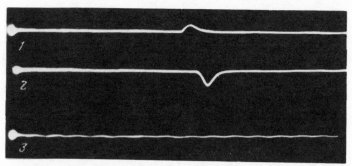

Fig. 16. Typical oscillograms of laser radiation incident
(1) and reflected (2) from a plasma. 3) Calibration sinusoid with period 10 nsec.

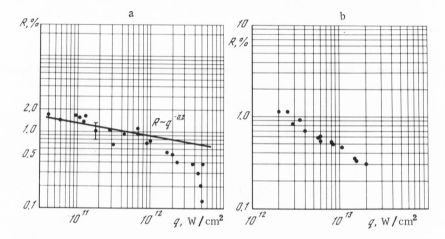

Fig. 17. Experimental dependence of the reflection coefficient R,
integrated over the pulse, of laser radiation into the solid angle of
the focusing lens, on the flux density q of the radiation incident on
an aluminum target, for two series of experiments in different flux-
density regions. a) From the data of [51]; b) from the data of [52].

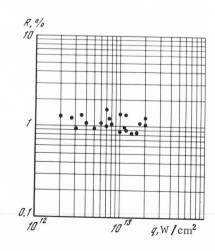

Fig. 18. Experimental dependence of laser-
radiation reflection coefficient R, integrated
over the pulse, on the flux density q of the
neodymium-laser radiation incident on a polyethylene target.

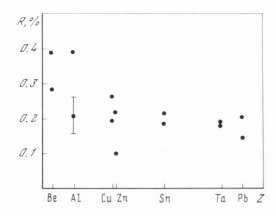

Fig. 19. Experimental dependence of the reflection coefficient R integrated over the pulse on the atomic number of the target material at a heating-radiation density $q = 5 \cdot 10^{12}$ W/cm².

periments in different flux-density regions [51, 52]. The radiation flux q, as usual, is determined by the ratio of the energy E_{las} to the duration of the laser pulse τ and the area S of the focal spot:

$$q = E_{las}/\tau S.$$

It is seen that the coefficient of reflection for aluminum decreases monotonically in the interval $q \sim 10^{10}$-10^{12} W/cm² from 1.5% to 1% ($R \sim q^{-0.2}$), and in the interval $q \sim 10^{12}$-$2 \cdot 10^{13}$ W/cm² the decrease of R is more abrupt, from 1 to 0.1-0.2%. Figure 18 shows a plot of R(q) for a polyethylene plasma. It is seen from the figure that the reflection coefficient remains constant in the interval $q \sim 10^{12}$-$2 \cdot 10^{13}$ W/cm². Figure 19 demonstrates the difference between the values of R for targets made of substances with different Z at a heating-radiation flux density $q = 5 \cdot 10^{12}$ W/cm². It is seen that the reflection coefficient is practically constant at $Z \geq 30$ ($R \simeq 0.2\%$) and is doubled for beryllium.

A low (R < 1%) reflection coefficient was obtained also in a series of experiments [53] at flux densities $q \sim 10^{14}$ W/cm². A detailed description of these investigations will be given in Sec. 5 of this chapter. The results [51-53] described above were the first to indicate that the coefficient R of reflection of laser radiation from a plasma with Z > 1 is small ($R \lesssim 1\%$) in a wide flux-density interval ($q \sim 10^{10}$-10^{14} W/cm²), and R can even decrease with increasing q (for aluminum). This is a very important fact, inasmuch as very serious doubts were raised, after the already mentioned experiments [9, 26] with solid hydrogen, whether laser heating of a plasma is feasible in principle, owing to the increased fraction of reflected energy with increasing flux densities. The results of our experiments have also made it necessary to review numerous data of nonsystematic experiments, in which also low values of R were obtained (for example, [34, 41, 42]).

It must be stated that the small absolute value obtained for the laser-radiation reflection coefficient and the decrease of the value of R with increasing flux density are difficult to explain within the framework of the ordinary (inverse-bremsstrahlung) absorption, i.e., in terms of Coulomb collisions of the plasma electrons with ions. It makes sense therefore to stop to discuss other possible mechanisms of absorption of an intense light wave in the plasma corona.

We discuss first effects of parametric resonance and present formulas for the threshold fluxes q_{thr} and the maximum growth increments γ_{max} of five most typical parametric instabilities in a homogeneous isotropic plasma [127-133] (see also [52]):

1) aperiodic instability (t $\rightarrow$ l + a):

$$q_{thr} = 6.7 \cdot 10^{12}(Z + T_i/T_e)T_e^{-1/2}\Lambda_1, \qquad \gamma_{max} = 2.4 \cdot 10^{-7}(Z + T_i/T_e)^{-1/2}(AT_e)^{-1/2}q^{1/2};$$

2) decay into an ion-acoustic wave and plasma wave (t → l + s)

$$q_{thr} = 1.9 \cdot 10^{11} Z^{3/2} (AT_e)^{-1/2} \Lambda_1, \qquad \gamma_{max} = 10^{-6} (Z/A)^{1/4} T_e^{-1/2} q^{1/2};$$

3) decay into two plasmons (t → l + l)

$$q_{thr} = 1.2 \cdot 10^{11} Z^2 T_e^{-3} \Lambda_1^2, \qquad \gamma_{max} = 2.4 \cdot 10^{-7} q^{1/2};$$

4) SMBS (t → t + s)

$$q_{thr} = 2.3 \cdot 10^{-10} N_e Z^{3/2} (AT_e)^{-1/2} \Lambda_2, \qquad \gamma_{max} = 2.6 \cdot 10^{-8} (Z/A)^{1/4} T_e^{1/4} q^{1/2} (1 - 10^{-21} N_e)^{3/4};$$

5) SRS (t → t + l)

$$q_{thr} = 1.6 \cdot 10^{-9} Z^2 N_e T_e^{-3} F(x) \Lambda_2^2; \qquad \gamma_{max} = 0.9 \cdot 10^{-17} N_e^{1/2} q^{1/2} (x - 1) F^{-1/2}(x).$$

Here

$$F = (x - 1)^{-1} [(x^2 - 1)^{1/2} + [x(x - 2)]^{1/2}]^{-2}; \qquad x = 3.16 \cdot 10^{10} N_e^{-1/2};$$

$$\Lambda_1 = 1 + 0.3 \log T_e; \qquad \Lambda_2 = 1 + 0.3 \log(10^{10} N_e^{-1/2} T_e),$$

the temperature T_e is in kiloelectron volts, q_{thr} in W/cm², and γ_{max} in nsec⁻¹.

Figure 20 shows for these parametric instabilities the plots of q_{thr} against the electron temperature T_e for an aluminum and polyethylene plasma. The dashed line corresponds to the relation $T_e \sim q^{4/9}$ which is typical of a laser plasma [124, 134, 135]. It is obvious that the range of values of the laser-plasma parameters in which any of the described five instabilities can be excited is located on the dashed line to the right of the point of intersection with the plot of the instability threshold against the temperature.

The thresholds of the first two instabilities (t → l + a) and (t → l + s) were calculated for the critical density $N_e = 10^{21}$ cm⁻³. The instability (t → l + l) of the type of decay of a light

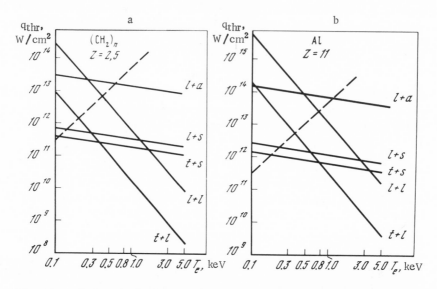

Fig. 20. Theoretical plots [52] of the threshold light fluxes q_{thr} needed to excite the five parametric instabilities against the electron temperature of a homogeneous polyethylene (a) and aluminum (b) plasma (solid curves). The dashed line corresponds to the relation $T_e \sim q^{4/9}$.

wave into two electron Langmuir oscillations develops at $N_e = 2.5 \cdot 10^{21}$ cm^{-3}, which is one-quarter the critical density. The temperature dependences of the threshold fluxes of the instabilities $(t \rightarrow t + s)$ and $(t \rightarrow l + l)$, excited in a wide range of plasma densities, were plotted for the densities $N_e \simeq 5 \cdot 10^{20}$ cm^{-3} and $N_e = 10^{20}$ cm^{-3} respectively.

The first three instabilities correspond to parametric generation, by the light wave, of longitudinal (potential) plasma oscillations that are absorbed by the particles (electrons and ions) not only as a result of Coulomb collisions but also because of the inverse Cerenkov effect, and are incapable of penetrating the plasma volume directly. It is these instabilities which can ensure absorption of the light wave.

The two other instabilities (SMBS and SRS), in contrast to the first three, are of scattering character, since they give rise not only to potential oscillations (sound and a plasmon, respectively), but also to a light wave that leaves freely the volume of the plasma that is transparent to Coulomb collisions. The appearance of any of the five described parametric instabilities leads to an anomalous interaction of the laser radiation with the plasma corona, with the first three leading to an anomalous absorption of the light by the plasma and the last two to anomalous scattering (reflection).

As seen from the plots of Fig. 20, starting with a flux density $q \simeq 10^{14}$ W/cm^2, all five parametric instabilities can in fact develop in both polyethylene and aluminum laser plasma.

It should be noted that the theoretical threshold fluxes presented above were obtained for a homogeneous unbounded plasma. Allowance for the finite volume of the real laser plasma and its inhomogeneities does not change the picture described above for the three instabilities that lead to anomalous absorption. Indeed (see [128], page 179), the inhomogeneity of a plasma with characteristic dimension a increases the threshold of such parametric instabilities if a is much less than the electron mean free path with respect to Coulomb collisions with ions, multiplied by the logarithm of the ratio of the light frequency to the collision frequency ν_{ei}. At a plasma electron temperature $T_e \sim 1$ keV and electron density $N_e = 10^{21}$ cm^{-3}, the thermal velocity is $v_{Te} \simeq 10^9$ cm/sec, $\nu_{ei} \simeq 10^{12}$ sec^{-1}, and the mean free path is $\sim 10^{-3}$ cm, i.e., at $a \simeq 10^{-2}$ cm the influence of the inhomogeneity on the threshold is negligible, but if $a \simeq 10^{-3}$ cm, then the inhomogeneity leads to an increase of the threshold. (More detailed conditions for neglecting the influence of the inhomogeneity are discussed in [128, 133] and are summarized in a table in [52].) On the other hand, instabilities of the SMBS and SRS type, inasmuch as the unstable (scattered) light waves can escape unimpeded from the inhomogeneous laser plasma, are susceptible to the much more appreciable influence of the inhomogeneity that increases their threshold to values $q_{thr} \sim 10^{15}$–10^{16} W/cm^2 (see Chapter I), i.e., exceeding the flux density in the experiment, although in a homogeneous plasma the threshold fluxes are lower for these two instabilities than for the remaining ones (see Fig. 20).

In addition to the anomalous nonlinear interaction mechanism discussed above, we should consider here the effect of linear transformation of a p-polarized light wave with frequency close to the plasma frequency into a potential electron Langmuir oscillation [136–138], since this phenomenon also leads to an additional absorption of the radiation in the plasma corona. Assume that the plasma is inhomogeneous along the z axis, so that the critical density is reached at $z = 0$ and $N(z) < N_{cr}$ at $z < 0$. We choose the xz plane of a rectangular coordinate system as the incidence plane of the light wave. We emphasize that transformation is possible only in the case of oblique incidence of the p-polarized waves, when the angle θ between the wave vector k and the z axis differs from zero. The electric field E lies then in the incidence plane, and the magnetic field B has a single component B_y perpendicular to the plane of incidence. The essence of the transformation of light into a Langmuir oscillation of an inhomogeneous plasma consists in an abrupt increase of the longitudinal component of the field E_z in the region of the critical density $z = 0$ at a smoother variation of E_x and B_y. The transformation coefficient, and consequently the coefficient of absorption of laser radiation for this mechanism, is independent of the temperature and is equal to [139]:

$$K_{1.t.} = \pi \frac{a\omega}{c} \frac{\sin^2 \theta}{|\cos \theta|} |B_y(0)|^2.$$

Inasmuch as the condition $a\omega/c \gg 1$ is satisfied in laser experiments, we can use the results of [138], where the quantity $B_y(0)$ was calculated in the geometric-optics approximation. It turns out that the absorption coefficient $K_{1.t.}$ has a narrow maximum near $\sin\theta = 0.5(a\omega/c)^{-1/3}$ and decreases very strongly if it deviates from the optimal incidence angle (there is no transformation at all at normal incidence). Therefore by choosing the method of illuminating the target we can obtain an appreciable ($K_{1.t.} \sim 50\%$) increase of the fraction of laser energy absorbed in the plasma.

3. Dependence of the Reflection Coefficient on the Time; Plasma Probing by Ruby-Laser Radiation

In the preceding section we have discussed the results of investigations of the time-integrated reflection coefficient R. Oscillographic measurements, however, have made it possible to determine also the dependence of R on the time during the laser pulse. Figure 21 shows typical incident and reflected light pulses and the function R(t). It is seen that the reflected signal is shortened somewhat (by approximately 2-3 nsec), and the duration of the leading front becomes shorter and approaches the temporal resolution of the recording apparatus (10^{-9} sec). The largest reflection occurs at the start of the laser pulse, and subsequently the reflection

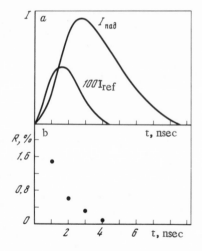

Fig. 21. Typical incident and reflected pulses ($q = 5 \cdot 10^{12}$ W/cm^2) (a) and dependence of the reflection coefficient on the time during the laser pulse (b).

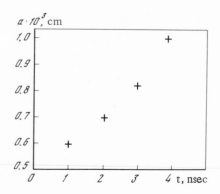

Fig. 22. Behavior of the characteristic plasma dimension $a = [(1/N_e)(dN_e/dx)]^{-1}$ at $N_e = 10^{21}$ cm^{-3} during the laser pulse (assuming a constant temperature).

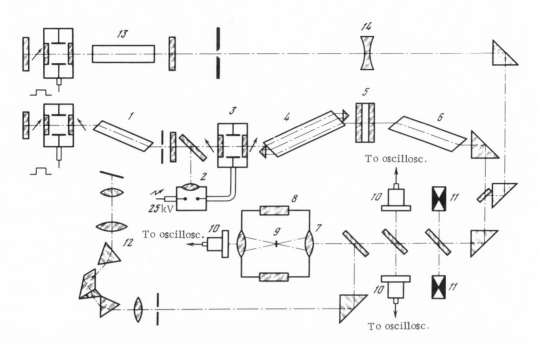

Fig. 23. Diagrams of experimental setup. 1) Neodymium laser Q-switched with a Kerr cell; 2) laser-ignited discharge gap; 3) shaping Kerr shutter; 4) first amplifier with three passes; 5) cell with saturable dye; 6) second amplifier; 7) focusing lens; 8) vacuum chamber; 9) targets; 10) coaxial photocell; 11) calorimeters; 12) prism spectrograph; 13) ruby laser Q-switched with Kerr cell; 14) negative lens to compensate for the chromatic aberration of the focusing lens.

coefficient decreases rapidly (in analogy with [32]). If it is assumed that the bremsstrahlung mechanism plays the principal role in the absorption of the light energy by the plasma corona (a sufficiently correct assumption at not too high flux densities $q < 5 \cdot 10^{12}$ W/cm^2) then it is possible, knowing the reflection coefficient R(t), to estimate the characteristic dimension of the plasma inhomogeneity $a = [(1/N_e)/(dN_e/dx)]^{-1}$ and its evolution during the laser pulse [see Chapter I, Fig. 2, and Eq. (2)].

The change of the characteristic inhomogeneity dimension a with time at a flux density $q = 5 \cdot 10^{12}$ W/cm^2 is shown in Fig. 22. It has been assumed in these estimates that the temperature T_e is ~300 eV [124] and is constant during the laser pulse [32, 140] and along the absorption region. The degree of ionization was calculated by an approximate method [97].

We note that measurements of the gradient of the electron density of a laser plasma near the critical density by other methods are apparently impossible at the present time.

Great interest attaches to a determination of the electron-density gradient at $N_e > 10^{21}$ cm^{-3}, i.e., in regions in which the the heating radiation of the neodymium laser does not penetrate. Therefore, to probe the plasma and to investigate the reflection coefficient at a different wavelength during the laser pulse, a ruby laser is used ($\lambda = 0.69$ μm), the critical density for the radiation of which is $N_{cr} = 2.4 \cdot 10^{21}$ cm^{-3}.

The setup for this experiment is illustrated in Fig. 23. The ruby-laser radiation was guided with the aid of a system of prisms and mirrors to the axis of the focusing system and was focused onto the target surface by a lens with $f = 50$ mm into the same focal spot as the neodymium-laser radiation. To cancel the chromatic aberration of the focusing lens and to equalize the dimensions of the focal spots, a negative lens and diaphragms were placed in the

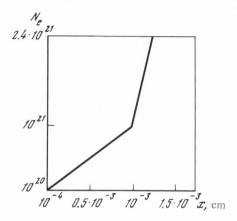

Fig. 24. Density profile $N_e(x)$ in a laser flare at the third nanosecond at $q = 5 \cdot 10^{12}$ W/cm^2.

path of the ruby-laser beam. The ruby laser was synchronized with the neodymium laser so that the neodymium-laser radiation pulse occurred at the maximum of the ruby-laser probing pulse. The flux density of the sounding radiation on the target was $q < 5 \cdot 10^9$ W/cm^2, and the ratio of the probing and heating signals was less than 10^{-3}, while the presence of the probing radiation did not alter the temporal parameters of the reflected radiation at the neodymium-laser frequency. The incident and reflected ruby-laser pulses were also diverted by a plane-parallel plane to coaxial photocells (not shown in the diagram), the signals from which were registered on 6LOR-2 and I2-7 oscilloscopes.

The experiments have shown that the reflection coefficient for the ruby laser R(t) is several times larger than for the fundamental radiation. Thus, the reflection coefficient of the ruby laser during the third nanosecond of the heating pulse was ~1% (as against 0.3-0.4% for a neodymium laser) and decreased towards the end of the pulse. Again using (2) of Chapter I, we find that, at a flux density $q = 5 \cdot 10^{12}$ W/cm^2, we obtain at a point with $N_e = 2.4 \cdot 10^{21}$ cm^{-3} an electron-density gradient $dN_e/dx = 9 \cdot 10^{24}$ cm^{-4} (at the third nanosecond), i.e., dN_e/dx increases sharply (by approximately 7-8 times) at densities above 10^{21} cm^{-3}. Figure 24 shows the profile of the electron density for the third nanosecond at $q = 5 \cdot 10^{12}$ W/cm^2.

In concluding this section we note once more that the estimates presented above for the electron-density gradients are meaningful only if the bremsstrahlung mechanism plays the principal role in the absorption of the laser radiation by the plasma corona.

4. Oscillations of Reflected Radiation with Time

Starting with flux densities $q > 10^{12}$-$5 \cdot 10^{12}$ W/cm^2, the radiation reflected from a plasma was observed on several oscillograms to be amplitude-modulated with a period $\tau \simeq 10^{-9}$ (Fig.

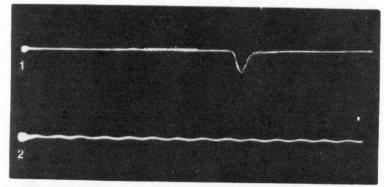

Fig. 25. Oscillogram of the reflected pulse showing the modulation of the light flux (1) and calibration sinusoid with period 10 nsec (2).

25). This fact by itself, however, is not irrefutable proof of modulation of the reflected radiation intensity in the plasma, since when a light signal with a leading front shorter than the time of resolution of the apparatus is applied to a coaxial photocell, "ringing" of the recording system can take place and lead to modulation of the electric signal (it was indicated in the preceding section that the leading front of the reflected radiation becomes shorter).

To attain a larger time of resolution of the measurements, a photoelectronic streak camera of the FÉR-2 type was used with an image converter UMI-92, which ensured a time of resolution better than 10^{-10} sec at a sweep duration 32 nsec and at a slit width 100 μm. This is higher by approximately one order of magnitude than the time of resolution obtained with oscilloscopes.

The reflected and incident radiation was focused with a cylindrical lens on the photochromator entrance slit, on which a nine-step attenuator was placed. The incident radiation was delayed optically by approximately 10 nsec relative to the reflected radiation. The presence of a step attenuator made it possible to operate in the range of normal photographic density in practically every shot. Thus, the incident and reflected signals were registered with a single sweep. The measurements were performed mainly with aluminum target [52]. The characteristic time sweep is shown in Fig. 26a. The streak photographs were reduced with a microphotometer. Figures 26b and 26c show plots of the incident and radiation flux signals on the time for the streak photographs of Fig. 26a.

The reduction of a large number of streak photographs yielded the following results: In all the flashes there was observed a shortening of the leading front of the reflected signal by approximately 2-3 times in comparison with the leading front of the incident signal (with approximate duration 1.5 nsec). In all the flashes, beats of the radiation flux are seen, with a modulation depth up to 30-50% of the maximum value in some of the flashes. The period of these beats was on the average 0.7-0.8 nsec. In 40% of the incident signals, individual much weaker flux bursts were also observed, either coinciding with the maxima (10% of the cases), or not coinciding with any of the maxima of the reflected signals (30% of the cases).

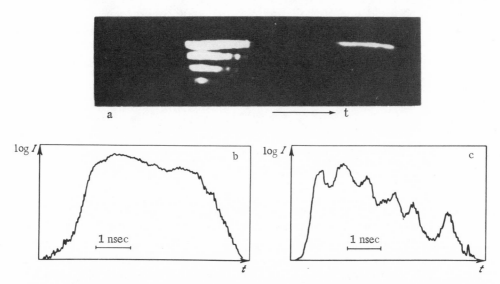

Fig. 26. Characteristic streak photography of incident (right) and reflected (left) laser radiation (a), obtained with a step attenuator, and dependence of the intensity of the incident (b) and reflected (c) signals on the time for the streak photograph given in the upper figure. Aluminum target, $q = 7 \cdot 10^{12}$ W/cm^2.

The experimental results obtained in this manner suggest that the deep modulation of the reflected light pulse is a physical effect due to the nonlinear interaction of the intense light fluxes with the laser plasma. The nonlinearity of the effect manifests itself clearly primarily in the fact that it has a threshold: the time oscillations of the reflected signal are observed only at a flux density $q > 10^{12}$ W/cm^2. A possible explanation of the oscillatory character of the reflected radiation may be the onset of relaxation oscillations of the nonstationary turbulent parametric noise that sets in as a result of the development of one of the potential parametric instabilities in the plasma [52, 141-143]. The time oscillations of the energy density of the parametric turbulence and the inverse transformation of the turbulent noise can lead to experimentally observable oscillations of the intensity of the reflected (reradiated) signal [52]. On the other hand, if this reradiation of the light by the turbulence can be neglected, one can speak again of modulation of the signal reflected by the plasma, with the modulation period of the turbulent noise, since a turbulence level of passage with time leads to modulation of the anomalous absorption of the intense optical illumination, meaning also to modulation of the reflection coefficient. An estimate of the oscillation period for the instabilities $(t \rightarrow l + l)$ [52, 53, 142, 143] and $(t \rightarrow l + s)$ [52, 137] under the conditions of our experiment yield values close to those observed ($\tau \sim 10^{-9}$ sec). On the other hand, to identify the actual instability that causes the oscillations of the reflected radiation, more complete experimental data are needed concerning the dependence of the period τ on the excess of the flux of the intense light wave over the threshold value, on the plasma electron temperature, on the ion masses, etc. On the other hand, satisfactory quantitative agreement between the estimate of the period of the oscillations of the radiation reflected by the plasma and the noise turbulence level for different instabilities does not exclude the possibility of other interpretations [144, 145] of the observed effect. It is possible, for example, that the oscillations of the reflected radiation may be due to amplification, in the laser plasma, of weak intensity oscillations of the incident light pulse, which cannot be observed with our apparatus (with a modulation depth less than $\sim 10\%$).

5. Directivity of Reflected Radiation

Great interest attaches to the investigation of the directivity of the reflected radiation for both normal and oblique incidence of the high-intensity radiation on the plasma layer. The reason is that the laser radiation is incident on the spherical target at different angles and a clarification of the character of reflection for inclined incidence is of practical interest. On the other hand, knowledge of the directivity pattern of the reflected radiation extends our knowledge of the processes occurring in the plasma corona.

The experiments that will be discussed in the present section [53] were performed at laser energies up to 30 J, pulse durations $3 \cdot 10^{-9}$ sec at half-energy level, a radiation contrast better than 10^5, and a divergence $5 \cdot 10^{-4}$ rad. The laser radiation was focused with a lens of focal length 10 cm onto the surface of a target placed at the center of the vacuum chamber. In this series of experiments the targets were samples of ordinary $(CH_2)_n$ and deuterated $(CD_2)_n$ polyethylene and of aluminum. The flux density reached values $q \sim 10^{14}$-$2 \cdot 10^{14}$ W/cm^2.

A specially developed vacuum chamber (see Fig. 15), the cross section of which in the plane perpendicular to the polarization vector of the heating radiation was an irregular heptagon, made it possible to investigate the parameters of the radiation scattered by the plasma in directions lying in this plane and making angles 0, 45, 60, 120, 135, 180° with the laser-beam axis, and also in a direction parallel to the incident-radiation polarization vector. This made it possible to investigate the spatial distribution of the radiation reflected and scattered by the plasma at various orientations of the target plane relative to the laser-beam axis, and also to investigate the temporal structure of the scattered radiation along all the indicated directions.

An investigation of the spatial distribution of the radiation reflected from the plasma at different angles of incidence of the laser beam on the surface of the flat target was carried out with the aid of calibrated coaxial photocells.

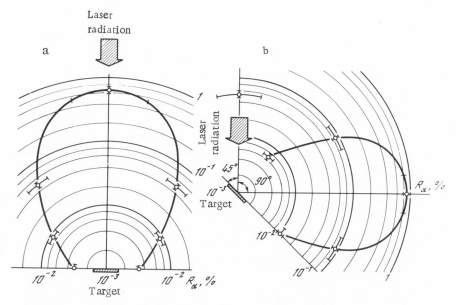

Fig. 27. Directivity patterns of the reflected radiation in a plane perpendicular to the laser-radiation polarization vector, for the case of normal incidence of light (q = 6 · 10^{13} W/cm^2) (a), and of the reflected and scattered radiations when the target was inclined 45° to the laser beam (the polarization vector lies in the plane of the target, q = 8 · 10^{13} W/cm^2) (b).

Figure 27a shows a typical directivity pattern of the reflected radiation in a plane perpendicular to the laser-radiation polarization vector, for the case of normal incidence of light on the target at a flux density q ≃ 6 · 10^{13} W/cm^2. The coefficients of reflection along different directions are referred in this diagram to the aperture of the focusing lens, which in our experiment amounted to ~0.16 sr.

The coefficient of reflection into the aperture of the focusing lens was ~0.65%, and when the observation angle (i.e., the angle between the investigated direction and the laser beam) was increased from 0° to 90° the reflection coefficient decreased rapidly and amounted to ~0.32 · 10^{-2}% at an observation angle 90° for an aperture ~0.16 sr.

The observation of the reflected radiation at 90° to the direction parallel to the polarization vector yielded approximately the same value of the scattered energy (it is of interest to compare the results with [146]); this indicates that the laser radiation is scattered isotropically by the plasma at 90° to the laser beam axis, at least at the flux densities of this particular experiment.

Even though the directivity pattern of the reflected radiation is strongly elongated in the direction of the focusing length and the energy of the back-reflected radiation is much higher than for other directions, allowance for the radiation scattered at all angles leads to a noticeable increase of the fraction of the energy reflected from the plasma. For example, for the case of Fig. 27a, integration of the directivity pattern over the half-space in front of the target leads to a value ~3% for the total reflection coefficient into a solid angle of 2π sr. It should be noted that in practically all the experiments in which the coefficient of reflection from a plasma was investigated (see Chapter I), this coefficient was measured only in the solid angle of the focusing system, so that only the lower bound of the reflected energy was obtained; this can result in a substantial error in the determination of the fraction the laser energy absorbed in the plasma.

An investigation of the temporal structure of the reflected radiation has shown that, just as at low light fluxes (see Sec. 4 of this chapter), the intensity of the reflected pulse is modulated, not only at the observation angle 0°, but also at other angles. The period of the oscillations was the same for the different observation angles, $\tau \geq 1$ nsec, and the depth of modulation was apparently maximal for the angle 45° (more than 50% is some flashes).

Another important factor observed in this experiment is the "specular" character of the reflection for oblique incidence of the radiation on a flat target. Figure 27b shows the characteristic directivity pattern of the reflected and scattered radiations for the case when the aluminum target is mounted at an angle 45° to the laser beam, and the polarization vector lies in the plane of the target. The diagram is elongated in the direction of 90° relative to the laser beam, and the integrated reflection coefficient was equal, accurate to a factor of 2, to the reflection coefficient in the case of normal incidence. The energy fraction reflected into the solid angle of the focusing system was less than the parasitic reflection from the latter, i.e., less than $10^{-1}\%$ of the laser energy.

This result contradicts in general the results of [26], where it was observed that the laser radiation is reflected not "specularly," i.e., each ray is reflected from the plasma along the direction of incidence on the target surface. It must be borne in mind, however, that the experiments of [26] were performed for a D_2 target and with normal incidence of the radiation on the target.

CHAPTER IV

GENERATION OF HARMONICS OF THE HEATING-RADIATION FREQUENCY IN A LASER PLASMA

In the next two chapters we describe the results of spectral investigations of the reflected and scattered radiation, and also of an investigation of the continuous x-ray emission from a laser plasma in the case of sharp focusing. Providing undisputed evidence of the anomalous character of the processes occurring in the plasma corona, the described results make it possible in a number of cases to identify exactly the mechanism of the interaction of high-power radiation with a plasma.

1. Investigation of the Generation of the Second Harmonic of the Heating Radiation in a Laser Plasma; Dependence on the Flux Density; Variation with Time

The spectral composition of the radiation reflected into the solid angle of the focusing lens was investigated in the visible region with the aid of an ISP-51 prism spectrograph, which was used with a UF-89 camera. Starting with flux densities $q \geq 10^{11}$ W/cm^2, for all the targets investigated in the experiment, the spectrum of the back-reflected radiation revealed a line with wavelength $\lambda = 5300$ Å corresponding to the second harmonic of the radiation incident on the target, with $\lambda = 10.600$ Å [51, 72]. This appears to be the first such observation for targets with Z > 1. It must be noted here that the appearance of the second harmonic in the reflected-radiation spectrum is not restricted to a laser plasma. Second harmonic generation was observed in the reflection of radio waves from the ionosphere; it was observed also in the investigation of the spectral composition of the irradiation of the solar corona during certain types of flares that the shape of the spectrum is duplicated at double the frequency, etc. [147].

Figure 28 shows a spectrogram [52] containing the line with wavelength $\lambda/2 = 5300$ Å for several flashes at different flux densities q.

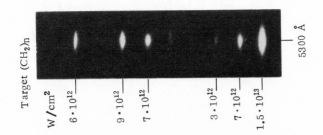

Fig. 28. Photograph of the second-harmonic line
$\lambda/2 = 5300$ Å in laser radiation reflected from a
plasma at different flux densities of the laser radiation with $\lambda = 10,600$ Å incident on the plasma.

The photographic density curve of the film during an exposure time equal to the duration
of the laser pulse (i.e., of the order of the time of second-harmonic generation) was plotted
in the following way. In one of the rungs of the flashes the spectrograph was tuned in such a
way that the spectral region with $\lambda = 5300$ Å had the worse dispersion (to decrease the width
of the line $\lambda = 5300$ Å on the photographic film), while the spectrograph slit, on which a stepped
attenuator was placed, was illuminated with second-harmonic radiation with the aid of a cylindrical lens. Photometry of the lines with $\lambda = 5300$ Å has shown that the second-harmonic energy $E_{2\omega}$ is connected with the flux density q (or with the laser energy E_{las}) by the power-law
relation $E_{2\omega} \sim q^{\alpha}$ with exponent $\alpha = 1.4 \pm 0.1$. The experimental dependence of the second-
harmonic energy $E_{2\omega}$ on the flux density q for an aluminum target is shown in Fig. 29. Unity
in this figure represents the second-harmonic energy at a heating-radiation flux density
$q \simeq 10^{13}$ W/cm^2, and amounts to approximately 10^{-5}-10^{-4} of the laser radiation. An investigation of the function $E_{2\omega}(q)$ was carried out also at lower flux densities [51]. It showed that
$\alpha \simeq 1.4$ also at $q \sim 10^{11}$-10^{12} W/cm^2. There were no differences between the characteristics
of the second-harmonic radiation for the different targets. It should only be noted that the line
intensity decreased slightly with increasing atomic number.

Theoretically, second-harmonic generation by a laser plasma can be interpreted as a
result of the coalescence of the potential electron Langmuir osillations excited in the plasma
by the incident light with the light wave reflected from the critical density ($t + l \rightarrow t$), or else
as a result of coalescence of two electron Langmuir oscillations ($l + l \rightarrow t$). The mechanisms
whereby the incident light excites the electron Lagmuir oscillations may vary. Sufficiently
small light fluxes, which do not exceed the threshold value q_{thr} needed for parametric buildup

Fig. 29. Experimental dependence of the second-harmonic energy $E_{2\omega}$ passing in the solid
angle of the lens on the flux density q of the
light incident on the plasma.

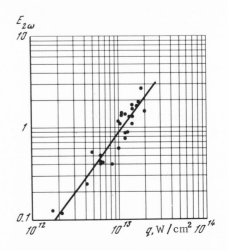

of the oscillations, lead to an increase of the potential electric field in the vicinity of the critical density of the inhomogeneous plasma, owing to the already discussed linear transformation phenomenon [136]. This is precisely the plasma-wave excitation mechanism discussed in [71, 148-150], where the theory of second-harmonic generation is treated.

According to the results of [71], the laser-plasma radiation flux at the second-harmonic frequency is determined by different relations, depending on the type of coalescing waves $(t + l \rightarrow t$ or $l + l \rightarrow t)$ and on the incidence angle θ of the radiation on the plasma. Assuming the relation between the plasma electron temperature T_e and the flux to be $T_e \sim q^{4/9}$, we obtain for the different types of coalescence

$$
\begin{array}{lll}
t + l \rightarrow t, & q_{2\omega} \sim q^{46/27} \simeq q^{1.7}, & \theta > \theta_0, \\
t + l \rightarrow t, & q_{2\omega} \sim q^{40/27} \simeq q^{1.48}, & \theta < \theta_0, \\
l + l \rightarrow t; & q_{2\omega} \sim q^{26/27} \simeq q^{0.96}.
\end{array}
$$

Here θ_0 is a certain critical value of the angle θ of incidence of the radiation on the plasma, and is determined by the plasma temperature (for details see [71]).

It is seen from these equations that the theoretical value of the flux exponents agrees satisfactorily with the obtained ones, the second of which being closest.

So good an agreement between the results of theory and experiment confirms the possibility of interpreting our results within the framework of the linear-transformation theory, at any rate at not too large flux densities q. Favoring the linear transformation is also the fact that the emission of the second-harmonic line into the solid angle of the focusing system was registered by starting with heating-radiation flux densities $q \sim 10^{11}$ W/cm^2, which is much lower (by not less than one order of magnitude) than q_{thr} for any of the parametric instabilities discussed above. The fact that in experiment no differences were noted in the characteristics of the second-harmonic radiation for different targets indicates that the mechanism of generation of the second harmonic is purely electronic, which again does not contradict the results in [71, 148-150]. It is interesting to note that with this generation mechanism and under optimal conditions [71], 1% of the laser energy can be converted into the second harmonic at $q = 2 \cdot 10^{13}$ W/cm^2.

At light fluxes exceeding the threshold for the buildup of parametric instabilities $q > q_{thr}$, the second harmonic can be generated by the plasma as the result of parametric instabilities of the type $t \rightarrow l + a$, $t \rightarrow l + s$, which develop in the vicinity of the plasma profile with critical density. The fact that the directions of the optimal development of parametric turbulence are singled out relative to the polarization vector of the incident light wave and its vector makes it possible to identify the directions of predominant second-harmonic generation due to the development of parametric instability.

If the second harmonic is generated by the mechanism discussed above, on the plasma electron density gradient, the second-harmonic radiation will propagate mainly backward [71] into the focusing system (if refraction is neglected and no account is taken of the complex configuration of the surface of the plasma with critical density). On the other hand, generation of the second harmonic, due to the development of parametric instability, takes place mainly along the polarization vector of the incident light wave (at $q/q_{thr} \sim 1$).

At large excesses above the threshold ($q \gg q_{thr}$), a line of wavelength $\lambda = 5300$ Å should be observed in the entire half-space in front of the target.

In the described experiments, the characteristics of the second-harmonic radiation were investigated mainly in the solid angle of the focusing system. Several dozen flashes were produced, however, to investigate second-generation perpendicular to the laser beam and to the polarization vector of the incident radiation at flux densities $q \sim 10^{12}$-$5 \cdot 10^{12}$ W/cm^2 (i.e., near the theoretical threshold of the potential parametric instabilities that develop at the point of the profile with critical density.

At normal incidence of the laser radiation on the target, none of these flashes produced lines with wavelength $\lambda \simeq 5300$ Å in the spectrum of the radiation scattered at 90°. It must be stated, however, that it would be unwise to draw from this experimental fact any conclusions concerning the possible mechanism of second harmonic generation. The point is that in the case of "sharp" focusing the flux density is not uniform over the diameter of the focal spot, and decreases towards its edges. The temperature of the plasma regions that are peripheral in the focal spot therefore decreases in comparison with the central and hottest part of the spot. The inverse-bremsstrahlung coefficient $K \sim 1/T^{3/2}$ is inversely proportional to the plasma temperature and increases from the center of the focal spot towards its edges. Therefore, if second-harmonic generation takes place parallel to the surface of the target, its radiation can be absorbed in the peripheral regions of the plasma. Interest attaches, however, to the fact that when the target is inclined 45° to the incident radiation, an intense $\lambda = 5300$ Å line appears in the spectrum observed at 90°, thus indicating a unique "specularity" (cf. Sec. 5, Chapter 3) of the second harmonic generation in the case of oblique incidence on the target in the flux-density region $q \sim 10^{12}$-$5 \cdot 10^{12}$ W/cm^2.

A study was made also of the temporal structure of the second harmonic propagating into the solid angle of the focusing lens (only for flux densities $q < 10^{13}$ W/cm^2). To this end, radiation with wavelength $\lambda \simeq 5300$ Å was focused either through an interference filter or from the exit slit of the spectrograph, on the entrance slit of the FÉR-2 streak camera, with the aid of which a time sweep was effected with a resolution not worse than 10^{-10} sec. The reduction of several dozen ($\sim$30) streak photographs has led to the conclusion that the waveform of the plasma irradiation pulse at the second-harmonic wavelength duplicates approximately the envelope of the reflected pulse at the fundamental frequency, but in none of the flashes was modulation of the second-harmonic intensity observed. It appears that this circumstance makes it difficult to attribute the oscillatory character of the reflected radiation at the fundamental frequency to energy-density oscillations of the parametric turbulence, which sets in as a result of excitation in the plasma of aperiodic instability ($t \rightarrow l + a$) and instability of the type of decay into an ion-acoustic and a plasma wave ($t \rightarrow l + s$), which develop at the point with critical density.

2. Generation of $^3/_2\omega_0$ Line

At flux densities $q \simeq 10^{13}$ W/cm^2 on the surface of a polyethylene (CH$_2$)$_n$ target, a line with wavelength $\lambda \simeq 7066$ Å was registered in the spectrum of the light scattered into the solid angle of the focusing lens; this line corresponded to the laser radiation frequency multiplied by a factor of 1.5. The signal/noise ratio for this line was much less than for the second harmonic ($\lambda \simeq 5300$ Å). We note that in the experiments [74] in which the $^3/_2\omega_0$ line was first observed for a hydrogen target, its intensity exceeded the $2\omega_0$ intensity, but the observation was carried out at an angle of 45° to the laser beam.

A possible explanation of the appearance of the $^3/_2\omega_0$ line in the spectrum of the scattered radiation may be the development of parametric instability of the type of the decay of light waves into two plasmons ($t \rightarrow l + l$). The point is that this instability develops at a profile point with density equal to one-quarter the critical value, and excites plasma waves with frequencies equal to $\omega_0/2$. The coalescence $\omega_0/2 + \omega_0$ of the potential plasma oscillation with the light wave reflected from the plasma region with critical density can lead to the appearance of the $^3/_2\omega_0$ line in the spectrum of the scattered radiation.

CHAPTER **V**

ANISOTROPY OF X RAYS FROM A LASER PLASMA

1. Procedure of Multichannel Measurement

of Continuous X Radiation

In this study, besides the reflected radiation of flux densities q $\simeq 10^{13}$-2 $\cdot$ 10^{14} W/cm^2, we investigate the spatial distribution of the intensity of the x rays passing through beryllium and aluminum filters of various thickness. A feature of this experiment (Fig. 30) was the fact that the x rays were registered with x-ray film in a sensitivity region down to 1 Å (UF-R or UF-VR), placed in a special cassette, at the entrance to which were mounted absorbing filters with different transmission curves. We used beryllium filters from 100 to 2500 μm and aluminum filters from 20 to 200 μm. The cassettes were oriented at various angles to the investigated plasma, and two or three cassettes were placed along each direction at different distances (from 2 to 10 cm) (a typical arrangement of the cassettes in the experiment is shown in Fig. 30). This has made it possible to operate in each flash in the region of normal photographic density, and consequently, obtain information on the distribution of the x-ray intensity in the given direction in a wide spectral interval. In addition, this made it possible to obtain in each flash the photographic density curve for each filter (since the number of quanta striking the photographic film was N $\sim 1/l^2$, where l is the distance from the cassettes to the target), information needed for the determination of the relative intensity of the x rays propagating along different directions in the investigated spectral region. The photographic film in each cassette measured 1 $\times$ 1 cm and was covered with 14 filters of various thicknesses. Thus, each filter covered $\sim$7 mm^2 of photographic film, and this permitted a correct reproduction of the results. This method of registration of continuous x rays does not, generally speaking, give the temporal resolution.

However, for the case of a laser plasma whose temperature decreases rapidly after the end of the laser pulse (T$_e$ $\sim$ $^1/r^2$, where r is the plasma dimension), the power, integrated over the spectrum, of both the bremsstrahlung [151] $\varepsilon_B = 4.86 \cdot 10^{-31} N_e N_Z Z^2 T_e^{1/2}$ W/cm^2 (where T$_e$ is in keV) and of the recombination radiation [152] $\varepsilon_R = 1.58 \cdot 10^{-32} N_e N_{Z+1} Z^4 T_e^{-1/2}$ W/cm^2, de-

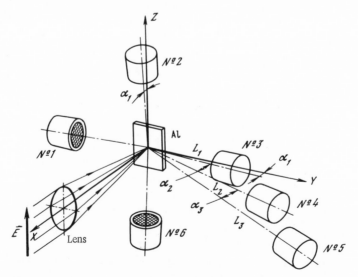

Fig. 30. Arrangement of the multichannel x-ray detectors Nos. 1-6 relative to the target. $\alpha_1 = 5°$; $\alpha_2 = \alpha_3 = 10°$.

creases as a result of the decrease in temperature and the recombination-induced decrease of the effective charge of the plasma ions Z. Inasmuch as the x rays registered in the experiments are from the short-wave end of the spectrum, the rapid decrease of the intensity of the registered signal after the end of the laser pulse is due also to a shift of the emission spectrum towards longer wavelengths. Thus, the overwhelming number of quanta registered by the x-ray film are emitted by the plasma during the time of the action of the laser pulse. On the other hand, the registration method widely used in laser-plasma diagnostics, wherein a scintillator converts the x rays into visible light registered with a photomultiplier, does not provide a temporal resolution better than 3-4 nsec, owing to both the finite scintillation time and to the parameters of the existing photomultipliers. The registration method using photomultipliers therefore offers no advantages over the registration of x rays with quantum energy $h\nu < 10$ keV, and becomes necessary only when quanta of higher energy are registered, i.e., outside the limits of the sensitivity of the presently existing x-ray films.

On the other hand, the method of registration with the aid of x ray films has a number of most important advantages, such as the high reliability of the registration process and the compactness of the recording detectors, so that measurements can be performed in a practically unlimited number of channels. The number of quanta per detector can be varied in a wide range by installing the detectors along one direction at different distances from the investigated plasma (at the geometry of the described experiment, an attenuation by two orders of magnitude was possible), and the dynamic range of the procedure can be greatly expanded by using films with different sensitivity. All this had made it possible for us to operate in practically each flash in the region of normal density of the films and to have information on the spectral distribution of the x rays at a sufficiently wide range without prior calibration of the detectors.

2. Investigation of the Directivity of the X Rays

It was observed that starting with a flux density $q \simeq 2 \cdot 10^{13}$ W/cm^2 a noticeable anisotropy* of the x rays sets in and manifests itself in the fact that in the plane of the target, in a direction perpendicular to the polarization vector of the incident light wave E_0, the plasma emits a noticeably larger number of quanta than in a direction parallel to E_0. Anisotropy was observed in 118 out of 259 flashes at $q \geq 2 \cdot 10^{13}$ W/cm^2, which amounts to approximately 43% of the cases, and in none of the flashes was the number of quanta in the parallel direction larger than the number of quanta in the perpendicular direction, even within the limits of the experimental error.

In fact, the percentage of flashes with anisotropy was apparently higher, since in some of the cases, which we assumed to be isotropic, only films with filters of thickness larger than 2000 μm were used in the region of normal density; in this spectral range, the radiation was isotropic even in the flashes with anisotropy (see Fig. 32). Figure 31 shows photographs of films that register the numbers of quanta passing through filters of various thicknesses in two directions. The results of the photometry for a typical shot with anisotropy of the x rays is shown in Fig. 32.

Figure 33 shows the dependence of the degree of anisotropy, defined by the relation $\xi = dN_1/dN_2$, where dN_1 and dN_2 is the number of quanta emitted by the plasma in the solid angle $d\Omega$ along directions perpendicular and parallel to the vector E_0, on the thickness δ of the beryllium filter (or on the end-point energy E_{ep} of the filter transmission corresponding to an attenuation of the radiation intensity by the filter by two orders of magnitude) for the flash of

---

* In earlier studies of x rays from a laser plasma, e.g., [153], no anisotropy of the x rays was observed. In [57, 154], however, polarization of the x rays was observed, thus pointing to an anomalous character of the interaction of the laser radiation with the plasma.

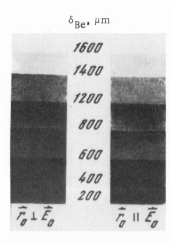

δ_{Be}, μm

Fig. 31. Photographs of films recording the x rays passing through beryllium filters of various thickness δ, in two directions.

Fig. 32, It is seen that the degree of anisotropy is maximal for thin filters and tends to unity with increasing δ (or E_{ep}).

We now discuss the possible causes of the anisotropy of the x rays. First, we note that generally speaking in an incompletely ionized plasma the intensity of the line spectrum can be greatly higher than the spectral intensity of the continuous radiation [152]. The wavelength of the spectral lines due to the bound−bound transitions of hydrogenlike ions of a multiply charged plasma are given by the expression

$$\lambda \simeq \frac{9.12 \cdot 10^2 n^2 n'^2}{Z^2 (n^2 - n'^2)} \text{Å}.$$

For the shortest-wavelength line of the Lyman series (at n' = 1, n = ∞) we then have λ_{min} = 9.12 · $10^2/Z^2$, or for fully ionized aluminum (Z = 13) λ_{min} = 5.4 Å. However, the energy of the radiation quanta with this wavelength (hν = 2.3 keV) is lower than the end-point transmission energy E_{ep} of a beryllium filter 600 μm thick, and it can be concluded that there is no line spectrum in the spectral region investigated in the given experiment. It is therefore meaningless to discuss here the possible mechanisms that lead to anisotropy of the line spectrum.

The effects of anisotropy of the angular distribution of the number of x ray quanta of a laser plasma, as well as other experimental results discussed in Chapters III and IV, can be

Fig. 32. Plot of the dependence of the photographic density S of the films on the filter thickness δ for a characteristic shot with anisotropy of the x rays. Aluminum target, beryllium filters, q $\simeq$ 5 · 10^{13} W/cm². 1) $\perp E_0$; 2) $\| E_0$.

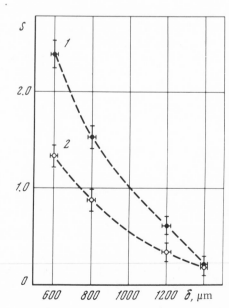

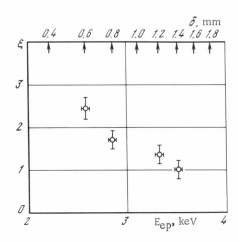

Fig. 33. Dependence of the degree of anisotropy ξ of the x rays from an aluminum plasma on the end-point energy E_{ep} of the transmission of a beryllium filter (corresponds to attenuation of the radiation intensity by a filter of thickness δ by a factor 10^2) ($q \simeq 5 \cdot 10^{13}$ W/cm^2).

interpreted within the framework of the theory of parametric resonance. The threshold fluxes (see Chapter III) for the development of periodic and aperiodic potential instabilities in an aluminum plasma with density close to critical range from 10^{12} to 10^{13} W/cm^2, depending on the type of the instability and on the relative contribution of the effects of dissipation and inhomogeneity of the laser plasma. Under the conditions of our experiment, the fastest to grow is the amplitude of the plasma oscillations, which propagate in a direction parallel to $\mathbf{E}_0$. This growth, in conjunction with one of the mechanisms of the linear, quasilinear, or nonlinear saturation of the instability [128], brings the plasma to a turbulent state with sufficiently high level of plasma oscillations, and in accordance with the Cerenkov effect, quasilinear deformation of the electron distribution function in the region of epithermal velocities $v \gg v_T$, equal in order of magnitude to the phase velocity of the high-frequency oscillations. As a result, the electron distribution function in this velocity range becomes non-Maxwellian and close to the distribution function of the electron beams propagating along $\mathbf{E}_0$ [155]. The relative number of such fast electrons increases in proportion to the laser radiation flux, and under the conditions of our experiment it is of the order of a fraction of 1% [155] of the total number of electrons in the region of the critical density of the plasma. For an instability wherein the light wave decays into electron Langmuir and ion-acoustic oscillations, the distribution function has the power law form $\delta f \sim v^{-\alpha}$. Integration of the cross section for the bremsstrahlung [156, 157] of fast electrons with such a distribution function, when these electrons are stopped by aluminum ions [53, 158], yields for the x-ray quanta of energy 1.3–12 keV a much larger number than the number of recombination-radiation quanta, which is approximately two orders of magnitude larger than the number of the bremsstrahlung quanta in the case of a Maxwellian electron distribution. This circumstance is decisive for the appearance of anisotropy of the angular distribution of the x rays, which is typical of a nonrelativistic electron beam. The calculated degree of anisotropy ξ agrees with the results of the experiment (at $\alpha = 4$ and in the fast-electron velocity interval from 7.5 v_T to 25 v_T).

3. Possibility of Measuring the Electron "Temperature" of a Laser Plasma by the "Absorber" Method

It should be noted that if we know the relative number $R_j^i = N\delta_i/N\delta_j$ of the quanta passing through filters of various thicknesses we can determine the electron temperature T_e of the plasma by comparing these ratios with the theoretical ones calculated for continuous radiation from a plasma with a Maxwellian electron velocity distribution. This method (of absorbers or filters) [159] was first proposed for the measurement of the electron temperature in a "theta pinch," and has been extensively used of late in the diagnostics of laser plasmas [28, 46, 160, 161].

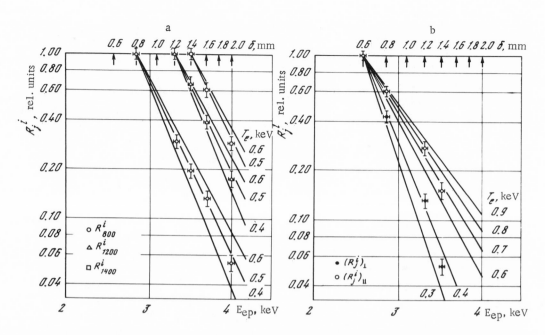

Fig. 34. Dependence of the ratio of the number of quanta passing through beryllium filters of various thicknesses δ on the end-point transmission energy E_{ep}. a) In the case of isotropic radiation ($q \simeq 8 \cdot 10^{13}$ W/cm^2); b) for a flash with emission anisotropy ($q \simeq 5 \cdot 10^{13}$ W/cm^2). Aluminum target.

Figure 34a shows the dependence of the ratio $R = N_{\delta_i}/N_{\delta_j}$ of the number of quanta passing through beryllium filters of various thicknesses, for a flash in which no anisotropy was observed (the change of the spectral sensitivity of the film was taken into account in accordance with the data of [162]). It turns out that for all combinations of filters, in the investigated spectral range, the ratios R_j^i correspond to a temperature $T_e \simeq 0.5 \pm 10\%$ keV for a plasma with a Maxwellian electron distribution. The "temperature" T_e determined in this manner in flashes with absence of anisotropy amounted to $T_e \simeq 0.5$–0.8 keV at flux densities $q \simeq 5 \cdot 10^{13}$–10^{14} W/cm^2.

A similar dependence for a flash with anisotropy is shown in Fig. 34b. In this case, since the ratios R_j^i are different in the directions parallel and perpendicular to the vector $\mathbf{E}_0$, the plasma electron temperatures T_e determined in this manner for these directions are also different. The experimental points are located close to the calculated ones for a Maxwellian distribution with temperature $T_e \simeq 0.7$ keV for a direction parallel to $\mathbf{E}_0$, and $T_e \simeq 0.4$ keV for the perpendicular direction, but the scatter of the "temperature" for each direction is increased in comparison with the isotropic case, amounting to approximately 25%.

We emphasize in this connection that it is necessary to approach with caution the interpretation of the experimental data on the spectral distribution of x rays from a nonequilibrium laser plasma, obtained with the aid of the absorber method. If no account is taken of the anisotropic character of the x rays and no account is taken of the disposition of the x-ray detectors when determining the ratios R_j, we can obtain, for one and the same flash, a variety of different distribution functions, none of which agrees with the real one.

Notice should also be taken of another important circumstance. As shown by calculations, even if the bulk of the plasma electrons have a Maxwellian velocity distribution and only an insignificant fraction of the electrons (less than 1%) in the region of epithermal velocities $v \gg v_T$ has a distribution function close to that of the electron beam, a very strong distortion of the spectral and spatial distributions of the x rays from the laser plasma can take place. All this allows us to conclude that the absorber method may be unsuitable for the measurement of the electron temperature of the laser plasma, even if an extremely insignificant fraction of the electrons has a velocity distribution function different from Maxwellian.

In conclusion the authors thank Academician N. G. Basov for constant interest in the work, V. P. Silin for valuable remarks, Yu. A. Afanas'ev, V. A. Gribkov, V. V. Pustovalov, and S. I. Fedotov for useful discussions, and Yu. A. Zakharchenkov, Yu. A. Mikhailov, and A. A. Rupasov for help with the experiments and useful discussions.

LITERATURE CITED

1. N. G. Basov and O. N. Krokhin, Proc. Conf. on Quantum Electron., Paris (1963).
2. N. G. Basov and O. N. Krokhin, Zh. Éksp. Teor. Fiz., 46:171 (1974).
3. N. G. Basov and O. N. Krokhin, Vestn. Akad. Nauk SSSR, No. 6:55 (1970).
4. N. G. Basov, Yu. S. Ivanov, O. N. Krokhin, Yu. A. Mikhailov, G. V. Sklizkov, and A. S. Shikanov, Pis'ma Zh. Éksp. Teor. Fiz., 15:589 (1972).
5. N. G. Basov, Yu. A. Zakharenkov, O. N. Krokhin, Yu. A. Mikhailov, G. V. Sklizkov, and S. I. Fedotov, Kvant. Élektron., 1(9):2069 (1974).
6. N. G. Basov, E. G. Gamalii, O. N. Krokhin, Yu. A. Mikhailov, G. V. Sklizkov, and S. I. Fedotov, Preprint No. 15, FIAN (1974).
7. J. Nuckolls, L. Wood, A. Thiessen, and G. Zimmerman, Nature, 239:139 (1972).
8. G. V. Sklizkov, Doctoral Dissertation, Physics Institute, Academy of Sciences of the USSR, (1973).
9. F. Floux, J. E. Benard, D. Cognard, and A. Saleres, Laser Interaction and Related Plasma Phenomena, Vol. 2, Plenum Press, New York (1972); F. Floux, Nucl. Fusion, 11:635 (1971).
10. J. M. McMahon and J. L. Emmett, Paper presented at Gordon Conf. B. D. (1971).
11. A. Haught, D. Polk, and W. Fader, Phys. Fluids, 13:2841 (1970).
12. F. Floux, D. Cognard, L. G. Denoed, G. Piar, D. Parisot, J. L. Bobin, F. Dolobeau, and C. Fauguignon, Phys. Rev., A1:821 (1970).
13. M. Lubin, J. Soures, E. Goldman, T. Bristow, and W. Leising, Laser Interaction and Related Plasma Phenomena, Vol. 2, Plenum Press, New York (1972).
14. J. Soures, S. Kumpan, and J. Hoose, Appl. Opt., 13:2081 (1974).
15. N. G. Basov, O. N. Krokhin, G. V. Sklizkov, S. I. Fedotov, and A. S. Shikanov, Zh. Éksp. Teor. Fiz., 62:203 (1972).
16. H. Hora, Laser Interaction and Related Plasma Phenomena, Vol. 1, Plenum Press, New York (1971).
17. Yu. V. Afanas'ev, N. G. Basov, P. P. Volosevich, O. N. Krokhin, E. I. Levanov, V. B. Rozanov, and A. A. Samarskii, Preprint No. 66. FIAN (1972).
18. K. A. Brueckner, Preprint KMS Fusion, Inc. KMSF-NRS (1972).
19. J. S. Clarke, H. N. Fisher, and R. J. Mason, Phys. Rev. Lett., 30:249 (1973).
20. A. Thiessen, G. Zimmerman, J. Weaver, J. Emmett, J. Nuckolls, and L. Wood, Proc. Sixth Eur. Conf. on Controlled Fusion and Plasma Physics, Vol. 2 (1973), p. 227.
21. J. Nuckolls, J. Lindl, W. Mead, A. Thiessen, L. Wood, and G. Zimmerman, Paper, Fifth IAEA Conf. on Plasma Phys. and Controlled Nuclear Fusion Res., Tokyo (1974).
22. G. S. Fraley, W. P. Gula, D. B. Henderson, R. L. McCrory, R. C. Malone, R. J. Mason, and R. L. Morse, Paper, Fifth IAEA Conf. on Plasma Phys. and Controlled Nuclear Fusion Res., Tokyo (1974).
23. N. G. Basov, V. A. Boiko, O. N. Krokhin, O. G. Semenov, and G. V. Sklizkov, Zh. Tekh. Fiz., 34(11) (1968).

24. N. G. Basov, V. A. Boiko, O. N. Krokhin, O. G. Semenov, and G. V. Sklizkov, Preprint No. 33, FIAN (1968)

25. A. M. Bonch-Bruevich, Ya. A. Imas, M. N. Libenson, G. S. Romanov, and L. N. Mal'tsev, Zh. Tekh. Fiz., No. 5 (1968).

26. K. Eidmann and R. Sigel, Internal Report No. IPPIV 46, Max-Planck Inst. Plasmaphysik (1972).

27. A. Saleres, F. Floux, D. Cognard, and J. L. Bobin, Phys. Lett., 45A:451 (1973).

28. F. Floux, D. Cognard, A. Saleres, and D. Redon, Phys. Lett., 45A:483 (1973).

29. J. Dawson, P. Kaw, and B. Green, Phys. Fluids, 11:875 (1969).

30. J. Dawson, in: Advances of Plasma Physics (A. Simon and W. Thompson, eds.), Interscience, New York (1968), p. 1.

31. S. Witkowski, Proc. Japan−U. S. Seminar on Laser Interaction with Matter, September 24-29 (1972).

32. M. Waki, T. Yamanaka, H. B. Kang, K. Yoshida, and C. Yamanaka, Jpn. J. Appl. Phys., 11:420-421 (1972)

33. C. Yamanaka, T. Yamanaka, H. B. Kang, K. Yoshida, M. Waki and K. Shimamura, in: Annual Review (April 1971-March 1972), Nagoya University.

34. H. B. Kang, T. Yamanaka, M. Waki, K. Yoshida, Y. Sakagami, and C. Yamanaka, IPPJ-142, Nagoya University (1972).

35. C. Yamanaka, Laser Interaction and Related Plasma Phenomena, Vol. 2, Plenum Press, New York (1972).

36. C. Yamanaka, T. Yamanaka, T. Sasaki, K. Yoshida, M. Waki, and H. B. Kang, Phys. Rev., A, 6:2335-2342 (1972).

37. C. Yamanaka, T. Yamanaka, T. Sasaki, M. Waki, and K. Shimamura, Proc. Japan−U. S. Seminar on Laser Interaction with Matter, Sept. 24-29 (1972).

38. C. Yamanaka, T. Yamanaka, and H. B. Kang, Laser Interaction and Related Plasma Phenomena, Vol. 3, Plenum Press, New York (1974).

39. F. C. Jahoda, E. M. Little, W. E. Quinn, G. A. Sawyer, and T. F. Stratton, Phys. Rev., 119:843 (1960).

40. K. Nishikawa, J. Phys. Soc. Jpn. 24:916, 1152 (1968).

41. E. D. Jones, G. W. Gobeli, and J. N. Olsen, Laser Interaction and Related Plasma Phenomena, Vol. 2, Plenum Press, New York (1972).

42. J. W. Shearer, S. W. Mead, J. Pettruzzi, F. Rainer, and J. E. Swain, Phys. Rev. A, 6:764-769 (1972).

43. K. Büchl, K. Eidmann, P. Mulser, and H. Salzmann, Report N IPPN/28, Max-Planck Inst. Plasmaphysik (1971).

44. W. L. Kruer and J. M. Dawson, Phys. Rev. Lett., 25:1174 (1970).

45. A. Carion, J. DeMetz, and A. Saleres, Phys. Lett., 45A:439 (1973).

46. R. Sigel, S. Witkowski, H. Baumhacker, K. Büchl, K. Eidmann, H. Hora, H. Mennicke P. Mulser, D. Pfirsch, and H. Salzmann, Kvant. Élektron., No. 8:37 (1972).

47. K. Büchl, K. Eidmann, and R. Sigel, Laser Interaction and Related Plasma Phenomena, Vol. 2, Plenum Press, New York (1972).

48. K. Büchl, K. Eidmann, A. Salzman, and R. Sigel, Appl. Phys. Lett., 20:231 (1972).

49. D. Beland, C. DeMichelis, M. Mattioli, and R. Papular, Appl. Phys. Lett., 21:31-33 (1972).

50. V. A. Boiko, Yu. A. Drozhbin, S. M. Zakharov, O. N. Krokhin, V. Ya. Nikulin, S. A. Pikuz, G. V. Sklizkov, A. Ya. Faenov, Yu. V. Chertov, and V. A. Yakovlev, Preprint No. 77, FIAN (1973).

51. A. A. Rupasov, G. V. Sklizkov, V. P. Tsapenko, and A. S. Shikanov, Zh. Éksp. Teor. Fiz., 65:1888-1904 (1973); Preprint No. 53, FIAN (1973).

52. N. G. Basov, O. N. Krokhin, V. V. Pustovalov, A. A. Rupasov, V. P. Silin, G. V. Sklizkov, V. T. Tikhonchuk, and A. S. Shikanov, Zh. Éksp. Teor. Fiz., 67(7) (1974); Preprint No. 17, FIAN (1974).

53. O. N. Krokhin, Yu. A. Mikhailov, V. V. Pustovalov, A. A. Rupasov, V. P. Silin, G. V. Sklizkov, and A. S. Shikanov, Preprint No. 22, FIAN (1975); Zh. Éksp. Teor. Fiz., 69: 206-220 (1975).

54. N. G. Basov, A. R. Zaritskii, S. D. Zakharvo, O. P. Krokhin, P. G. Kryukov, Yu. A. Matveets, Yu. V. Senatskii, and A. I. Fedosimov, Kvant. Élektron., No. 5:63-71 (1972).

55. A. Caruso, A. DeAngelis, G. Gatt, et al., Phys. Lett., 33A:320-321 (1970).

56. A. J. Andrews, T. A. Hall, and T. P. Hughes, Proc. Fifth Eur. Conf. on Controlled Fusion and Plasma Phys., Grenoble, August 21-25 (1972), Vol. 1, p. 62: Vol. 2, p. 237.

57. K. Boyer, LASL Report, LA-5251−PR (1973).

58. H. Salzmann, J. Appl. Phys., 44:113 (1973).

59. V. D. Dyatlov, R. N. Medvedeva, V. N. Sizov, and A. D. Starikov, Pis'ma Zh. Éksp. Teor. Fiz., 19:125 (1974).

60. L. V. Dubovoi, V. D. Dyatlov, V. L. Kryzhanovskii, A. A. Mak, R. N. Medvedev, A. N. Popytaev, V. A. Serebryakov, V. N. Sizov, and A. D. Starikov, Zh. Tekh. Fiz., 44(11) (1974).

61. L. M. Goldman, J. Soures, and M. J. Lubin, Phys. Rev. Lett., BI:1184-1187 (1973).

62. J. Soures, L. M. Goldman, and M. Lubin, Nucl. Fusion, 13:829-838 (1973).

63. C. S. Liu and M. N. Rosenbluth, Report NO C11 3237-11, Princeton Institute for Advanced Study (1972).

64. I. K. Krasyuk, P. P. Pashinin, and A. M. Prokhorov, Pis'ma Zh. Éksp. Teor. Fiz., 17:130 (1973).

65. A. E. Kazakov, I. K. Krasyuuk, P. P. Pashinin, and A. M. Prokhorov, Pis'ma Zh. Éksp. Teor. Fiz., 14:416 (1971).

66. P. Belland, C. DeMichelis, M. Mattioli, and R. Popular, Appl. Phys. Lett., 18:542 (1971).

67. N. G. Basov, V. A. Boiko, V. A. Gribkov, S. M. Zakharov, O. N. Krokhin, and G. V. Sklizkov, Zh. Éksp. Teor. Fiz., 61 (1970).

68. P. Mulser and S. Witkowski, Phys. Lett., 28A, 703 (1969).

69. A. Caruso, A. DeAngelis, G. Gatti, T. Gratton, and S. Martelucci, Phys. Lett., 33A:29 (1970).

70. G. Piar, B. Meyer, and M. Decroisette, Tenth Internat. Conf. on Phenomena in Ionized Gases, Oxford, Sept., 13-18 (1971).

71. A. V. Vinogradov and V. V. Pustovalov, Zh. Éksp. Teor. Fiz., 63:940 (1972).

72. A. A. Rupasov, V. P. Tsapenko, and A. S. Shikanov, Preprint No. 94, FIAN (1972).

73. C. Yamanaka, T. Yamanaka, H. B. Kang, M. Waki, and K. Shimamura, in: Annual Review (April 1972-March 1973), Nagoya University.

74. J. L. Bobin, M. Decroisette, B. Meyer, Y. Yitel, Phys. Rev. Lett., 30:594 (1973).

75. M. Decroisette, B. Meyer, and Y. Yitel, Phys. Lett., 45A:443 (1973).

76. A. A. Gorokhov, V. D. Dyatlov, V. B. Ivanov, R. N. Medvedev, V. P. Poponin, A. N. Popytaev, and A. D. Starikov, Abstracts of Proceedings of the Seventh All-Union Conference on Coherent and Nonlinear Optics, Tashkent [in Russian], Izd. Mosk. Gos. Univ., May 10-13 (1974).

77. J. Scott Hildum, Preprint KMS Fusion, Inc. KMSF-VI21 (1973).

78. H. Mennicke, Phys. Lett., 36A:127 (1971).

79. Yu. V. Afanas'ev, O. N. Krokhin, and G. V. Sklizkov, IEEE J. Quant. Electron. QE-2:483 (1966).

80. A. Caruso and R. Gratton, Plasma Phys., 10:867 (1968).

81. P. Mulser, Z. Naturforsch., 25A:282 (1970).

82. A. Caruso and R. Gratton, Phys. Lett., 27A:49 (1968).

83. P. Mulser and S. Witkowski, Phys. Lett., 28A:151 (1968).

84. P. Mulser and S. Witkowski, Phys. Lett., 28A, 703 (1968).

85. W. J. Linlor, Phys. Rev. Lett., 12:383 (1964).

86. W. G. Griffon and J. Schlüter, Phys. Lett., 26A(6):241 (1968).

87.   R. Sigel, Z. Naturforsch., 25a:488 (1970).

88.   J. Brunetead et al., Phys. Fluids, 13:1795 (1970).

89.   E. Fabre and P. Vasseur, J. Phys., 29C:123 (1968).

90.   R. Sigel et al., Phys. Lett., 26A:498 (1968).

91.   E. G. Gamalii, A. I. Isakov, Yu. A. Merkul'ev, A. I. Nikitenko, E. R. Rychkova, and G. V. Sklizkov, Kvant. Élektron., 2:1043 (1975).

92.   V. A. Gribkov, G. V. Sklizkov, S. I. Fedotov, and A. S. Shikanov, Prib. Tekh. Éksp., No. 44:213 (1970).

93.   N. G. Basov, V. A. Gribkov, O. N. Krokhin, and G. V. Sklizkov, Zh. Éksp. Teor. Fiz., 54:1073 (1968).

94.   N. G. Basov, O. N. Krokhin, and G. V. Sklizkov, Pis'ma Zh. Éksp. Teor. Fiz., 6:638 (1967).

95.   G. V. Sklizkov, S. I. Fedotov, and A. S. Shikanov, Preprint No. 45, FIAN (1972).

96.   G. V. Sklizkov, Candidate's Dissertation, Physics Institute, Academy of Sciences of the USSR (1967).

97.   Ya. B. Zel'dovich and Yu. P. Raizer, Elements of Gas Dynamics and the Classical Theory of Shock Waves, Academic Press, New York (1968).

98.   V. P. Korobeinikov, N. S. Mel'nikov, and E. V. Rozanov, Theory of Point Detonations [in Russian], Fizmatgiz, Moscow (1961).

99.   Yu. A. Zakharenkov and A. S. Shikanov, Prib. Tekh. Éksp. No. 5:166 (1974).

100. L. Foucault, Ann. Observ. Imp. Paris, 5:203 (1853).

101. L. A. Vasil'ev, Shadow Cathodes [in Russian], Nauka, Moscow (1968).

102. A. Toepler, Beobachtungen nach einer Neuen Optischen Methode, Bonn (1864).

103. Physical Measurements in Gas Dynamics and Combustion (R. W. Ladenburg et al., eds.), Princeton University Press (1954).

104. R. Huddlestone and S. Leonard (editors), Plasma Diagnostics Techniques, Academic Press, New York (1965).

105. L. A. Dushin and O. S. Pavlichenko, Plasma Research Using Lasers [in Russian], Atomizdat, Moscow (1968).

106. N. N. Zorev, G. V. Sklizkov, S. I. Fedotov, and A. S. Shikanov, Preprint No. 56, FIAN (1971).

107. G. V. Sklizkov and S. I. Fedotov, Prib. Tekh. Éksp., No. 2:176 (1972).

108. Yu. A. Zakharenkov, N. N. Zorev, A. A. Kologrivov, P. A. Konoplev, G. V. Sklizkov, and S. I. Fedotov, Preprint No. 121, FIAN (1973).

109. V. Ascoli-Bartoli, A. DeAngelis, and S. Martelluci, Nuovo Cimento, 18:1116–1137 (1967).

110. G. V. Sklizkov, in: Laser Handbook, Vol. 2, North Holland Publ. Co., Amsterdam, pp. 1545–1576.

111. Yu. V. Afanas'ev, V. M. Krol', O. N. Krokhin, and I. V. Nemchinov, Prikl. Mat. Mekh., 30(6) (1966).

112. J. L. Bobin, Y. A. Durand, P. P. Langer, and G. Tonon, J. Appl. Phys., 39:4184 (1968).

113. D. C. Emmony and J. Irving, Eighth Inter. Conf. on Ionization Phenomena in Gases, Vienna (1967).

114. N. G. Basov, O. N. Krokhin, and G. V. Sklizkov, Laser Interaction and Related Plasma Phenomena, Vol. 2, Plenum Press, New York (1972), p. 389; Preprint No. 132, FIAN (1971).

115. N. G. Basov, O. N. Krokhin, and G. V. Sklizkov, IEEE J. Quant. Electron., QE-4:988 (1968).

116. A. C. Kolb, Phys. Rev., 107:345 (1957).

117. N. G. Basov, O. N. Krokhin, and G. V. Sklizkov, Trudy FIAN, 52:118 (1970).

118. V. A. Gribkov, V. Ya. Nikulin, G. V. Sklizkov, Preprint No. 153, FIAN (1970).

119. Yu. V. Afanas'ev, V. A. Gribkov, O. N. Krokhin, V. Ya. Nikulin, G. V. Sklizkov, and M. A. Sultanov, Preprint No. 87, FIAN (1973).

120. V. A. Gribkov, O. N. Krokhin, and G. V. Sklizkov, Preprint No. 136, FIAN (1973).

121. R. N. Arkhipov, B. L. Vasin, I. A. Dubovik, M. I. Fedyanina, and A. S. Shikanov, Preprint No. 70, FIAN (1975).

122. D. Herriot, H. Kogelnik, and R. Kompfner, Appl. Opt., 3:523 (1964).

123. D. Herriott and H. Schalte, Appl. Opt., 4:883 (1965).

124. V. A. Boiko, O. N. Krokhin, and G. V. Sklizkov, Preprint No. 121, FIAN (1972).

125. V. S. Imshennik, Astron. Zh., 38:652 (1961).

126. J. M. Walsh, M. H. Rice, R. G. Queen, and F. L. Yarger, Phys. Rev., 108:196 (1957).

127. V. P. Silin, Zh. Éksp. Teor. Fiz., 48:1679 (1965); Usp. Fiz. Nauk, 108:625 (1972); Preprint [in Russian], No. 62:84, FIAN (1973).

128. V. P. Silin, Parametric Effects of High-Power Radiation on a Plasma [in Russian], Nauka, Moscow (1973).

129. N. E. Andreev, Yu. A. Kirii, and V. P. Silin, Zh. Éksp. Teor. Fiz., 57:1024 (1969).

130. L. M. Gorbunov, Zh. Éksp. Teor. Fiz., 55:2298 (1968).

131. N. E. Andreev, Zh. Éksp. Teor. Fiz., 53:2168 (1967).

132. R. R. Ramazashvili, Zh. Éksp. Teor. Fiz., 53:2168 (1967).

133. V. P. Silin and A. N. Starodub, Zh. Éksp. Teor. Fiz., 66:176 (1974); Preprint No. 124, FIAN (1973).

134. O. N. Krokhin, "High-temperature and plasma phenomena induced by laser radiation," in: Physics of High-Energy Density, KLVIII, Corso (1971).

135. G. Tonon, D. Schirmann, and M. Rabeau, Tenth Intern. Conf. on Phenomena in Ionized Gases, Oxford (1971), p. 225.

136. V. L. Ginzburg, Propagation of Electromagnetic Waves in Plasmas, 2nd. rev. ed., Pergamon Press (1971).

137. A. G. Denisov, Zh. Éksp. Teor. Fiz., 31:609 (1956).

138. A. D. Piliya, Zh. Tekh. Fiz., 36:818 (1966).

139. A. V. Vinogradov and V. V. Pustovalov, Pis'ma Zh. Éksp. Teor. Fiz., 13:317 (1971).

140. N. G. Basov, V. A. Boiko, V. A. Gribkov, S. M. Zakharov, and G. V. Sklizkov, Preprint No. 111, FIAN (1968).

141. V. V. Pustovalov and V. P. Silin, Zh. Tekh. Fiz., 45:2472 (1975).

142. N. E. Andreev, V. V. Pustovalov, V. P. Silin, and V. T. Tikhonchuk, Pis'ma Zh. Éksp. Teor. Fiz., 18:624 (1973); Kvant. Élektron., 1:1099 (1974).

143. N. E. Andreev, V. V. Pustovalov, V. P. Silin, and V. T. Tikhonchuk, Kratk. Soobshch. Fiz., No. 1:31 (1974).

144. V. I. Emel'yanov and Yu. L. Klimontovich, Abstracts of Proceedings of the Second International Conf. on Plasma Theory [in Russian], Kiev, October 28-November 1 (1974).

145. S. Wolschke, Abstracts of Proceedings of the Second International Conf. on Plasma Theory [in Russian], Kiev, October 28-November 1 (1974).

146. A. A. Galeev, G. V. Laval, T. O. Neil, M. Rosenbluth, and R. Z. Sagdeev, Zh. Éksp. Teor. Fiz., 65:973 (1973).

147. V. V. Zheleznyakov, Radioemission of the Sun and Planets [in Russian], Nauka, Moscow (1964).

148. N. S. Erokhin, V. E. Zakharov, and S. S. Moiseev, Zh. Éksp. Teor. Fiz., 56:179 (1969).

149. N. S. Erokhin and S. S. Moiseev, in: Reviews of Plasma Physics, Vol. 7, Consultants Bureau, New York (in press).

150. N. S. Erukhimov and S. S. Moiseev, Zh. Tekh. Fiz., 40:1144 (1970).

151. L. Spitzer, Physics of Fully Ionized Gases, Wiley (1962).

152. I. I. Sobel'man, Introduction to the Theory of Atomic Spectra, Pergamon Press (1972).

153. K. Eidmann and R. Sigel, Contributed Papers, Sixth Eur. Conf. on Controlled Fusion and Plasma Physics, Moscow, Vol. 1, July 30-August 4 (1973), p. 435.

154. R. P. Godwin, J. E. Kephart, and G. H. McCall, Bull. Am. Phys. Soc., Ser. II, 17:971 (1972).

155. V. V. Pustovalov, V. P. Silin, and V. T. Tikhonchuk, Pis'ma Zh. Éksp. Teor. Fiz., 17:120 (1973).

156. V. B. Berestetskii, E. M. Lifshits, and L. P. Pitaevskii, Relativistic Quantum Theory, Course in Theoretical Physics [in Russian], Vol. 4, Part 1, Nauka, Moscow (1968).

157. I. B. Borovskii, Physical Principles of X-Ray Spectral Research [in Russian], Mosk. Gos. Univ., Moscow (1956).

158. O. N. Krokhin, Yu. A. Mikhailov, V. V. Pustovalov, A. A. Rupasov, V. P. Silin, G. V. Sklizkov, and A. S. Shikanov, Pis'ma Zh. Éksp. Teor. Fiz., 20:239 (1974).

159. F. C. Jahoda, E. M. Little, W. E. Quinn, G. A. Sawyer, and T. E. Stratton, Phys. Rev., 119:843 (1960).

160. S. B. Segall, G. Charatis, R. R. Johnson, and E. J. Mayer, Report KMSF-U120, KMS Fusion, Inc. (1973).

161. E. J. Mayer and G. Charatis, Report KMSF-U116, KMS Fusion, Inc. (1973).

162. V. G. Movchev, A. N. Ryabtsev, and N. K. Sukhodrev, Zh. Prikl. Spektrosk., 12:274 (1970).

EXPERIMENTAL STUDY OF CUMULATIVE PHENOMENA
IN A PLASMA FOCUS AND IN A LASER PLASMA

V. A. Gribkov, O. N. Krokhin, G. V. Sklizkov, N. V. Filippov, and T. I. Filippova*

High-speed laser interferometry at two wavelengths, high-speed multiframe shadow photography, x-ray and neutron-radiation studies, as well as a probe procedure are used to show that the processes of cumulation in a plasma focus and in a laser plasma of special geometry proceed in two stages — hydrodynamic and kinetic. In the first stage, both in the case of MHD cumulation (plasma focus) and in the case of hydrodynamic cumulation (laser plasma), the main processes are determined by the formation of a cumulative jet. In the second stage, an intense relativistic electron beam is produced in the plasma focus and interacts strongly with the plasma, whereas in the cumulative laser plasma there is observed generation of high-power Langmuir noise that leads to the formation of a group (up to 1% of the total number) of fast electrons having an energy 5-10 times larger than the energy of the bulk of the electrons.

INTRODUCTION

Great interest is attached, in the present-day research on controlled thermonuclear fusion, to the investigation of plasma obtained in installations that make use of the cumulative effect. The gist of this method is that the energy initially accumulated in the large mass of the plasma or in the magnetic field is transferred to a relatively small number of particles. This is usually done by collapsing plasma clusters that are accelerated with a magnetic field or by gas-dynamic motion [1, 2]. A characteristic feature of cumulative processes is usually that an appreciable fraction of the particles of the large plasma mass flows out of the cumulation region, so that a high temperature can be attained in the remaining plasma (see, e.g., [3]).

The cumulative effect can be obtained also by another method, wherein compression without shock waves is produced when a spherical target is exposed to a high-power radiation pulse [4-6] of suitably chosen form. High density and temperature are then reached at the center of the target.

The reason for the interest in the cumulation installations is that the theoretical studies of the last decade have demonstrated the possibility of producing in them controlled thermonuclear fusion in the relatively near future (laser plasma, relativistic electron beam). In addition, a number of experimental setups already in existence and based on the cumulative effect have unprecedented neutron yields (for example, the plasma focus), so that investigations with the aid of these setups yield much valuable information on the behavior of the plasma under conditions close to thermonuclear.

The cumulation effect was discovered in the second half of the past century, when a sharp increase was observed in the piercing ability of shells in the presence of a pit in the explosive

*I. V. Kurchatov Institute of Atomic Energy.

charge. The first theoretical studies were made in the forties by a group of American scientists headed by Taylor [7]. In our country, a theory of cumulation was developed by Lavrent'ev [8]. It was shown in the theoretical papers that when a conical shell with angle 2α is collapsed under the influence of the explosive, a cumulative jet is produced at the apex of the cone, on the cone axis, and this jet propels approximately 10% of the collapsed mass at a velocity given by the equation

$$v_j = \frac{v_0}{\tan(\alpha/2)}, \tag{1}$$

where v_0 is the rate of collapse of the cone elements. It is seen from (1) that the jet velocity is larger the smaller the angle α. The idea of cumulation came subsequently into use in plasma physics. Since modern electric-discharge devices and lasers produce higher collapse velocities than explosives, an appreciably larger velocity of the cumulative jet can be obtained. The greatest progress in this direction was attained by now with installations of the plasma-focus type and in a laser plasma.

The present paper is devoted to an experimental investigation of the physical processes that take place in the course of magnetic cumulation of a plasma (a "plasma-focus" setup) and in the case of gas-dynamic cumulation (laser plasma obtained by using a special target geometry). The principal experimental plasma investigations were made with interferometers, x rays, and probes. Particular attention was paid in the investigations to the final stage of plasma cumulation in which the highest parameters are reached as a rule.

Despite the larger number of experimental and theoretical studies performed to date on magnetic cumulation, on the phenomena accompanying it in general, and on the plasma focus in particular [9-34], there is still no clear idea concerning the operation of this installation. The least understood are precisely the most interesting phenomena that occur in the plasma focus, namely the generation of high-power hard x rays and neutron pulses. The studies of cumulative laser plasma are only in their initial stage, and information contained in the articles published on this subject [36-39] are quite skimpy.

We proceed now to describe a procedure especially developed by us, using high-speed photography in a laser beam, and note that all other diagnostic methods used by us are slight modifications of classical procedures [40, 41].

CHAPTER I

PROCEDURE OF HIGH-SPEED INTERFEROMETRIC INVESTIGATION OF A NONSTATIONARY DENSE PLASMA

1. The Maximum Information Obtained by Optical

Laser Research Methods

A question encountered in laser interferometry of fast-moving objects is that of the volume of information obtainable as the result of the experiment.

When an object moving with velocity v is photographed, the blurring of the image in the object plane is $v\tau$, where τ is the exposure time. Accordingly the resolution is $N_1 = (v\tau)^{-1}$. On the other hand in the case of an object that is at rest, the resolution of the interference pattern is determined by the width of the laser line, which is uniquely connected with the exposure time

$$\tau \simeq k/\Delta\nu = k\lambda^2/c\Delta\lambda \tag{2}$$

(k is a coefficient that characterizes the degree of temporal coherence of the laser pulse, and $c = 3 \cdot 10^{10}$ cm/sec is the speed of light).

Given the dimension l of the interference field, the number N_2 of fringes on this field, and the accuracy of $(N_3)^{-1}$ which is required to reduce each fringe, we find that the order of the interference $p = \lambda / \Delta\lambda$ should be

$$p = N_2 N_3, \tag{3}$$

and the necessary resolution of a static object should be

$$N_4 = N_2 N_3 / l. \tag{4}$$

It is clear that the optimal solution of a moving object will be reached when $N_1 = N_4$. Combining the foregoing relations, we obtain an expression for the optimal exposure time for the maximum possible resolution:

$$\tau_{opt} = (k\lambda l/vc)^{1/2}. \tag{5}$$

With increasing exposure time, the resolution is determined initially by the velocity of the object, and after decreasing to $\tau < \tau_{opt}$ it is determined by the width of the laser emission line. For example, at $\lambda = 0.7\ \mu m$, $l = 10$ cm, $v = 10^7$ cm/sec, $\Delta\lambda = 0.1$ Å, and $k \simeq 10$, we have

$$\tau_{opt} = 1.5 \cdot 10^{-10}\ \text{sec.}$$

In this case $N_{opt} = 650$ lines/mm.

On the other hand, at $\tau = 10^{-9}$ sec and under the conditions indicated above we have $N_1 = (v\tau)^{-1} = 10$ lines/mm.

It must be borne in mind, however, that the plasma density frequently varies relatively slowly in plasma formations whose boundaries move with large velocity. The resolution in these regions is then determined by the coherence of the laser, and in the case of a ruby laser and $l = 10$ cm it amounts to

$$N_4 = p/l = \lambda/l\Delta\lambda \simeq 10^3\ \text{lines/mm.}$$

We proceed now to a description of the high-speed method used by us for the investigation of the plasma-focus plasma and cumulative configurations of laser plasma, using a ruby laser as the illumination source and employing in some of the experiments a nonlinear KDP crystal to obtain the second harmonic with $\lambda = 0.35\ \mu m$. In light of all the foregoing, this choice of the illumination source is optimal in our case.

2. High-Speed Laser Setup for Interferometric Investigations of a Plasma Focus and Cumulative Laser-Plasma Configurations

It was found in the course of this study that the plasma focus moves from shot to shot in an uncontrolled manner in a range of several millimeters about the center of the electrode. Consequently the only feasible method of optically investigating a plasma focus and cumulative plasma configurations is to use frame-by-frame shadow, Schlieren, and interferometric investigations. It is desirable to have a probing pulse of ~1 nsec duration and with good spatial and temporal coherence. To obtain such a short pulse we used an external shutter to cut out a spike from an ordinary Q-switched laser pulse [42]. A diagram of the installation used by us is shown in Fig. 1. Rough synchronization of the electric pulses on both Kerr cells was carried out by adjusting the lengths of the discharge cables. The fine synchronization was adjusted by means of the pump level of the ruby rod. This system made it possible to obtain a light pulse of 1-2 nsec duration.

By special measures, the stability of the laser operation was made practically equal to 100%.

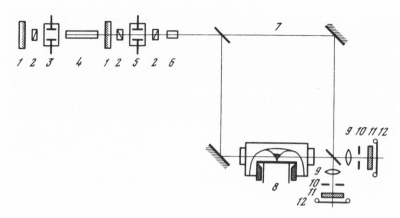

Fig. 1. Optical system of the experiment. 1) Mirrors; 2) polarizers; 3) Kerr cell; 4) ruby crystal; 5) pulse-sharpening Kerr cell; 6) KDP crystal; 7) Mach−Zehnder interferometer; 8) discharge chamber; 9) lengths; 10) diaphragms; 11) filters; 12) photographic film.

To improve the spatial coherence, we used a diaphragm of 3 mm diameter, which exposed a small section of the end face of the crystal, with subsequent collimation of the beam.

It is known [43] that the best temporal coherence is possessed by the leading front of a giant Q-switched pulse. We have taken special measures to obtain our pulse from precisely this part of the giant pulse. When working near the lasing threshold, the coherence length of our laser was of the order of several centimeters.

3. Synchronization Methods

One of the complicated questions encountered in the experiment was that of synchronizing the plasma focus with the laser. The sequence of synchropulses was chosen as follows. The laser lamps were first turned on. Approximately 500 μsec later, a synchronizing pulse was applied and triggered the plasma-focus installation. From the derivative of the discharge current at the instant when its sign was reversed, a synchronizing pulse was picked off and triggered, via a delay line, the thyratron of the high-voltage pulse generator controlling the Kerr cells.

Notice should be taken of the significant synchronization difficulties connected with the efforts to combat the noise, which was quite large in our case at discharge currents ∼1 mA. The greatest complication in the synchronization was that the peak on the oscillogram of the current derivative, and consequently also the instant of the collapse, could "float" relative to the start of the discharge (and relative to the point where the current derivative reverses sign) in a range of 2 μsec at a total discharge duration of 5 μsec.

Our investigations have revealed that the time of the appearance of the "singularity" depends on the discharge parameters; this helped produce the laser pulse at the required instant of cumulation. A two-beam oscilloscope was used to record the hard x-ray and neutrino radiation pulses, which were separated in time because of the different flight velocities, as well as the laser pulse. All the delays were compensated by cable segments. A typical synchronization oscillogram is shown in Fig. 2.

The synchronization of the lasers with one another and with the electron optical converter raised no difficulty and was affected by varying the lengths of the cables of the five-channel modulator for the Kerr cells.

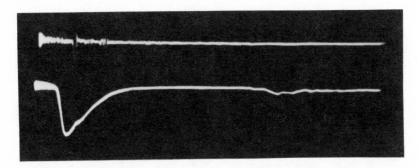

Fig. 2. Oscillograms of synchronization of plasma focus with laser. The break in the upper trace represents a short laser pulse. At the start of the lower trace one can see the large pulse of hard x radiation of the plasma focus, and at the end of the trace one can see the neutron pulse of the plasma focus. Total sweep duration is 2 μsec. The distance from the chamber to the multiplier is 15.22 m.

4. Discussion of the Applicability of Laser Interferometry and Interpretation of the Interference Patterns

Inasmuch as the targets used in various laser-plasma experiments had high values of Z, it was necessary to carry out plasma interferometry at two wavelengths in order to exclude the contribution of the ions to the plasma refractive index.

Without going into a detailed description of two-beam interferometry [40], we present a practical formula for the determination of the electron density in the case when the transmitted light is the first or second harmonic of a ruby laser with λ = 0.694 and 0.347 μm:

$$N_e = 6.166 \cdot 10^{21} \, [(n-1)_{\lambda_2} - (n-1)_{\lambda_1}], \tag{6}$$

where N_e is the electron density, and $(n-1)_{\lambda_i} = \nu_i$ is the polarizability of the plasma at the corresponding wavelength. This equation is valid for the case of a sufficiently hot plasma and under the condition that the plasma frequency is much lower than the frequency of the transmitted radiation. It must be borne in mind, however, that, generally speaking, situations can be realized in which this equation cannot be used.

We write the equation for the total polarizability of the plasma [44]:

$$(n-1)_{pl} = -\, 4.49 \cdot 10^{-14} \lambda^2 N_e + 4.49 \cdot 10^{-14} \sum_\alpha \nu_\alpha, \tag{7}$$

where ν_α is the polarizability of the ions:

$$\nu_\alpha = \sum_i \sum_k \frac{\lambda^2 \lambda_{ik}^{\alpha^2}}{\lambda^2 - \lambda_{ik}^{\alpha^2}} f_{ik}^\alpha N_i^\alpha, \tag{8}$$

λ is the wavelength of the electromagnetic field (in cm); N_i^α is the density of the ions in state i and with ionization multiplicity α: $\lambda_{ik}^\alpha, f_{ik}$ are the radiation wavelengths and the oscillator strength corresponding to the transition from the i-th to the k-th level.

It is seen from (8) that the weak dependence of the ion polarizability on the wavelength, which is used in the derivation of (6), is possible only when $\lambda_{ik}^2 \ll \lambda^2$. In the opposite case its dependence on the wavelength will be the same as for the electron polarizability. It is clear therefore that interferometry measurements should be preceded in general by spectroscopic measurements that yield information on the number of ions that are in "undesirable" states.

It should be noted that we have referred here to the phase refractive index, with which one deals in the case of interferometry in monochromatic light: $n_{ph} = c/v_{ph}$, where $v_{ph} = \omega/|\text{grad } f|$ is the phase velocity, that is, the propagation velocity of each constant-phase surface $f(\mathbf{r}) = \text{const}$ of the wave:

$$V(\mathbf{r}, t) = a(\mathbf{r}) \cos[\omega t - f(\mathbf{r})]. \tag{9}$$

The connection between the phase and the group refractive indices is given by the equation

$$n_{gr} = n_{ph} - \lambda \frac{dn_{ph}}{d\lambda}, \tag{10}$$

where n_{ph} pertains to the average frequency $\bar{\omega}$ of the wave packet.

It is known that the phase velocity v_{ph} of a wave having a more complicated form than a plane monochromatic wave differs in the general case from $c/(\varepsilon\mu)^{1/2}$ and varies from point to point even in a homogeneous medium. In an inhomogeneous medium, on the other hand, it is necessary to take into account in the wave equation the term with $\text{grad } \ln \varepsilon$. Furthermore, the large plasma dispersion $(n - 1 \sim \lambda^2)$ can lead to a strong spreading of the wave packet. This means that in the case of laser interferometry with ultrashort exposures ($\sim 10^{-10}$–10^{-11} sec), in a plasma with abrupt electron-density gradients at large optical path lengths, the distortions of the wave packet can play a significant role and call for a special analysis in each individual case.

It is thus desirable, by virtue of the mentioned difficulties, to work with a monochromatic light source in high-speed plasma interferometry where, as can be easily seen, n_{gr} differs strongly from n_{ph}, since interferometry in white light becomes meaningless. We point out also that when choosing the wavelength of the probing beam it is necessary to verify whether the emission of the investigated plasma contains a line with a frequency equal to that of the sounding radiation, for in this case even a relatively small number of ions in a given state can, as seen from Eq. (8), distort the picture significantly.

The absence of impurity emission lines with wavelengths that coincide with the wavelength of the sounding radiation was checked by us against the tables of [45]. Nor were there in the plasma focus and in our laser plasma any intense lines with wavelengths exceeding the laser wavelength.

Since the plasma focus has sharp density gradients and at the same time large absolute values of the density, it can be asked whether the refraction influences the interference pattern. For the angle at which the sounding ray emerges from the plasma we obtain from the eikonal equation [46]

$$\tan \alpha \simeq \int \frac{\partial}{\partial y} \ln n(x, y, z)\, dx, \tag{11}$$

where n is the refractive index of the plasma.

An expression for the displacement of the interference fringes resulting from refraction and not compensated for by the objective is given in [47]:

$$\delta \simeq \alpha^2 x/12\lambda, \tag{12}$$

where α is the angle at which the ray leaves the inhomogeneity, x is the path length in the plasma, and λ is the wavelength of the sounding radiation. It is easy to see from this expression that when the interference pattern is analyzed accurate to one-tenth of a fringe and at a path length in the plasma ~ 1 mm the distortion of the interference pattern due to refraction lies within the measurement accuracy if the image-producing optical system gathers only rays with deflection angles smaller than 2°. The geometry of our experiments satisfied this condition.

Diaphragms placed in the foci of the image-producing objectives, besides cutting off the plasma radiation, helped satisfy the condition described above.

We chose for the analysis the interference patterns with the best cylindrical symmetry.

If it is assumed that geometrical optics is valid and that the refraction of the beam is negligibly small, then the shift of the interference fringe for axially symmetrical inhomogeneities is determined by the Abel integral [48]. Several methods for numerically solving these equations are described in detail in [49]. All these methods employ for the sought function $\nu\,(r) = n(r) - n(0)$ an approximation linear in r in the annular zones into which the cross section of the inhomogeneity breaks up: $0 < r_1 < r_2 < \dots < r_N = R$ (Fig. 3b). The gist of the method proposed by us [50] consists in approximating the sought function $\nu\,(r) = n(r) - n(0)$ in each band $(r_i,\ r_{i+2})$ by a Lagrange parabolic interpolation polynomial.

ALGOL programs for the reduction of the interference patterns by the known Van Voorhis method and by the method proposed by us are given in [36, 51].

Inasmuch as in the solution of the Abel equation the integration is replaced by a finite sum, it is necessary to consider the resultant approximation errors. This was done in [36]. The error for the exponential density profile $\nu\,(r) = \nu\,(0)\exp(-r/r_0)$ turned out to be

$$\frac{\Delta\nu\,(r_i)}{\nu\,(r_i)} \leqslant 0.25\sqrt{\frac{r_N}{r_i + 0.25w}}\left[1 - \exp\left(-\frac{r_N - r_i}{r_0}\right)\right]\left(\frac{r_N}{r_0}\right)^2\frac{1}{N^{5/2}},\qquad(13)$$

where $w = r_{i+1} - r_i$ is the width of the subdivision band and N is the number of the subdivision bands. It is seen that with increasing N the approximation error decreases in proportion to $N^{-5/2}$. With the increasing number of bands, however, the error is increased because of the inaccurate determination of the band shift (measurement error). Consequently, there exists an optimal number of bands at which the total error is minimal. Equating the expressions for the measurement errors and for the approximation errors, we obtain the optimal band number N_{opt} for the parabola method in the case of an exponential density profile [36]

$$N_{opt} = 0.63\,(r_N/r_0)/\sqrt[3]{\Delta\delta/\delta}.\qquad(14)$$

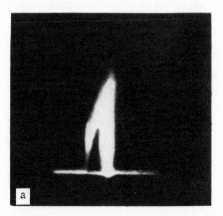 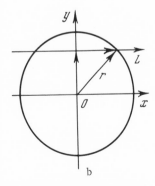

Fig. 3. Pinpoint photographs of the plasma focus and of the soft x-ray region of the spectrum (a) and ray paths through the investigated inhomogeneity (b). a) The straight long column corresponds to the first contraction of the plasma focus, the shorter formation to the left corresponds to the second "contraction"; b) O, center through which the symmetry axis z of the object, perpendicular to the plane of the figure, passes; L, probing ray.

Thus, for example, at $r_N/r_0 = 5$ and $\Delta\delta/\delta = 1/100$ the optimal number of bands is 15. The total error for ν_0 is then 7%.

The optimal number of bands for the trapezoid method is equal to $N_{opt.} = (0.25 r_N/r_0)(\Delta\delta/\delta)$. If $r_N/r_0 = 5$ and $\Delta\delta/\delta = 1/100$, then the optimal number of the bands is 125, in which case the error in the zero band is 50%.

The main difficulty lies in the reduction of the interference pattern of the discontinuities of the electron density, of the type of shock waves and of the boundary between the current sheath and the vacuum, when the change of the electron density occurs over lengths of the order of the particle mean free path in the plasma. So far, this question has not been sufficiently well dealt with. The methods proposed in the literature (see, e.g., [49]) are as a rule not suitable for plasma objects with large absolute values of N_e. Fortunately, in our case, owing to a certain penetration of the magnetic field into the current sheath [15] and owing to the radiant thermal conductivity in the shock wave front [52], the gradient on the corresponding density discontinuities are not too large, so that they can be reduced in the usual manner. The values of the gradients can be checked (as was indeed done by us) against the diffraction pattern near the discontinuity by using shadow photographs of the object.

CHAPTER II

INVESTIGATION OF CUMULATIVE STAGE OF PLASMA FOCUS

1. Parameters of the "Plasma Focus" Installation

The plasma-focus experiments were performed with an installation having the following parameters: diameter of internal electrode d = 400 mm, diameter of external electrode D = 700 mm, height of internal electrode l = 120 mm, rating of capacitor bank C = 180 μF, anode working voltage V = 24 kV, external inductance (specially chosen to fit the discharge parameters) L = 30 cm, initial deuterium pressure ρ_0 = 1 Torr, and energy stored in capacitors W = 52 kJ.

It is known [22] that the discharge parameters are greatly improved by addition of special heavy gases (~1-2%). In our case the experiments were performed with addition to the chamber of nitrogen or xenon amounting to ~1-2% of the total deuterium mass. The neutron yield of the installation range from 10^9 to $5 \cdot 10^9$. Simultaneously with registration of the neutrons, the hard x rays, current and voltage oscillograms, and photographs of the pinch in the laser beam, we photographed also the discharge glow with the aid of a pinpoint camera in the soft x-ray (SX) region (Fig. 3a).

2. Results of Reduction of the Interference Patterns

of the First Contraction of the Plasma Focus

Figure 4 shows typical interference patterns of two stages of cylindrical contraction of the current sheath to the discharge axis, obtained by photographing the discharge in the central region. The time t = 0 was assumed to be the start of the hard x-ray pulse. The following conclusions can be drawn from the interference patterns and from the results of their reduction.

After the current sheath with strong skin concentration has been produced, in accordance with the theoretical "snow plow" model [15], the sweeping of the gas towards the axis of the focus begins, and a shock wave is produced in front of the funnel-shaped current sheath. The convergence of the shock wave towards the axis can be easily determined from several interference patterns: $v \sim 3 \cdot 10^7$ cm/sec.

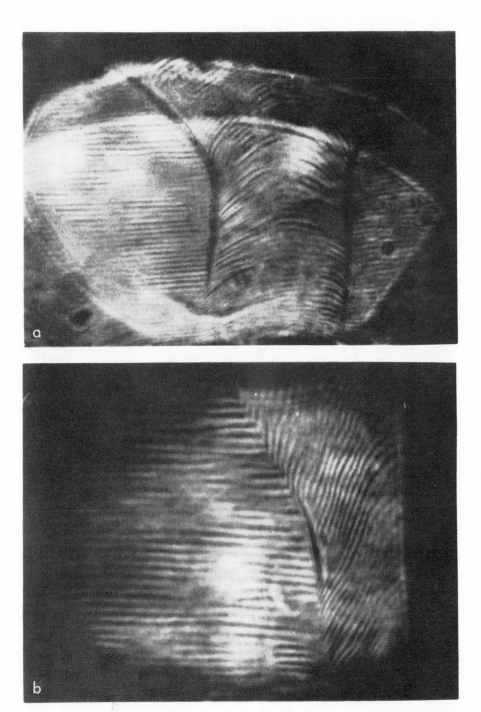

Fig. 4. Interference patterns of plasma focus prior to the first contraction with appreciable scattering in the skin layer (a) and at the instant directly preceding the first contraction (b). There is no scattering in the skin layer ($\lambda^7 < \lambda_0/2$).

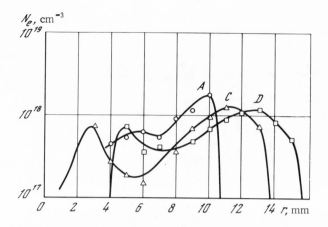

Fig. 5. Reduction of interference pattern of convergence of a shock wave in a plasma focus in three sections z. Section A corresponds to the reflected shock wave and sections D and C correspond to successive stages of cumulation of the shock wave on the z axis.

Figure 5 shows a reduction of the interference pattern obtained at the instant $\tau = -110$ nsec in three sections: below (curve A), where the collapse of the shock wave has already occurred, and above (curves C and D), prior to the collapse. It is seen that the compression in the shock wave (curve C) is > 4.

The lack of exact data on the axis at the location where the shock waves converge is due to the fact that the velocity of the convergence point of the shock waves directed upwards along the z axis is so large that it is impossible to resolve the change of the electron density. Estimates of this velocity from the size of the region with the smeared bands (2-4 mm) yield values

$$v_z \simeq 1\text{-}2 \cdot 10^8 \text{ cm/sec.}$$

The cross sections for which the interference pattern was processed were chosen such that two of them (D and C) constitute a converging shock wave, while one (A) is reflected. It is seen that the reflected shock wave is weak. The compression in it is very small.

From the reduction of the successive stages of the convergence of the current sheath itself towards the z axis it is seen that as the convergence progresses the density at the center of the pinch increases and reaches a value ~10^{20} cm^{-3} by the instant of time t = −40 nsec (Fig. 6a), after which the pinch expands slightly and the density again decreases. The convergence

Fig. 6. Reduction of interference pattern of plasma focus during the first compression stage (a), and also in the "neck" of the constriction, indicating diffusion of the heavy impurity to the z axis (b).

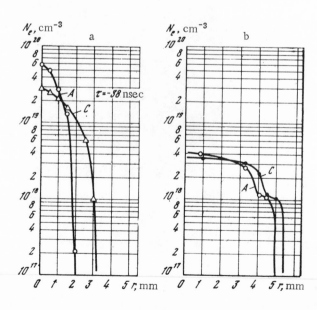

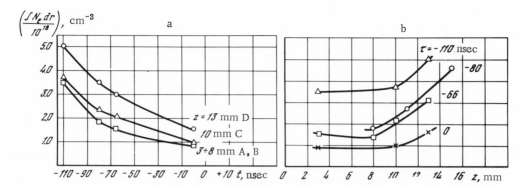

Fig. 7. Plots of the running mass in the plasma focus against the time in three sections z (a) and as a function of the distance to the anode at different instants of time (b).

rate of the current sheath is approximately equal to the shock wave velocity, that is, ~3 · 10⁷ cm/sec. From the obtained diagrams we can plot the time evolution of the amount of plasma in the different cross sections, as well as the variations of the amount of plasma along the z axis. From these curves (Fig. 7) it is possible to estimate what part of the total gas mass plowed by the magnetic field remains in the focal region. This quantity is of the order of 10^{18} cm^{-1}, that is, only ~1/100 of the initial mass of matter remains in the region of the focus as a result of the strong outflow of the plasma along the z axis.

As will be shown in the second half of this paper, in which the second "compression" is described, the current sheath begins to bend strongly after the instant of time t = −40 nsec, and exhibits discontinuities, so that during that period the plasma focus contains no plasma formations of high density and of sufficient length. Thus, it is precisely at the instant t = −40 nsec that the pinpoint camera (Fig. 3) shows a picture of the pinch in soft x rays in the form of a straight relatively long column.

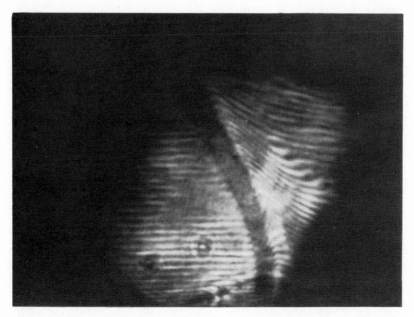

Fig. 8. Complete scattering of probing radiation in the skin layer of a plasma focus (xenon additive).

We note that in the plasma focus one frequently observes the following interesting phenomenon. Following the instant of the first contraction, and also particularly when the amount of the additive increases to 2-3%, the distribution of the electron density over the radius of the pinch took the form shown in Fig. 6b. It is seen that the distribution has a "step" and by comparing this form of the distribution with the usual form it can be stated that for nitrogen the height of this step is approximately double the average ordinary electron density.

In a certain (rather large) part of the experiments, we observed that during the time preceding the first compression (Fig. 4a) the interference fringes become indistinguishable near the boundary of the current sheath. This phenomenon cannot be attributed either to smearing of the fringes because of the rapid motion of the sheath (it is practically immobile during that time), or to absorption of the sounding radiation (the density is not high enough and the temperature is high), or else to classical Thompson scattering of light by the electrons, or to refraction of the light by the density discontinuity (this scattering is not initiated at all at the plasma boundary, where the refraction is maximal, but at a certain distance from it, that is, where the path of the sounding radiation through the skin layer is the longest). Consequently, anomalous scattering of the probing laser radiation takes place in that location in the pinch.

A discharge in a plasma-focus installation filled with deuterium with xenon additive (~1-3%) has certain important features. First, as seen from Fig. 8, the quasi-cylindrical shock wave traveling through the unperturbed gas in front of the current sheath remains practically undetached from the latter, and is frequently difficult to identify. The maximum detachment of the wave from the sheath, which was observed during the course of the experiments, was 2.5 mm, whereas in a discharge with nitrogen additive, as seen from Fig. 5, a typical value of the corresponding separation is 8-10 mm. In addition, it is seen that the scattering of the probing radiation in the current sheath, referred to in the preceding section, is observed also in this case, and is furthermore much more strongly pronounced.

By using several interference patterns it is easy to plot the dependence of the radius of the current sheath on the time, and to determine from this plot also the dependence of the rate of convergence of the sheath to the axis. The corresponding curves for discharges with nitrogen additive are shown in Fig. 9.

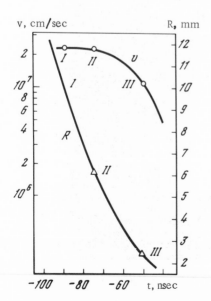

Fig. 9. Plots of the radius and velocity of the plasma-focus sheath against time.

3. Intermediate Phase

We proceed now to consider processes that occur after the instant of the maximum compression, prior to the main neutron pulse. As seen from the interference patterns of Fig. 10, obtained at the approximate instant $t \lessgtr 0$, the lateral surface of the plasma focus begins to be perturbed after the instant of the maximum compression, and these perturbations grow quite rapidly with time. The characteristic time of development of this instability turned out to be of the order of 30 nsec, and the typical wavelength is of the order of 1 cm. As seen from the interference patterns, the pinch distortions usually developed in the form of constrictions with m = 0 [18] (Fig. 10a) and inflections (Fig. 10b), m = 1, but constrictions were most frequently encountered, and their growth rates exceeded the corresponding values for the inflections.

During the development of this instability, the current sheath becomes discontinuous. Examples of such discontinuities are shown in Fig. 11. Most frequently the discontinuity is produced in the manner shown in Fig. 11a. It is interesting that the region of the discontinuity does not coincide with the narrowest point of the pinch, as was assumed in [31]. To the con-

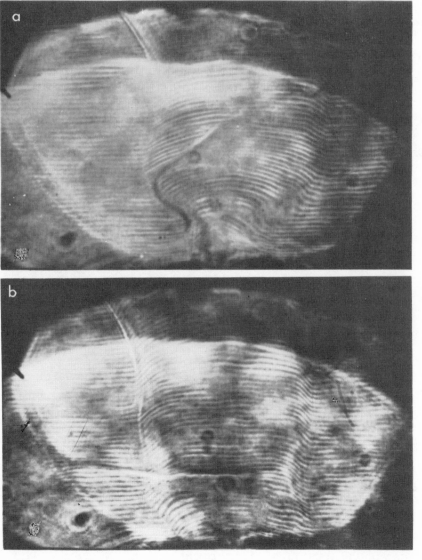

Fig. 10. Rayleigh—Taylor instability in plasma focus. a) Mode m = 0;
b) mode m = 1.

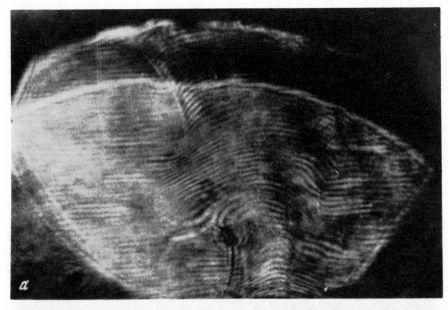

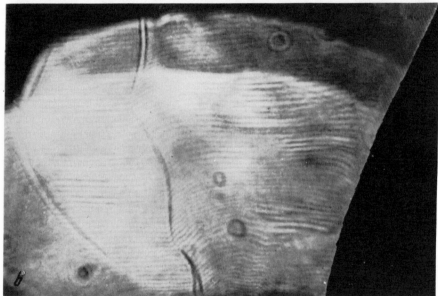

Fig. 11. Discontinuities of plasma-focus sheath after the instant $\tau = 0$ in the case of a strongly (a) and weakly (b) developed Rayleigh—Taylor instability.

trary, the pinch is relatively stable in this place, and the plasma streams emerge from the sheath in regions directly adjacent to the constriction or even in a place where the pinch is thickened.

The disintegration of the current sheath takes place either simultaneously over the radius, or else at a single point, but then it propagates over the entire periphery of the pinch very rapidly, within a time of the order of 1 nsec.

As revealed by the synchronization oscillograms, the instant of the discontinuity $t = 0$ usually coincides with the start of the hard x-ray pulse (the front duration of which is quite short in this case and is determined by the resolution of the apparatus).

4. Second "Contraction" of Plasma Focus

Traditionally, the process occurring at the instant of time corresponding to the most powerful neutron pulse is called the second "contraction." This terminology was adopted in the plasma-focus literature in papers dealing with cylindrical z pinches, where several successive contractions of the plasma were in fact observed. For convenience in the exposition, we shall continue to use this universal term for the time being.

Thus, after the break of the sheath (t = 0), the plasma begins to flow out of the region of the pinch. This process continues for approximately 70 nsec. As shown by the interference patterns (Fig. 12), the neck of the constriction continues to exist during this entire time. By the 70th nsec, the distribution of the density in the neck takes the form as shown in Fig. 12b, which was obtained after the reduction of the interference pattern. Usually this tubular formation manages to be displaced within this time from its initial position by 2-3 mm, and undergoes inclinations relative to the initial z axis.

Next, after about 100-120 nsec, this neck also disintegrates, and the field of the interference pattern represents, during 120-140 nsec, random inflections of the fringes each about one fringe in depth, together with breaks and smearing of the fringes. An estimate of the electron density yields a value $\sim 10^{16}$ cm^{-3} for the greater part of the plasma situated in this region. However, in addition to this irregular cloud, there exists also relatively regular formations in the form of filaments, with diameter of the order of 1 mm, having a tubular structure and directed toward the anode, but not reaching the latter (Figs. 13a and 13b). It is seen that these filaments are bent and at the lower ends there is a region in which the interference fringes can be traced. Such filaments were observed in some of the discharges also in a region adjacent to the pinch (Fig. 13c).

Very interesting distinguishing features mark the phase of the second "contraction" in the case of a discharge in deuterium with xenon additive. The interference patterns show that in this case filaments 1-2 mm thick develop first along the pinch axis, and then practically over the entire cross section of the pinch. No interference fringes are traced in the filaments, and the radiation of the probing beam does not reach the photographic film at all. In some discharges (Fig. 14) several filaments exist simultaneously, they are strongly bent, and ter-

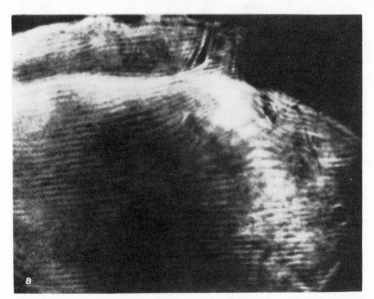
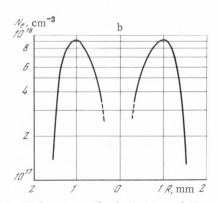

Fig. 12. Interference pattern of "frozen" neck of constriction ("plasma cathode") (a) and its reduction (b). τ = +120 nsec.

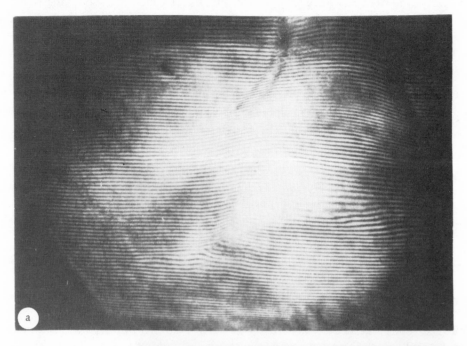

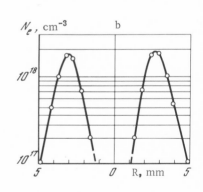

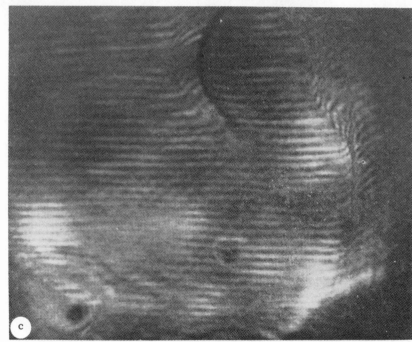

Fig. 13. Typical interference pattern of the magnetic channel of the electron beam in a plasma focus with one focusing point (the additive to the deuterium is nitrogen)(a), and its reduction (b), as well as a typical beam produced alongside the pinch (the additive is xenon) (c).

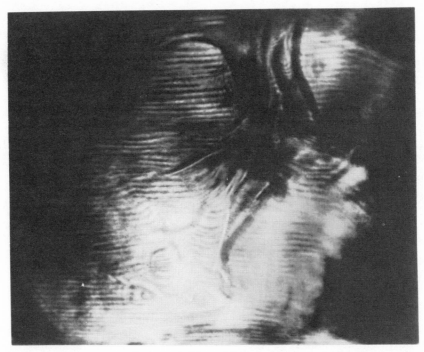

Fig. 14. Hose instability of beam in plasma focus.

minate before they reach the anode. In others (Fig. 15), however, there is a sufficiently thick filament that bears against the anode.

5. Concluding Stage

By the instant of time t ~ 200 nsec, the field of the interference pattern constitutes rather even fringes without distortions, and in the lower part, near the anode, a broad cloud appears, which is absolutely opaque to the laser radiation. From the different interference patterns, it is easy to estimate the speed of the boundary of this cloud, which obviously is made up of anode-material vapor. This speed turns out to be of the order of 10^5 cm/sec.

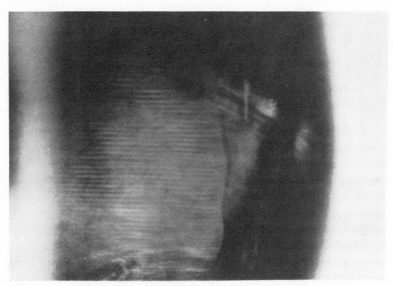

Fig. 15. Good "transportation" of beam in plasma focus.

CHAPTER III

DISCUSSION OF RESULTS OF EXPERIMENTS WITH PLASMA FOCUS

1. First "Contraction"

The authors of all the plasma-focus numerical calculations performed to date start from the magnetohydrodynamic (MHD) model of the object. It should be noted that this model, as already noted by many investigators (see e.g., [15]), has a limited region of applicability. First, during the final stages of the cumulation the average particle mean free path in the plasma turns out to be of the order of the characteristic dimension of the object — the radius of the pinch. Second, under certain conditions the time of the ion—ion collisions in the plasma focus becomes comparable with the ion cyclotron time, and at small pinch radii the viscous-stress tensor used in these calculations turns out to be indeterminate. In addition, the satisfaction of the well-known Dreicer condition for "runaway" electrons in the plasma focus [18] by the instant of time t = 0 ($N_e = 10^{18}$ cm^{-3}, $T_0 = 10^{7}$°K),

$$E_k = 10^{-8} \frac{N_e Z^2}{T_e^0} \simeq 10^3 \text{ [V/cm]} \lesssim \frac{U}{l} \sim 2.4 \cdot 10^3 \text{ [V/cm]}, \qquad (15)$$

where $l \sim 10$ cm is the distance from the anode to the liner and U is a voltage on the order of that used in the plasma focus, together with other conditions, leads to the conclusion that the law governing the effective connection of the energy and the pressure in the pinch deviates from adiabatic for a fraction of the particles. However, as we shall see, the magnetohydro-dynamic model describes quite well a number of processes up to the first contraction.

The presently most complete formulation of the magnetohydrodynamic problem for a non-cylindrical z-pinch is given in [15, 16, 53]. It should be noted, however, that all these calculations pertain to a discharge in pure deuterium.

Before we proceed to compare the experiment with the theory, let us dwell on several special problems. We consider first a question connected with the azimuthal Rayleigh—Taylor instability, which is caused by the acceleration of the separation boundary during the time of the sheath convergence. A comparison of the experimental data on the acceleration with the theoretical calculation of the critical acceleration [22] shows that the stability condition may not be satisfied, although only with a slight margin. An experiment, however, reveals no azimuthal modes (Chapter II). The step that appears on the electron-density distribution in certain discharges during the later stages of the cumulation (Fig. 6b) cannot serve as evidence of such perturbations, since its width is exceedingly narrow (this would indicate small wave-lengths of the perturbations, that is, large m) and no subsequent "steps" have been observed on the plot of the density distribution.

It is shown in [15] that during the concluding stage of the convergence of the sheath of the plasma focus towards the axis, the current velocity in it exceeds the ion-acoustic velocity by more than one order of magnitude, and consequently turns out be of the order of the electron thermal velocity. In addition, for a converging sheath of ($N_e \sim 10^{17}$-10^{18} cm^{-3}, $T_e \sim 100$-1000 eV) the Dreicer criterion (15) is satisfied, and a small fraction of the electrons is dragged through. This means that ion-acoustic and Langmuir oscillations should build up in the current layer. It is known [54] that when a weak electromagnetic wave passes through a plasma in which there exists sufficiently intense pulsations, such a wave can undergo scattering, with a cross section greatly exceeding the Thomson value. Such a scattering is possible if the wave-length of the electromagnetic radiation is smaller than double the wavelength of the turbulent pulsations. The characteristic wavelengths of the two-stream instability and of the instability of the Buneman type [55] are

$$\lambda = 2\pi/k = 2\pi V/\omega_{pl} \geqslant 5 \cdot 10^{-5} \text{cm} > \lambda_0/2 = 3.5 \cdot 10^{-5} \text{cm}, \tag{16}$$

where V is the current velocity of the electrons or the velocity of the beam electrons.

A wavelength of similar magnitude is obtained also for high-frequency ion-sound oscillations when $v_{T_i} \ll V < v_{T_e}$ [55]:

$$\lambda = 2\pi v_{T_e}/\omega_{pl} \simeq 5 \cdot 10^{-5} \text{ cm}. \tag{17}$$

For small scattering angles we can write

$$< \theta^2 > = \sigma^* L/2c, \tag{18}$$

where θ is the scattering angle, L is the length of the path of the beam in the plasma, and $\sigma = \sigma^*/cN_e$ (σ^* is a quantity playing the role of the cross section for scattering by the pulsations). In our case, the scattered ray will not strike the film at angles of the order of several degrees. On the other hand, as seen from Figs. 4a and 8, the vanishing of the interference fringes and the slight illumination of the film are observed only in the region near the edge of the pinch, when the path ray in the plasma is of the order of 1 cm, and are not seen at the center, where this ray crosses the sheath in a transverse direction (L $\simeq$ 0.1 cm). Substitution of the indicated quantities in (18) yields a cross section $\sigma^* = 6 \cdot 10^8 \text{ sec}^{-1}$. On the other hand, the cross section for the scattering by Langmuir pulsations [54] is proportional to $(M_i/m_e)^{1/2}$, whereas the cross section for the scattering by ion-acoustic oscillations is proportional to $(m_e/M_i)^{1/2}$. From a comparison of Figs. 4a and 8 it is seen that in the case when xenon is added the scattering cross section is strongly increased, that is, the scattering is by the Langmuir oscillations.

From the equations of [54] we can attempt to estimate roughly the ratio W/nT, where W is the density of the Langmuir oscillations. It turns out to be of the order of 10^{-1}. Such values are perfectly realistic. For example, measurements of this ratio, carried out with an installation in which turbulent heating is used [56] and the heating is produced by methods that are essentially analogous to the mechanism existing in the plasma focus, yielded a similar result.

At the instant of the maximum contraction, the wavelength of the Langmuir oscillations, calculated from Eq. (16), turns out to be of the order of 10^{-5} cm < $\lambda_0/2$, making it impossible to observe this instability. This is confirmed by experiment (see Fig. 4b). However, experiments [57] offer evidence that the noise power at the plasma frequency increases by six orders of magnitude at that time.

From work on pinches [58] it is known that a high-power Langmuir noise is always accompanied by ion-acoustic noise. Thus, the arguments presented above can apparently explain why there are no azimuthal perturbations of the pinch when the current sheath of the plasma focus converges to the axis, since the perturbations are dissipated by the anomalously high resistance of the skin-layer plasma.

We proceed now to compare experiment with theory.

A. Shock wave. The difference between the density ratios in the front and behind the front in the sections C and D, which was mentioned in Chapter II, is in good qualitative agreement with [53]. This phenomenon is explained by the fact that the sections C and D actually constitute so to speak successive stages of the convergence of the shock wave to the axis, and the cumulation of a cylindrical shock wave on the axis already begins "at a later instant of time" (section C). It should be noted that the observed value of the "contraction" by a factor of more than 4 could be explained in this case as being due to an increased content of highly ionized impurities.

The "disengagement" of the shock-wave front from the current sheath amounts (in dimensionless units x/R_0, where R_0 is the pinch radius and x is the front coordinate reckoned from the axis) to approximately 0.2. This is double the value obtained in [53] for pure deuterium. As noted in the preceding section, in the case of a discharge in deuterium with xenon, the shock wave is hardly detached at all from the sheath, and the corresponding quantity is 0.8-1.0. This phenomenon can apparently be explained in the following manner.

The resultant force acting on the gas compressed between the shock wave and the current sheath is equal to the difference between the pressures exerted by the "piston" and by the unperturbed medium. The pressure of the "piston" is the resultant of the magnetic-field force and the inertial forces. The only difference between the theoretical calculations and experiment is that no account is taken of the heavy additives to the deuterium. But the amount of the additive is very small. Consequently, it can be assumed that when the sheath converges this additive accumulates in some way in the region behind the skin layer. Then, owing to the increase of the sheath mass, the inertia increases and the pressure exerted by the "piston" increases, the plasma behind the shock wave becomes more strongly compressed, and the ratio x/R_0 increases. At the same time, the impurity increases the radiation from the plasma that is compressed between the shock wave and the sheath, leading to a cooling of the plasma and again to a decrease of the detachment of the shock wave from the sheath. What kind of mechanism, obviously connected with the diffusion processes, can cause the accumulation of the impurity in the sheath?

It must first be stated that this accumulation must not be too large. Indeed, since the ratio of the nitrogen molecule mass to the deuterium molecule mass is 7 (and reaches even 66 for xenon), an increase of its concentration to 15% doubles the inertial force (for xenon, 2% concentration suffices). Further, we have two zones in which the most intense separation of the components by diffusion is possible. These are the skin layer and the shock wave. The expressions for the thermal diffusion and barodiffusion coefficients [52] are too cumbersome to present here, but it is important that from the form of these coefficients we can draw the following conclusions: First, when the condition $\nabla P/P \sim \nabla T/T$ is satisfied (this takes place in our strong shock wave which propagates in front of the sheath) the role of thermal diffusion is small and can be neglected. Second, barodiffusion leads to a concentration of the lighter component in the shock-wave front. This is clear also from simple physical considerations. Indeed, the light-gas molecules in the region heated behind the shock wave front have a larger thermal velocity than the molecules of the heavier gas, by a factor $(m_2/m_1)^{1/2}$, and, so to speak, "break away" forward. It is shown in the book [52] that in this case the width of the region in which the lighter component is concentrated is larger by $(m_2/m_1)^{1/2}$ times than the viscous shock discontinuity, and its excess in this zone, in the case of a strong shock wave and a large difference between the component molecule masses, turns out to be of the order of the concentration behind the front itself.

The situation is more complicated in the skin-layer region. Here, on the one hand, the rate of penetration of the plasma into the magnetic field should be slowed down by a factor $(\nu_{ei}/\omega_{H_e})^2$, owing to allowance for the external force of the magnetic field for the case of classical plasma conductivity. On the other hand, the analysis presented above for the skin-layer region demonstrates the important role of the turbulent pulsations in the layer. The results of [59] for a two-component fully ionized plasma cannot be used here directly, since no account is taken in [59] of the inertial forces, which in our case are large. It is qualitatively clear, however, that since the component ions with $m_1/Z_1 \sim m_2/Z_2$ can penetrate into the magnetic field approximately equally, and the inertial forces of the heavier component are larger than of the lighter one, it follows that the edge of the skin layer should also be oversaturated with the lighter component.

It is shown in [16] that when the sheath converges to the axis, the outflow of the plasma from the cumulation region takes place on the shock-wave front, while the authors of [53] note that a numerical calculation has shown a predominant outflow of the plasma from the rarefied region adjacent to the outer boundary of the current sheath. These phenomena were not discussed in the cited papers, but it is clear that this outflow should be most effective from both points where the external forces are applied (the magnetic field from the outside and the unperturbed gas inside) towards the plasma that is compressed between the shock wave and the magnetic field. This is confirmed by the increased electron density in the front of the shock wave and in the maximum of the skin layer, in comparison with the numerical calculation, and also by the sag of the distribution of the electron density behind the shock wave front. These phenomena point to an abrupt increase in the shell of the concentration of the ions with the larger charge.

This explains also the results connected with the degree of "disengagement" of the shock wave from the sheaths.

B. The skin layer. This has been discussed above. We can add that during the earlier stages its value amounts to $\sim 5\%$ of the radius, in agreement with the numerical calculation. During the later stages, however, when the sheath is practically immobile and the temperature does not change strongly, its size should increase towards the instant $\tau = +60$ nsec in the case of classical plasma conductivity, by an amount $\delta^* \sim c\sqrt{\tau}/\sqrt{2\pi\sigma} \lesssim 100\ \mu m$. Actually, by that time the size of the skin layer, at a plasma radius 5 mm, turns out to be of the order of 2 mm. This discrepancy can be resolved by assuming that the conductivity of the plasma is determined by turbulent processes and is smaller by two orders of magnitude than the classical value. This result is an independent confirmation of the foregoing analysis.

C. The first contraction. This sets in at the instant of time $\tau \simeq -40$ nsec. The principal disparity between the numerical calculation and the experiment lies in the value of the mass remaining in the region of the focus, which turns out in the theory to be larger by one order of magnitude. This difference can be attributed to the fact that, as seen from the preceding arguments, in the final stage of cumulation we are actually dealing with the compression of the deuterium by a heavy sheath. In this case, as is well known [7, 8], the cumulation is much more effective, and this explains the described effect, as well as the increase of the neutron yield in the discharge with impurity as compared with cumulation in pure deuterium. The same considerations explain also the estimate in Chapter II of the velocity of the cumulative jet, produced at the instant of the first contraction, which turns out to be almost twice as large as that obtained in a numerical calculation [53].

An interesting question is that of the subsequent behavior of the impurities in the pinch at the instant of the maximum compression and for some time later. The Enskog−Chapman approximation for the calculation of the diffusion is no longer applicable in this case, since the role of the viscosity increases very strongly at that instant of time and it is necessary to take into account the viscous momentum transfer. In addition, by virtue of the temperature rise, the collisions become less frequent, and caution must be exercised when solving the kinetic equation. Grad's method is convenient in this case. Calculations of the diffusion in a shock wave by this method, and simply in the presence of gradients of macroscopic quantities, were carried out in [60] on the basis of the 13-moment approximation. The results coincide in the main with the preceding analysis in the Enskog−Chapman approximation for our case $m_1/m_2 \ll 1$ and $n_1/n_2 > 1$. This means that the heavy component should go off in the direction of the increasing pressure, that is, towards the center of the pinch. This agrees with arguments connected with inertia, and also with the theory of [59], which becomes valid after the collapse "stops." The foregoing arguments explain the difference between the pinch diameter as determined from interferometry and from pinpoint cameras obtained in the soft x-ray region of the spectrum. This difference, which reaches values of the order of 3-5, can now be understood from the fact that the integrated intensity of the bremsstrahlung is proportional to Z^2 [52]. In addition, this phenomenon explains also the "step" in the distribution of the electron density (Fig. 6b).

We proceed now to consider the intermediate phase, when, as shown in Chapter II, the decisive process is the development of a Rayleigh−Taylor instability on the surface of the pinch.

2. Intermediate Phase

The theory of the Rayleigh−Taylor instability is developed in the book [55]. The mechanism of its appearance in our case is connected with a difference between the drift direction of the electrons and ions across the force lines of the magnetic field. A drift of this type can be due to two causes: the acceleration transverse to the field, and curvature of the force lines of the field. Both can take place in the final stage of the cumulation, whereas the instability far from the axis can be caused only by acceleration.

We discuss our results within the framework of the linear theory. Up to the instant of time I (see Fig. 9) the acceleration of the sheath is directed into the plasma and should coincide in order of magnitude with the value measured at the point III (−50 nsec): ~$1.5 \cdot 10^{14}$ cm/sec^2. An acceleration of the same order will take place also in the intermediate phase, after the pinch has "diverged" after the first compression and started to return to the axis. On the other hand, in a magnetic field having cylindrical symmetry and having only an azimuthal component, the instability of the constriction type develops also in the absence of acceleration of the separation boundary [55]. The magnetic drift of the plasma particles can then be simulated by a gravitational field with an "acceleration" $g \simeq v_{T_i}^2/R$, where R is the curvature radius of the magnetic field. In our case, in the final cumulation phase, the value of this effective acceleration reaches an order of 10^{15} cm/sec^2. The growth time of the instability has turned out to be of the order of 15 nsec, in agreement with the experiment.

From the formula for the increment of the flute instability caused by the curvature of the force lines, $\gamma^{**} \simeq [(v_{T_i}^2/R)k]^{1/2}$, it is seen that this mechanism should be quite effective at the instant of cumulation, and particularly if account is taken of the arguments presented above concerning the diffusion of the heavy component. Indeed, after the sheath converges to the axis, the heavy impurity, which has "stabilized" the sheath as a result of the low ionic thermal velocity, goes off to the center, and what appears on the surface of the pinch is in the main deuterium with a large value of v_{T_i}, after which the instability develops quite rapidly. Recognizing that the pinch surface is unperturbed prior to the first "contraction" and taking into account the large value of the "acceleration," it must be admitted that the most likely mechanism responsible for the development of the instability in the intermediate phase is the effect of the curvature of the force lines.

Let us present some estimates. It is known that the rate of growth of the fundamental mode of the perturbations in the plasma focus should be of the order of magnitude of the speed of sound in the plasma. On the other hand, the plasma temperature after the instant of maximum compression, as already mentioned, decreases, as indicated by the "pause" in the soft x rays from the plasma focus. From a numerical calculation it follows that its values can decrease by approximately one order of magnitude [53]. Then assuming the average radius of curvature of the magnetic field at this instant to be 0.5 cm (Fig. 10a), we obtain

$$\tau = R \Big/ \sqrt{\gamma \frac{p}{\rho}} = R \Big/ \sqrt{\gamma \frac{kT}{M}} \simeq 30 \text{ nsec}, \qquad (19)$$

which is also in good agreement with experiment.

From the value of the wavelength of the instability we can draw a number of important conclusions concerning the dissipative processes that occur in the plasma focus in that period, and in particular, impose a lower bound on the wavelength of the developed perturbation.

To simplify the discussion that follows, we neglect first the effects due to the finite Larmor radius. This will enable us to regard such parameters as the conductivity and viscosity of the plasma as scalars. In this case, the principal effects that influence the development of the m = 0 mode of the Rayleigh−Taylor instability will be the following:

Inertia or curvature of the magnetic force lines with growth rate $\gamma_1^* = \sqrt{|gk|}$,

Viscosity with growth rate $\qquad\qquad\qquad\qquad\qquad\qquad \gamma_2^* \simeq -\nu\lambda^2$, $\qquad$ (20)

Conductivity with growth rate $\qquad\qquad\qquad\qquad\qquad \gamma_3^* = -k^2/4\pi\sigma$;

where ν is the kinematic viscosity and σ is the conductivity. The last two growth rates characterize the diffusion times of the momentum and of the magnetic field over the spectrum, thereby in fact determining the dissipative processes.

From the indicated quantities we can make up a dimensionless parameter, namely the ratio of the hydromagnetic and hydrodynamic Reynolds numbers, which will characterize the relative role of the indicated mechanisms in the dissipation of an instability with large wave numbers:

$$\mathcal{N} = \frac{R_{hm}}{R_{hd}} = 4\pi\sigma\nu = \frac{2\cdot10^{-3}\gamma_E(Z)\,T_e^{3/2}T_i^{5/2}}{A^{1/2}N_iZ^3\,(\ln\Lambda)_{T_e}\,(\ln\Lambda)_{T_i}}\,. \qquad (21)$$

Here σ is the conductivity in cgs emu units; $\ln\Lambda$ is the Coulomb logarithm. The dependence of $\mathcal{N}$ on the charge Z is in our case cubic, inasmuch as in an operating regime with small impurity content it is necessary to take into account only the collisions between the deuterons and the impurity nuclei, and the collisions between the impurity nuclei themselves can be neglected. Expressions for the viscosity and the conductivity were taken respectively from [53]. The temperatures T_e and T_i are in degrees Kelvin, and $\gamma_E(Z)$ is the Spitzer correction for the electron−electron collisions [53]. The meanings of the remaining parameters are the usual ones.

Substitution of the values for D−D collisions into this formula yields the value $\mathcal{N} = 15$. Thus, in this case the principal mechanism of the attenuation of the short-wave modes is viscosity. This conclusion must be verified by determining the time of the ion−ion collisons. It turns out in this case to be

$$\tau_{ii} = T_i^{3/2}A^{1/2}/0.6N_i \simeq 10^{-9}\,\sec \ll 1/\gamma_1^* = 3\cdot10^{-8}\,\sec, \qquad (22)$$

i.e., it is equal to the instability development time, thus confirming the result.

Equating the inertial growth rate to the viscosity growth rate, we obtain the minimum instability wavelength that can still develop in the given case:

$$\lambda_{min} = 2\pi g^{-1/3}\nu^{2/3} = 1.18\,\text{cm}, \qquad (23)$$

which agrees well with the experimental data.

We consider now the influence of the impurities and the instability. If it is assumed that the heavy ions do not diffuse towards the center from the skin layer, and take into account in (21) only the collisions of deuterons with nuclei having $Z \sim 10$, then the quantity $\mathcal{N}$ turns out to be of the order of $2 \cdot 10^2$, that is, in this case the dominant dissipation mechanism should become resistive. In this case the conductivity is equal to

$$\sigma = 2.6\cdot10^{-13}\frac{T_e^{0\,1/2}\gamma_E(Z)}{Z\ln\Lambda} = 2.5\cdot10^{-6}, \qquad (24)$$

that is, it is smaller by one order of magnitude than for the case of deuteron collisions.

The minimal wavelength determined by equating the inertial and resistive growth rates, however, turns out to be of the order of

$$\lambda_{min} = (\pi/2)^{1/3} g^{-1/3} 5^{-2/3} \simeq 0.1 \text{ cm,} \tag{25}$$

which is smaller by one order of magnitude than the experimentally observed value, that is, the presence of impurities in the skin layer cannot explain the dissipation. It is seen from (25) however, that if we assume the validity of the aforementioned reasoning concerning the anomalously low (by two orders of magnitude) plasma conductivity as a result of the turbulence in the skin layer, then the minimum wavelength will approach the experimentally obtained value. Thus, it can be stated that in the intermediate phase it appears that approximately equal roles are played by the two mechanisms of stabilization of the short-wave modes of the Rayleigh−Taylor instability, namely the viscosity mechanism and the mechanism due to the anomalous conductivity of the plasma.

We consider now the processes that occur at the instant of time $\tau = 0$, that is, at the instant when the hard x radiation pulse begins. In practice, the instant $\tau = 0$ corresponded always to interference patterns with discontinuities in the current sheath. At the present time it is impossible to indicate exactly the reason for the interruption of the current. However, as noted above, the decrease of the anomalous scattering of the laser radiation in the skin layer at that time cannot be connected with the decrease in the turbulence level in this layer. To the contrary, it seems to increase to values $(E_e^2/8\pi)/(N_e T_e) \sim 1$ ("strong turbulence"). Indeed, after the development of the Rayleigh−Taylor instability, as indicated, the plasma temperature at the place where the pinch becomes thicker decreases abruptly, owing to the pinch expansion and the "burning out" of the impurities that have gone off to the center. Consequently the electron–oscillation velocity amplitude in the longitudinal waves becomes much larger than v_{T_e} and can exceed the current velocity, leading to an interruption of the current in the skin layer. On the other hand, the current interruption can be brought about by a mechanism of the type indicated in [61], or else a parallel increase in the power of the ion sound to very large values. Nor can hydrodynamic instabilities be excluded from consideration.

Once the current layer is completely broken, the conduction current stops. The reason why the current does not drop to zero on the oscillograms at that instant is that usually the current is determined by measuring the magnetic field it produces (Rogowski loop) and the magnetic field is the same for the total current, consisting of the conduction current and the displacement current. Thus, during the course of a certain time the displacement current replaces completely the conduction current. A description of this process is contained in [31], and we can attempt to apply the results there to our case. Assume that at the instant of termination the current is $\mathcal{I}_0 = 10^6$ A, the skin layer diameter is 1 cm, and the time in which the constriction is broken is $\tau_0 = 10^{-9}$ sec. The electric field is then of the order of $3 \cdot 10^7$ V/cm. In this tremendous electric field, the energy of the electrons and of the ions reaches values of the order of 300 keV within a time of, say, 2 nsec. This rough estimate shows that this effect can explain fully the formation of the hard x rays, which will be emitted from the anode of the plasma focus when electrons are stopped in it, and also the appearance of the second neutron pulse. From the shift of the spectrum we can estimate the velocity acquired by the deuteron in this field. For the typical anisotropic part of the neutron spectrum, with a maximum near 3 MeV, the deuteron velocity turns out to be of the order of $4.5 \cdot 10^8$ cm/sec. As seen from the discussion of the first contraction, no such velocities can be reached in a cumulative jet, whereas the foregoing reasoning explains fully the presence of such deuteron velocities.

This acceleration mechanism provides also a good qualitative explanation for the parameters of the hard x rays from the plasma focus. Indeed, the power-law dependence of the energy of the accelerated electrons [27-29] reflects the existence of an accelerating mechanism of the x-ray-tube type, as is indeed the case in our situation. On the other hand, the impossibility of attributing the spectrum to one monoenergetic group of electrons is a reflection of the fact that the time of breaking of the skin layer is finite, and accordingly the field growth time is slow in comparison with (R_0/c). In addition, the characteristic energy that the electrons acquire in the same fields in which the ions have a velocity $4.5 \cdot 10^8$ cm/sec turns out to be of the order of 200 keV, a value usually obtained in experiments on hard x rays. Thus, the ion energy is always lower than the electron energy.

Consequently, in the intermediate phase, at the instant when the skin layer breaks, large electric fields appear and form near the system axis powerful deuterium electrons and ions that move in opposite directions. The electron beam, stopped in the anode, produces a flash of hard x rays, while the deuteron beam produces the second neutron pulse. It must be emphasized, however that to draw an unequivocal conclusion concerning the mechanism that generates fast particles in a plasma focus, further research is necessary, both experimental and theoretical, since attempts can be made, in general, to interpret the available experimental data in other ways.

3. Second "Contraction"

We proceed now to discuss the process of the so-called second "contraction," and examine first of all the evolution of the current sheath during that time. We have seen that after the first "contraction" the Rayleigh−Taylor instability (the mode m = 0) leads usually to the formation of two constrictions on the pinch − near the anode and above − and the sheath breaks at the location of the expansion. Approximately 100–120 nsec after the break, as noted in Chapter III, the pinch in effect disintegrates, but the upper constriction (obviously the plasma cathode) continues to exist (Fig. 12a). To discuss its subsequent behavior it is possible to resort in part to the results of a numerical calculation [17]. Analysis [62] shows that its decay can be due to flute instability (m = 1) and drift instability [55], which develop within a time of the order of 10^{-7} sec.

We consider now the behavior of the electron beam. For an electron beam of energy 200 keV, the known relativistic parameters $\beta = v_z/c$ and $\gamma = (1 - v_z^2/c^2)^{-1/2}$ are respectively 0.71 and 1.41. It is easier to determine for these parameters the Alfvén critical current, which turns out to be

$$I_A = \frac{mc^3}{e}\beta\gamma = 1.7 \cdot 10^4 \text{A} \ll 10^6 \text{ A},\tag{26}$$

the typical value of the plasma-focus current during the time of the second "contraction." This result is understandable. Indeed, in our case the beam is injected into the plasma. It is easy to determine the electron concentration in the beam at a beam diameter of the order of 1 mm:

$$N_e = j/ev_z = I/\pi r^2 e v_z \simeq 10^{16} \text{ cm}^{-3}.\tag{27}$$

In our experiments the plasma density at this point, measured from the interference patterns, is $\sim 10^{18}$ cm^{-3}. In a dense plasma, however, the restriction (26) is lifted, since current neutralization of the beam takes place in this plasma immediately. Therefore the current can exceed I_A. This current neutralization takes place over the "length" of the return current [63]:

$$l^* = v_z r^2 \sigma_{\text{cgsm}} \simeq 10^5 \text{ cm} \gg h_z,\tag{28}$$

where h_z is the distance from the plasma cathode to the plasma-focus anode. Thus, the beam should be completely cancelled over its entire path by the return current of the plasma. But at the same time it should spread out with thermal velocities owing to the neutralization of the beam's own field. Figure 13a demonstrates one more important feature: The plasma contains a configuration with tubular structure, which narrows down towards the anode. This structure arises in the case when the relativistic electron beam passing through a plasma having a finite conductivity does not lose completely its magnetic field and pushes out the return current from its channel [64]. When the beam passes through a plasma with finite conductivity, its self-field will focus the beam, and the field itself will be frozen in the surrounding plasma, which in this case is rapidly heated as a result of dissipation of the return current. Thus, a magnetic channel is produced for the beam in the plasma by the electrons that have passed earlier. It is possible to determine the focusing length [65], when the magnetic contraction force is balanced by the beam-pressure force:

$$l^{**} \simeq R \left(\sigma_0 p_0/j_0^2 t \right)^{1/5} \left(N_e m_e c^2/p_0 \right)^{1/2} \sim 3 \text{ mm}, \tag{29}$$

where σ_0 is the plasma conductivity; p_0 is the plasma pressure; N_e is the beam density; $\mathcal{I} = 10^6$ A is the beam current; $N_e = 10^{18}$ cm^{-3} is the plasma density; $R = 1.5$ mm is the initial radius of the beam (radius of the neck); $\tau \simeq 10^{-7}$ sec is the injection time. In our case (Fig. 13a) the focusing length is equal to 6 mm, in good agreement with the estimate (29).

It is seen from the figure that the beam is focused to dimensions of the order of 0.5 mm, after which there is no magnetic channel. From the dimensions of the minimal beam radius and the focusing length we can determine the average initial angular scatter of the electrons in the beam [66]:

$$\langle \theta_0 \rangle \sim r/l^{**} \sim 0.07 \text{ rad}. \tag{30}$$

It is seen from the same figure that the beam has been strongly bent and that its focal point has moved a long distance away from the axis. This effect is obviously due to the "hose" instability [66], wherein a beam moving along the curved trajectory is subjected to a centrifugal force that displaces it together with the magnetic channel and with the frozen-in plasma. The time of development of this instability is

$$\tau = \frac{1}{\gamma^+} \left(\frac{N_e' M}{N_e m} \right)^{1/2} \frac{r}{c \langle \theta \rangle} \sim 2.4 \cdot 10^{-8} \text{ sec}, \tag{31}$$

where $(N_e'/N_e) \sim 10^2$ is the ratio of the densities of the electrons in the plasma and in the beam. It is seen that with decreasing beam radius the instability develops more rapidly, so that the focused section of the beam is deflected to a larger distance. The order of magnitude of (31) agrees well with experiment.

When a beam passes through a plasma having a density larger than the beam density, the beam, in the absence of a longitudinal magnetic field, should produce turbulence in the plasma. The fastest to be excited are the longitudinal Langmuir oscillations [67]. We did not investigate in detail the beam relaxation process. It can be firmly stated, however, that the beam is completely stopped in the plasma only if the plasma manages to become heated to temperatures of the order of several kiloelectron volts [21]. During this time, the beam manages sometime to be focused in the plasma only once (Fig. 13a), and sometimes multiple ("banana") focusing is obtained. The relaxation length fluctuates in this case in the range 0.1-1 cm. It is interesting to estimate in our case the known parameter ν/γ. It turns out to be of the order of

$$\nu/\gamma \approx I/1.7 \cdot 10^4 \quad \beta\gamma = N_e r_0/\gamma \sim 60 \gg 1, \tag{32}$$

where r_0 is the classical radius of the electron, $\beta = v_b/c$, and $\gamma = (1 - \beta^2)^{-1/2}$.

Next, the beam relaxation process can proceed particularly effectively in the case when there are no mechanisms that pump out the waves from the regions of the unstable wave numbers. The most important of these mechanisms for the Langmuir oscillations is their induced scattering due to resonant interactions of the plasma ions with the beats of these oscillations, which are due to nonlinearities and have combination frequencies $(\omega_c - \omega_{c'})$ [67]. The stabilization condition can be written in the form

$$\delta \simeq 25 \frac{m_i}{m_e} \left(\frac{T_e}{T_i} + 1 \right)^2 \left(\frac{v_{T_e}}{v_z} \right)^6 \left(\frac{N_{pl}}{N_e} \right)^{1/3} \simeq 2 \cdot 10^6 \left(\frac{v_{T_e}}{v_z} \right)^6 > 1. \tag{33}$$

This condition is satisfied at a plasma temperature of the order of 1 keV. It is seen, however, that at an electron temperature 10 keV under our conditions the quasi-linear relaxation of the beam should proceed without obstacles.

Further, as already indicated, strong glow of the plasma appeared sometime at the point where the beam was focused. On the other hand, the pinpoint photographs of the plasma-focus soft x rays [22, 32] usually have, besides the principal straight pinch, small formations (see Fig. 4), which sometimes follow one another. All this, and also the determination of the plasma temperature at that time by measuring the soft x rays [21] and the neutron yield [20], indicates that strong beam-energy dissipation takes place in these regions and plasma heating is possible. Favoring these arguments are also experiments with injection of an external electron beam in the plasma focus [68] and a numerical calculation [64].

We note here also one distinguishing feature of the beam-formation process in the case of discharges in deuterium with xenon additive. From a comparison of figures 13a and 14 it is seen that the main differences between the beams propagating in the mixture of the deuterium with xenon from beams in nitrogen–deuterium discharges are the following: The beams are practically completely opaque to the ruby-laser light, the beam breaks up into several primary beams and, occasionally, there is no hose instability or self-focusing of the beam, that is, the beam is effectively transported to the anode of the chamber (Fig. 15). Taking into account the fact that the heavy impurity becomes accumulated in the pinch, it can be stated that even though the energy density of the Langmuir oscillations excited by the beam in the surrounding plasma is apparently of the same order as in the case when nitrogen is added, the scattering is much larger. The breakdown of the beam into filaments (the Weibel instability [69]) is the result of a local increase of the electron density, owing to which an uncompensated current is produced accompanied by a magnetic field that "unravels" the beam from the principal beam. This instability is described in [66, 69]. It is shown there that when beams propagate in a high-density plasma this instability should become stabilized. This is why sometimes, especially during the earlier stage, the beam constituted a single solid filament. The distinctive feature of its formation in this case consisted in the fact that, at least in the initial stage, the beam propagates inside the pinch, in which the current direction coincides with the beam propagation direction. As shown in [70], such a configuration makes it possible to guide the beams over considerable distances with very high efficiency, since the pinch plasma affects the current neutralization, and the external magnetic field of the pinch causes radial containment of the beam. This transportation method makes it possible to mix several beams, meaning also to suppress the indicated tendency of the beam to break up into individual filaments.

4. Neutron Emission from Plasma Focus

The synchronization oscillograms (Fig. 2) allow us to draw certain conclusions concerning the neutron emission from the plasma focus. Indeed, since the photomultiplier that registers the hard x rays and the neutron radiation was placed at an angle close to 90° to the z axis, the time intervals between the peaks of the neutron pulse on the oscillogram (that is, at the place where the photomultiplier was installed) should be equal to the corresponding intervals at the point of their production. Our multiplier was located 15 m away from the chamber axis. We registered simultaneously the neutron and hard x-ray emission, at a distance of 5 m from the plasma-focus axis, also at an angle of 90° to it. It is easy to calculate the velocity of the neutrons with energy 2.45 MeV (D–D reaction):

$$v = \sqrt{-\frac{1}{2}\left(\frac{2E}{m_0 c}\right)^2 + \sqrt{\frac{1}{4}\left(\frac{2E}{m_0 c}\right)^4 + \left(\frac{2E}{m_0}\right)^2}} = 21.6 \text{ m/}\mu\text{sec.} \tag{34}$$

Knowing the distances from the plasma-focus axis to the multipliers and the velocities of the neutrons and of the light (for the hard x rays) we can compare the peaks of the neutron pulses with the hard x-ray pulse and thus relate the data to the obtained interference patterns.

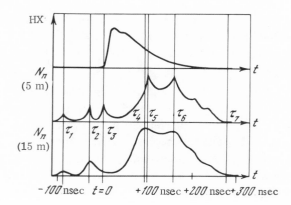

Fig. 16. Diagram of hard x-ray (HX) pulses and of the neutron radiation from a plasma focus on an absolute time scale.

It is seen from Fig. 16 that the first peak of the neutron pulse appears at the instant $\tau_1 \approx -100$ nsec. This instant corresponds, as we have seen, to cumulation of the shock wave on the axis. The instant of the appearance of the second peak ($\tau_2 \approx -30$ nsec) coincides with the first contraction; $\tau_3 = 0$ nsec is the instant when the sheath breaks. Frequently there was no peak of neutron radiation at that instant, as can be seen from the lower oscillogram. The peak that appeared at the instants $\tau_{4,5} = +100$ nsec was always unstable in amplitude.

The instant of the principal neutron pulse ($\tau_0 \approx +150$ nsec), which contains the overwhelming fraction of the neutrons, corresponds, on the one hand, to the end of the hard x-ray pulse, and on the other hand to the interference pattern with the magnetic channel of the beam of electrons focused in the interior of the plasma. This fact indicates that the principal role in the production of the plasma-focus neutrons is apparently played by the interaction of the electron beam with the plasma.

The foregoing explains the plasma-focus temperatures 10–20 keV in the second "contraction" obtained by measuring the soft x rays [21] or the neutron spectrum [20]. These values can be explained also in the framework of the Langmuir turbulence, which leads to heating of the plasma or to acceleration of the particles. If it is assumed that heating is realized then, starting from the dimensions of the region of the highest heating, that is, the region of the beam focus (0.5×1 mm), it is easy to estimate the neutron yield from one beam at the indicated parameters [18]

$$W \sim V\tau\langle\overline{\sigma v}\rangle_{\text{D-D}} \, n^2 \sim 2.5 \cdot 10^9 \text{ neutrons,} \tag{35}$$

where V is the volume of the hot plasma; $\tau \simeq 10^{-7}$ sec is the lifetime of the beam, and $n \simeq 10^{19}$ cm^{-3} is the plasma density in the region of the beam focus; $\langle\overline{\sigma v}\rangle_{\text{D-D}} \simeq 10^{-22}$ (T °K/10^7)^4, where $T \simeq 10^8$ °K is the plasma temperature. The quantity W is of the order of the experimental value of the neutron yield of the apparatus.

This concludes the description of the processes that occur in the plasma focus; we now proceed to another group of experiments, namely, investigations of the processes that occur in a cumulative laser plasma.

CHAPTER IV

INVESTIGATIONS OF CUMULATIVE LASER PLASMA

All the experiments on the cumulative laser plasma, described below, are based on collisions between expanding laser flares in one geometry or another. Important features of these processes are the absence of large external quasi-stationary magnetic fields and the presence of a high-power laser radiation that ensures plasma heating. The principal method of the investigation in this case was again high-speed frame-by-frame interferometry, but a number of experiments were performed using a five-frame shadow procedure of plasma photography, by determining the plasma temperature from soft x-ray measurements, by studying the plasma by the method of filters, by photography of the self-glow of the plasma with exposure of the order of 1 nsec, and by a probe procedure.

1. Experimental Setup

To heat the plasma we used a setup based on the neodymium-glass laser described in [42]. Its energy was raised to 100 J by additional amplification. Particular attention was paid to such important parameters as the divergence and the contrast of the radiation. A collimator reduced the maximum laser-beam divergence to $6 \cdot 10^{-4}$ rad, a value monitored with the aid of an image converter. To increase the radiation contrast (that is, the ratio of the Q-switched pulse energy to the energy of the "background" of the laser radiation, which consisted of the spontaneous generation of the entire volume of active elements of the laser and of the leading exponentially growing tail of the Q-switch pulse) we used an additional Kerr cell. This cell was placed at the output of the laser and was turned on at the instant of the maximum Q-switch pulse, making it possible to raise the contrast to $\gtrsim 10^3$ and to shorten the rise time of the pulse to 6 nsec. The energy flux density here turned out to be of the order of $5 \cdot 10^{13}$-10^{14} W/cm^2.

For the interferometric and shadow investigations we used a ruby-crystal setup analogous to that described in Sec. 2 of Chapter II. The time delay described in [42] was employed and made it possible to obtain a five-frame shadow photograph of the object with an exposure ~1 nsec for each frame and with arbitrary intervals between the frames. With the aid of this delay it was also possible to obtain one interferometry frame at the required instant of time relative to the maximum emission of the high-power laser.

To obtain the second harmonic of the ruby-laser emission we used a nonlinear element (KDP crystal) placed behind an optical pulse sharpener. From the KDP crystal, two beams corresponding to the ruby-laser emission and to its second harmonic were directed to an optical delay line and then to a Mach−Zehnder interferometer. The frequencies were separated directly in front of the photographic cameras with the aid of filters.

The plasma was photographed at the ruby-laser frequency with the aid of RO-2M and ERA-14 lenses with respective resolutions 180 and 625 lines/mm. Since the second harmonic of the ruby-laser emission lies in the ultraviolet, it was necessary to use quartz optics. We used a quartz lens with $f = 11$ cm and diameter 40 mm, corrected for spherical aberration. The resolution of the lens was ≈ 40 lines/mm in monochromatic light.

2. Collision of Two Laser Flares

Typical interference patterns in two wavelengths, of collisions of two laser flares from beryllium foils in a vacuum of $p \sim 10^{-5}$ Torr and at laser energies ~5 J in each of the two beams, are shown in Fig. 17. The reduction of most such interference patterns [62] obtained at various instants of time relative to the maximum of the heating pulse and for different distances between foils, and also a comparison of these interference patterns with similar ones for the free expansion of a laser flare (Fig. 18) have revealed the following characteristic features. When the distances between the foils are increased, the density at the point of collision decreases and tends to the value corresponding to the free expansion of the flare. At distances larger than 2-3 mm the density at the point of collision is approximately 1 mm higher than in the adjacent regions. The resultant cluster is obviously a "standing shock wave" with a density about 1.5-2 times larger than the plasma density in the incoming streams. This shock wave is therefore "weak," and the presence of the shock wave points to the efficiency of

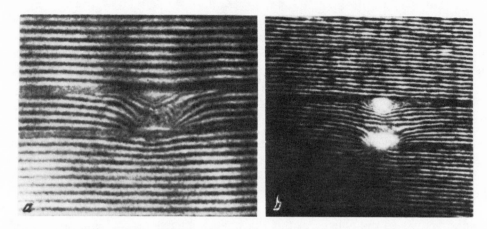

Fig. 17. Typical interference patterns of collisions of flares from two beryl-
lium foils (thickness 100 μm, τ_3 = 63 nsec). a) λ = 0.7 μm; b) λ = 0.35 μm.

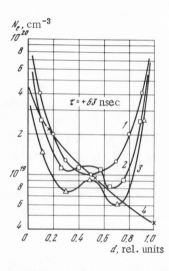

Fig. 18. Reduction of typical interference
pattern of collision of flares from two beryllium foils (thickness 100 μm) as a function of
the distance d between the foils. 1) d = 1
mm; 2) 3 mm; 3) 3.5 mm; 4) single flare (d =
1 mm).

the dissipative processes, that is, heating of the plasma in the collision zone. Experiments
carried out at laser energy 30 J and at τ_i = 15 nsec have shown that in the case when two flares
from beryllium foils collide, the "optimal" distance is of the order of 1-2 mm. There is time
in this case for the edge of the plasma to accelerate to supersonic velocity, but the density still
remains such as to permit effective conversion of this directional motion into thermal motion.

To explain the character of the gas-dynamic motion of the plasma in the collision of two
flares, we performed experiments on the collision of shock waves from two flares in atmospheres of H_2, He, and air at a pressure of the order of several Torr and at distances between foils up to 1 cm (see, e.g., Fig. 19). In this case the collision of the shock waves occurs
under practically stationary conditions [$\rho = \rho(x)$, $p = p(x)$ ahead of and behind the shock-wave
fronts]. The experiments have shown that the velocity of the shock waves decreases to approximately one-third, and in the zone behind the shock-wave front the material is in a compressed
state. It appears that in the case of flare collisions we are dealing with an analogous process,
but one occurring under essentially nonstationary conditions.

3.   Quasi-cylindrical Cumulation of Laser Plasma

The next stage consisted of experiments on quasi-cylindrical cumulation — focusing of
laser radiation into a hollow cone in a massive target with different vertex angles. The target
materials were Be, Li, and Pb.

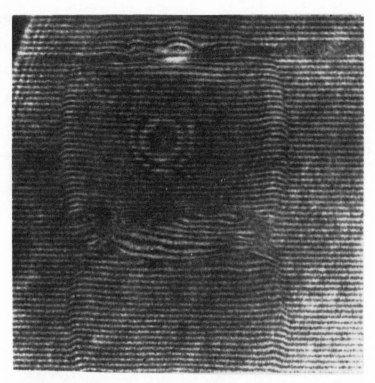

Fig. 19. Interference pattern of the collision of shock waves
from flares in a helium atmosphere (p $\simeq$ 1 Torr).

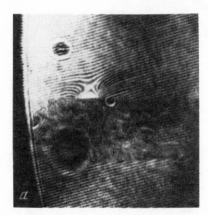

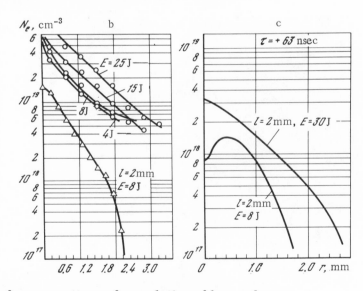

Fig. 20. Interference pattern of cumulation of laser plasma
when high-power laser radiation is focused into a hollow
cone with $\angle\, 2\alpha \sim 90°$ (target material, beryllium; $\tau_3 = 50$
nsec) (a), reduction of the interference patterns of the coni-
cal cumulation of the laser plasma for $\angle\, 2\alpha \lesssim 30°$ as a function of the energy of the high-power laser radiation (b), and
distribution of the electron density in a direction transverse
to the cone axis in the case of conical cumulation with $\angle\, 2\alpha =$
90° (c). In (b), the upper and lower curves show the electron density distributions along and across the cone axis.

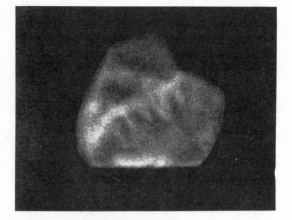

Fig. 21. Spontaneous glow of shock wave
from cumulative jet.

In this case (Fig. 20) it turned out that the parameters of the cumulative jet from the cone depend strongly on the angle at the vertex and turn out to be particularly high at $2\alpha \lesssim 30°$. At $2\alpha \lesssim 30°$, N_e is of the same order as in the region of cumulation of a "flat" collision of two flares. In addition, in this case the average gradient of the electron density along the beam axis turns out to be much less than in the case of ordinary flares or their collisions, and is of the order of $5 \cdot 10^{19}$ cm^{-4} (for a freely expanding flare this quantity amounts to $5 \cdot 10^{20}$ cm^{-4}).

When the angle at the apex of the cone was of the order of 90°, the distribution of the density took the form shown in Fig. 2c. It is seen that at a certain distance from the target surface the density falls off near the axis of the cone (of the laser beam), indicating a faster escape of the plasma along this axis. Photography of the spontaneous glow of the shock waves

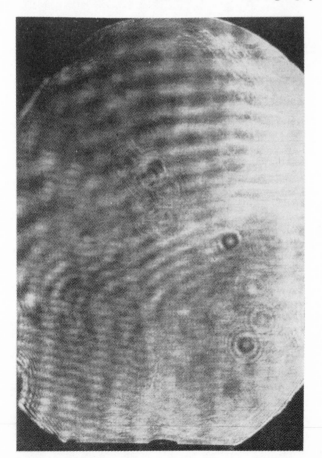

Fig. 22. Interference pattern of plasma in the case of a cone with $\angle 2\alpha \lesssim 30°$. Magnification $24 \times$ (the dimension of the entire interference field is 1 mm). Aperture 1:4.5. The spatial resolution of the objective was 800 lines/mm, and the spatial resolution of the interference patterns was of the order of 250 lines/mm. The laser beam is incident from above.

produced in this geometry when helium was admitted into the chamber (P ~ 10 mm Hg) has demonstrated the presence of this cumulative jet (Fig. 21).

Interference patterns taken at maximum laser energy have demonstrated an appreciable increase of the transparency band at the point of formation of the cumulative jet from the cone, if the angle at the vertex was ≲30°. Specially performed experiments with a magnification 24×, a spatial resolution of approximately 250 lines/mm, and large optical aperture of the projection system have shown (Fig. 22) that the character of this zone cannot be attributed either to classical scattering, or to refraction of the probing radiation by the density gradient, or else to insufficient resolution of the apparatus. Comparison of these interference patterns with the analogous ones for a flat target shows that the boundary of this zone occurs at densities ~$5 \cdot 10^{19}$ cm^{-3}, and the dimension of this zone was the same when in the light of the first and second harmonics of the ruby laser.

4. Investigation of X Rays from a Cumulative Laser

Measurements of the electron temperature of the plasma were made by using its soft x radiation and the well-known method of filters, with the laser beam focused into a hollow cone, in which case a greater part of the plasma participates in the cumulation, in contrast to collision of flares from foils.

The geometry of the experiment is such that the scintillator is protected from stray x rays by an aluminum screen 5 mm thick. We used a plastic scintillator of 3 mm thickness, based on polystyrene with p-terphenyl and POP as additives. A scintillator of this thickness is sufficient for a reliable absorption of x-ray photons of energy up to 5 keV [21].

To register the scintillator glow we used an ELU-FT photomultiplier with a gain ~10^7, a linear current ~2 A, and a time resolution of the order of 2 nsec. Typical oscillograms of the x-ray pulse are shown in Fig. 23.

It is seen from the oscillogram that for a target cone angle $2\alpha \lesssim 30°$, and at an initial temperature of the laser plasma produced on the target surface ≃280 eV, the corresponding x-ray pulse has in addition to a soft x-ray flash of ~15 nsec duration also a "hard" component with duration reaching 100 nsec in individual cases, and with randomly varying density. The employed pair of beryllium filters (18.48–55.43 mg/cm²) made it possible to establish only the lower limit of the corresponding electron temperature, which turned out to be of the order of 1 keV. (The upper temperature limit was determined by the scintillator thickness and amounted, as indicated above, to 5 keV). When the laser radiation was focused not on the apex but on the basis of the target cone, there was no soft x-ray pulse. It could be established from these oscillograms that the hard x-ray pulse begins at the instant of the laser-pulse maximum, whereas the soft x rays practically duplicate the waveform of the laser pulse, with a slightly decreased duration (by 3–5 nsec).

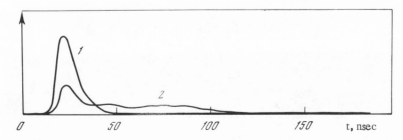

Fig. 23. Typical oscillogram of x-ray pulse of a cumulative jet from a cone with apex angle $2\alpha = 90°$ (1) and $2\alpha = 30°$ (2).

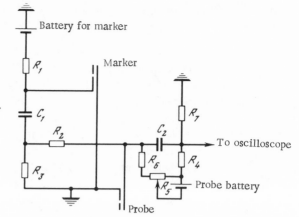

Fig. 24. Electric diagram of probe measurements in a cumulative laser plasma.

5. Probe Studies of Laser Plasma

The electric diagram of the probe procedure employed by us is shown in Fig. 24. The probe was a screened Faraday cup 8 mm long and 2 mm in diameter, insulated from the screen by a Teflon sleeve. It is clear that such a probe constitutes a certain modification of the double probe, one electrode being in our case the target and the metallic walls of the chamber, and the other the Faraday cup. A probe of this form makes it possible to reduce greatly the parasitic influence of the photoeffect due to the radiation of the plasma itself, as well as the influence of the secondary emission from the probe. Estimates and control experiments have shown that these effects are negligible in our case.

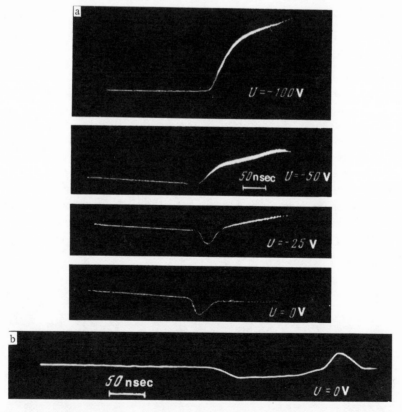

Fig. 25. Typical oscillograms of probe currents at different retarding potentials (a); the last pulse became longer in the case of focusing into a cone (b) (target material, copper).

Part of the laser beam was diverted and focused on the discharge gap, which became conducting at the start of the pulse. The electric pulse from this gap was fed through a capacitor to an oscilloscope and produced the zero time marker.

These experiments were performed on a flat target at a laser-pulse duration ~40 nsec at the base and at a radiation flux density ~$5 \cdot 10^{12}$ W/cm^2. To remove the impurity molecules (copper, carbon) absorbed by the target surface, the target was bombarded prior to the main laser pulse by defocused radiation. The vacuum in the chamber was not worse than 10^{-5} Torr.

Figure 25 shows typical oscillograms at different retarding potentials. It is seen that the bulk of the electrons in the tail of the expanding plasma has energies of the order of 100 eV, while the ion energy is of the order of 20 eV. In the case of a repeated shot on the same spot of the target, which obviously is equivalent to focusing into a cone with relatively small apex angle, an appreciable lengthening of the electron pulse at zero probe potential was observed (Fig. 25b). This pulse lengthening was likewise always observed in the case of a flat target when carbon and lithium targets were used. The pulse from the probe (at the zero probe potential) had a positive (ion) spike following the negative (electron) spike, thus obviously demonstrating that in these cases the electrons emitted from the plasma induce a positive potential on the plasma boundary.

CHAPTER V

DISCUSSION OF EXPERIMENTAL RESULTS

1. Collision of Flares

A complete theoretical analysis of the processes of collisions and cumulation of laser plasma is quite difficult. To ascertain the character of physical processes of this type it is therefore useful to employ a simplified model to analyze the problem of collision of laser flares. This problem was solved in [38]. Rather than present a detailed exposition of the results, we shall discuss here a question connected with the formation of a "standing" shock wave.

According to the theory developed in [38] for one-dimensional motion, the velocity of the shock wave produced after flare collision is always positive and not equal to zero in the laboratory reference frame. Under the experimental conditions, however, an appreciable role can be assumed by the fact that the motion is not one-dimensional (lateral expansion of the material), which leads to a faster damping of the shock wave than in the case of one-dimensional motion. As a result, a shock wave that is stationary in the laboratory frame can be produced in the experiments in question, as is apparently indeed observed in experiments on collisions of laser flares. We present numerical estimates of the parameters of the plasma produced in the region where two laser flares collide. Putting in the equations of [38] $n_0 = 5 \cdot 10^{18}$-$5 \cdot 10^{19}$ cm^{-3} ($\rho_0 = An_0/6 \cdot 10^{23}$ g/cm^3, A = 12 (C), $u_0 = 3 \cdot 10^7$ cm/sec), we obtain $p_{10} = 3 \cdot 10^5$-$3 \cdot 10^6$ atm, $T_{10} = 3$ keV, and $n_{10} = 10^{19}$-10^{20} cm^{-3}. These values are in good agreement with experiment.

We proceed now to discuss cone cumulation of a laser plasma. We shall consider mainly the experiments carried out when the laser radiation is focused on the base of a hollow cone in the target.

2. Cone Cumulation

In this case a theoretical analysis of the cumulation process becomes very complicated because the flow is not one-dimensional. For this reason, it is advantageous to present a qualitative description of the hydrodynamic picture of the phenomenon. Since the processes that take place in the case of cone cumulation of a laser plasma and in the case of convergence of the shock wave and sheath in the plasma focus are physically similar, the results of the analysis are applicable to both cases.

The symmetry of the problem enables us to start the analysis of the process with the planar case, replacing the symmetry axis by a rigid undeformable wall. Once the main characteristic features of the phenomenon are made clear, the results will be generalized to include the case of cylindrical symmetry.

It is known [71] that oblique collision between a supersonic stream (shock wave) and a rigid wall can produce two types of reflection, depending on the incidence angle (α) — regular (at $\alpha < \alpha_{cr}$) and irregular (at $\alpha \geq \alpha_{cr}$). A cumulative jet can be produced only in the second case. Expression (1) for the velocity v_j of the cumulative jet can be easily obtained from simple considerations [71].

It follows formally from (1) that by decreasing the angle (α) it is possible to attain arbitrarily large cumulative-jet velocities. However, as shown in [72], Eq. (1) is physically meaningful only at $\alpha \geq \alpha_{cr}$. In the case $\alpha < \alpha_{cr}$, however, the flow contains no jet. The latter circumstance is due to the formation, in the stream incident on the wall, of a shock wave whose front is "tied" at $\alpha < \alpha_{cr}$ to the rigid wall (to the symmetry plane), whereas at $\alpha > \alpha_{cr}$ this front is disengaged from the wall and the cumulative jet flows out through the produced "gap" [because $(\gamma + 1)/(\gamma - 1)$ is finite in the shock wave].

An expression for α_{cr} was obtained in [72] by numerically solving our problem for an ideal gas:

$$\tan \alpha_{cr} = (\gamma^2 - 1)^{-1/2}, \tag{36}$$

where $\gamma = C_p/C_V = 5/3$, i.e., in our case $\alpha_{cr} = 36°52'$. Consequently, the maximum velocity that the cumulative jet can attain in the case of a stationary planar flow is

$$v_{j_{max}} = v_0 (\gamma + \sqrt{\gamma^2 - 1}) = 3v_0 \simeq 9 \cdot 10^7 \text{ cm/sec} \tag{37}$$

for $v_0 = 3 \cdot 10^7$ cm/sec. Calculation [72] shows, however, that in the time interval corresponding to the transient process, the jet parameters are larger than the stationary values. Thus, for example, the maximum particles lost in the jet is

$$v_{j_{max}} = v_0 [\gamma + (2\gamma + 2)^{1/2}] \simeq 4v_0 = 1.2 \cdot 10^8 \text{ cm/sec}. \tag{38}$$

It is clear that this nonstationary collapse phase plays a very important role both in the plasma focus and in cone cumulation of a laser plasma, because the time during which the density varies in converging streams is of the order of the jet buildup time.

In the case of a large cone angle, the difference between conical geometry and the planar case is relatively small, so that the additional effect of the compression of the gas in converging conical streams (shock waves) is relatively small in this case by virtue of the rapid outflow of the plasma along the z axis. It is clear nevertheless that this effect leads to a decrease of the critical angle and to an increase of the velocity of the cumulative jet.

After emerging from the cone, the cumulative jet expands in accordance with the laws of gas dynamics, and its maximal parameters (density, pressure, temperature) are reached near the "focus" of the jet [72], i.e., at a point that separates the cumulative jet from its "trail." We see thus that our estimates and the general picture of the process are close to the experimental results.

Taking into account the values obtained above for the velocities, we can assume that the effect of the cumulative jet can indeed be responsible for the production of neutrons in the plasma focus at the instant of cumulation of the shock wave and of the sheath on the z axis.

More interesting, however, is the fact that the same effect can play an essential role also in the production of neutrons when a high-power laser beam is incident on a planar bulky target of $(CD_2)_n$ or D_2. Indeed, as noted by many workers, the production of neutrons in this case is accompanied by the following effects:

(a) Low-intensity hard x rays, which are not produced when targets of elements such as iron are irradiated [73];

(b) Fast ions with energy $\sim 10^4$ eV;

(c) Vanishing of the neutrons when gas is admitted into the chamber at a pressure of the order of 1 Torr [74];

(d) The neutron pulse is anomalously long (50-100 nsec), with clearly pronounced two maxima — one near the maxima of the laser pulse and the other much later; it was also established that the second maximum appears after several shots in the chamber, so that its character is governed by the collisions of the fast particles from the flare with the deuterium adsorbed on the chamber wall [73];

(e) A dip appears on the distribution of the electron density in the flare near the beam axis [75];

(f) In experiments with a long laser pulse ($\sim$50 nsec) and with heavy-element targets, the appearance of lines of multiply charged ions with maximum charge is observed at a certain distance from the target [76];

(g) The soft x rays are emitted from the plasma in this case only during the rise time of the heating pulse [77].

These facts are difficult to explain within the framework of the usual hydrodynamics of a laser plasma obtained from a flat target [42]. Their nature becomes clear, however, if it is assumed that as the target becomes heated the laser pulse produces in it a conical opening, so that the maximum of the laser pulse acts on a laser plasma obtained in conical geometry. It is indeed quite clear that this cone will be produced particularly effectively only in the case of light elements. It is shown in [78] that for a polyethylene target at a flux density $\sim 10^{12}$ W/cm^2 the depth of the hole burned in 10 nsec is of the order of the dimension of the focal spot, whereas for an aluminum target this depth is negligibly small. It is further obvious that after formation of the cumulative jet (which takes place in this case, since the depth of the "cone" turns out to be of the order of its diameter at the base, and consequently $\alpha > \alpha_{cr}$) the latter will propagate through a "ready-made" laser plasma with a decreasing density profile. The neutrons, in the case of targets with deuterium, can be produced when the deuterons of the jet $(v_j \sim 10^8$ cm/sec) collide either with the "ready-made" slow-moving plasma of the flare or with the deuterium absorbed by the chamber walls in the preceding shots. Ion lines with maximum ionization multiplicity will appear at the focal point of the jet, which at large $\angle \alpha$ may turn out to be located at a certain distance from the target. As seen from an examination of the irregular reflection of the flux (of the shock wave) from the symmetry axis, the plasma near the axis flows out along the axis with large velocity, whereas the plasma behind the reflected shock wave is compressed and moves more slowly. This effect should lead to the formation of a tubular conical structure in the distribution of the electron density. This is indicated also by a numerical calculation [72]. This phenomenon explains the dip in the distribution of the electron density near the laser-beam axis [75].

Finally, it should be noted that the indicated supersonic jet can produce in principle a shock wave (apparently a collisionless one [79]) in the ready-made plasma. Since this shock wave propagates along a decreasing density profile, energy will accumulate in it, as is the case when the shock wave emerges to the surface of a star [52]. This type of cumulative motion produces plasma parameters that reinforce all the effects listed above and add new ones, of the type of stochastic acceleration of particles to high energies.

We proceed now to consider cases when the angle at the vertex of the cone is small. Under this condition, first, the reflection becomes regular and the cumulative jet is not formed and, second, this case becomes close to cylindrically symmetrical. For the case of oblique collision of plane shock waves in a cavity made up of two converging planes and filled uniformly with gas, such a problem was solved in [71]. It was shown that at small angles almost the entire energy of the converging fluxes becomes potential, i.e., almost complete stagnation of the gas takes place. The velocity of its expansion along the z axis will be close to the velocity of sound c, and the largest value of this velocity for $\gamma = 5/3$ is 1.15c.

In the case of axisymmetric cumulation as $\alpha \to 0$, the additional effect of gas compression will be proportional to $[(\gamma + 1)/(\gamma - 1)]^2$ [80]. As seen from all the foregoing, in the case of a conical cavity with small apex angle in the target, the plasma on the axis is heated as a result of formation of a quasi-cylindrical shock wave. The first to be heated in this case are the ions. It is of interest to examine how the electrons are heated by these shock-heated ions. In [81] are given equations for the relaxation times of the ion and electron components, and also for the relaxation of the plasma as a whole. In our case, the initial temperatures are $T_i \sim 3$ keV and $T_e \sim 1$ eV. From the general relaxation formula

$$\tau_{\alpha\beta} = \frac{3}{8\sqrt{2\pi}} \frac{m_\beta}{\sqrt{m_\alpha}} \frac{\varepsilon_\alpha^{3/2}}{e_\alpha^2 e_\beta^2 n_\beta \Lambda}, \quad \text{where } \varepsilon_\alpha = T_\alpha + \frac{m_\alpha}{m_\beta} T_\beta, \tag{39}$$

it is easy to determine the quantity τ_{ie}.

Taking into account the fact that T_e varies with time, this quantity turns out to be (for $N_e \sim 10^{20}$ cm^{-3}, Be) of the order of 10^{-5} sec. During the time that the electrons are present in the region of shock compression, prior to the emergence of the plasma from the cone along the z axis, the temperature rises to values of the order of 100-200 eV, i.e., the ions do not have time to transfer their energy to the electrons by means of ordinary Coulomb collisions. To explain the hard x radiation obtained by us in the experiments with small aperture cones, we must therefore consider other mechanisms for the production of fast electrons. These mechanisms can in our case be the following:

(1) Plasma instabilities produced in collisions of plasma streams formed on the walls of the conical cavity

(2) The electrons of the Maxwellian "tail"

(3) The acceleration of the electrons due to stimulated Compton scattering of powerful laser radiation by the electrons of the cumulative plasma cluster

(4) Electron acceleration when parametric instability is developed in the cumulative plasma region

We now analyze the suggested mechanisms in succession.

1. When one plasma beam passes through another with velocity $V \simeq 3 \cdot 10^7$ cm/sec, the following instabilities can become excited [55]:

(a) Langmuir oscillations with wave number $k_{opt} = (\sqrt{3}/2)(\omega_{pl}/V) \simeq 2.5 \cdot 10^7$ cm^{-1} and growth rate $\gamma \simeq \omega_{pl}/2$. These oscillations will arise when $V \gg v_{Te}$. In our case, however, since $\tau_{ee} \ll 1/\gamma$, this condition is not satisfied.

(b) Buneman instability (resonant instability of drift oscillations of ions and of electron Langmuir oscillations) with wave number $k_\parallel \simeq \omega_{pl}/V$ and growth rate $\gamma = (\sqrt{3}/2^{4/3})\omega_{pl}(m_e/m_i)^{1/3}$. For this instability to build up, however, it is necessary to satisfy the condition $v_{Ti} \ll (m_e/m_i)^{1/3}V$. In our case, on the other hand, owing to shock compression we have $v_{Ti} \sim V$, i.e., this instability likewise does not develop.

(c) Instability of ion-sound and ion Langmuir oscillations are impossible, inasmuch as in our case $T_e \ll T_i$.

(d) Kinetic instability of hot ions is impossible since $v_{T_e} > V$.

(e) Buildup of ion-electron oscillations is also impossible in our case, since the condition $V < (T_e/m_i)^{1/2}$ required for these instabilities is not satisfied for either the ionic or the electronic component of the beam $[w_i > T_e, w_e > (m_e/m_i)T_e$, where $w_{i,e}$ is the energy of the ions and electrons in the beam].

2. From our experiments on x rays from a cumulative laser plasma we obtain for the ratio of the absolute intensities of the soft and hard components a value of the order of 10. As indicated above, at the ELU scintillator thickness used by us, the electron beam multipliers respond equally to equal radiation intensity of a plasma with energy 200 eV and 5 keV. At a plasma electron temperature in the cumulation region ~200 eV, the fraction of the radiation pertaining to the energies $h\nu_0 > 1$ keV is $I_d/I_d (\nu > \nu_0) \sim \exp(h\nu_0/kT_e) \sim 150$. Thus, the contribution of the Maxwellian-tail electrons is negligible. In addition, this mechanism does not explain the lengthening of the hard x-ray pulse in comparison with the soft one, nor the lengthening of the electron pulse from the probe at zero probe potential for the case of focus into a conical cavity with small angle 2α.

Using the formulas for the intensities of the bremsstrahlung and combination continuums, we can estimate the number of particles that emit hard x rays in the case of focusing into the base of a cone in a carbon target $[I_h \sim 0.1 \, I_s$, I_h is the intensity of bremsstrahlung and I_s is the intensity of the recombination radiation, $T_{e_h} \sim 10^3$ eV, $T_{e_s} \sim 10^2$ eV, $N_{e_s} \sim 10^{21}$ cm^{-3}, N_{e_h} (cumulative zone) $\sim 10^{20}$ cm$^{-3}]$. By equating the ratios of the absolute intensities of the soft and hard x rays to 0.1, we find that the particles that emit hard x rays occupy a region with dimension of the order of the focal spot of the focusing lens, while the number of particles emitting hard x rays is $N_e \sim N_{e_h} \cdot V_h \sim 10^{20} \cdot 10^{-6} = 10^{14}$ particles. This quantity agrees with the estimate of the number of fast electrons obtained from probe measurements when focused into a "cone" ($\mathscr{I}_{tot} \simeq 100$ A, $\tau_{fast \, el.} = 10^{-7}$ sec):

$$N_{tot} = N_e v \tau s = \frac{\mathscr{I}_{tot}\tau}{e} = 10^{14} \text{ particles.}$$

3. Stimulated Compton scattering of light is similar in its character to the parametric instabilities that will be described below, and can apparently be responsible for the formation of the hard x rays (fast electrons), although pure heating is apparently impossible, since for our parameters ($N_e \sim 10^{20}$ cm^{-3}, $(\Delta\omega/\omega_0)^2_{Nd} \simeq 10^{-5}$) the necessary laser radiation flux density is [82]

$$I > I_0 \simeq \frac{30\ln\Lambda}{11\pi^2} m\omega_0^3 \left(\frac{\omega_{pl}}{\Delta\omega}\right)^2 \frac{e_i}{|e|} \simeq 10^{16} \text{ W/cm}^2, \tag{40}$$

which is not satisfied in our case. We are thus left with only one possibility − the formation of fast electrons when parametric instabilities develop in the cumulation region as a result of the interaction of the laser radiation with the plasma of this cluster. To determine the type of instability that can take place in our case, we must take into account the following facts:

(1) The laser-radiation flux density in all our experiments was in the range $5 \cdot 10^{12}$-10^{14} W/cm^2.

(2) At the initial instant of time relative to the instability development we have $T_i \approx 10 \, T_e$ as a result of shock compression.

(3) The boundary of the opacity zone for the diagnostic laser radiation (i.e., the boundary of the turbulence zone) lies near a density of the order of $5 \cdot 10^{19}$ cm^{-3}.

(4) The maximum plasma density in the interaction region (obtained by extrapolation) is of the order of 2-$3 \cdot 10^{20}$ cm^{-3}.

As shown in [83], the growth rate of the SRS will exceed the growth rate of the SMBS at all transverse-wave field intensities if the condition $v_{T_e}/c > \omega L_i/2\omega_0$ is satisfied. In our case ($T_e = 100$ eV to 1 keV, $N_e = 7 \cdot 10^{19}$–$3 \cdot 10^{20}$ cm^{-3}) we obtain

$$\frac{v_{T_e}}{c} = 1.5 - 4 \cdot 10^{-2} \simeq \frac{5.4 \cdot 10^8 - 1.2 \cdot 10^9}{3 \cdot 10^{10}} > \frac{\omega_{L_i}}{2\omega_0} = \frac{(1.1 - 2.1) \cdot 10^{13}}{2.1 \cdot 8 \cdot 10^{15}} - 3 \cdot 10^{-3} - 5 \cdot 10^{-3},$$

i.e., the SMBS should be neglected. Thus, recognizing that the boundary of the turbulent zone lies near plasma densities typical of of SRS processes and of two-plasmon decay [84], we must verify for them the threshold fields and the growth rates. For the first process ($T_e = 200$ eV, $N_e \simeq 10^{20}$ cm^{-3}): $I_{thr} = 3 \cdot 10^{13}$ W/cm^2 and at $I \simeq 10^{14}$ W/cm^2 we have $\gamma \simeq 4 \cdot 10^{12}$ sec^{-1}. For the second process ($T_e \simeq 200$ eV, $N_e = 2.5 \cdot 10^{20}$ cm^{-3}): $I_{thr} \simeq 2 \cdot 10^{13}$ V/cm^2 and for $I = 5 \cdot 10^{13}$ V/cm^2 $\gamma \simeq 10^{12}$ sec^{-1}. It is seen that both processes can take place in our case, but the two-plasmon decay has a lower pump field threshold intensity.

Next, as is well known, the plasma density gradient can lead to an increase in the value of the threshold field. An investigation of its influence on the second process was carried out in [85]. However, substitution of the characteristic dimension of the inhomogeneity of our cumulative plasma (for which it is quite large in comparison with the case of a laser plasma from a flat target) in the formula obtained in our paper did not lead to an increase of the threshold.

Sometime after the start of its development, the instability should reach saturation. The problem of choosing the saturation mechanism is quite important, since it determines the solution of such processes as the plasma heating efficiency, the formation of fast particles, reflection (scattering) of the heating laser radiation, etc.

In our case these saturation mechanisms can be the following:

(a) Scattering of the laser radiation by Langmuir oscillations — "extraction" of the radiation from the plasma [86]

(b) Nonlinear interaction of parametrically excited oscillations and redistribution of the energy over the spectrum into the region of stable wave numbers [87, 88]

(c) An inverse Cerenkov process that leads to the appearance of a group of fast electrons [89]

(d) Effects of electron capture, when the energy of a plasma wave of large amplitude is transferred to the plasma electrons trapped by this wave, causing the electron component to be heated [90]

We have not investigated the reflected laser radiation, but the results on the hard x rays and the existence of a turbulent plasma that is opaque to $\lambda = 0.7$ μm indicates that the processes (c) and (d) can take place in any event, and the process (d) can obviously not account for the stretching of the hard x-ray pulse and the probe electron pulse.

Let us estimate the wavelength of the longitudinal Langmuir oscillations, recognizing that the hard x-ray measurements yield the energy of the fast particles in the range 1–5 keV. From the equation of [84] $\frac{3}{2}k^2 r_{D_e}^2 = (\omega_0 - 2\omega_{L_e})/2\omega_{L_e}$ we obtain for the densities and the temperatures

$$
\begin{aligned}
N_e &= 7 \cdot 10^{19} \text{cm}^{-3}, & T_e &= 5 \text{ keV}, & \lambda^l &= 5 \cdot 10^{-5} \text{cm} > \lambda_0/2 = 3.5 \cdot 10^{-5} \text{ cm}. \\
N_e &= 2.4 \cdot 10^{20} \text{cm}^{-3}, & T_e &= 500 \text{ eV}, & \lambda^l &= 6 \cdot 10^{-5} \text{cm} > \lambda_0/2, \\
N_e &= 2.4 \cdot 10^{20} \text{cm}^{-3}, & T_e &= 1 \text{keV}, & \lambda^l &= 1.2 \cdot 10^{-1} \text{cm} > \lambda_0/2.
\end{aligned}
$$

It is seen therefore that anomalous scattering of the transilluminating radiation by this turbulent plasma is indeed possible.

Recognizing further that the mean free path of electrons with energy of 1 keV is in our plasma of the order of 50 μm it is clear that hard x rays can be emitted from the plasma itself. On the other hand, the stretching of the hard x-ray pulse and of the probe electron pulse can now be explained for our case as follows. The damping of the high-frequency longitudinal plasma oscillations in our relatively cold plasma is a very slow process, so that few particles are at resonance with the phase velocity of these waves. Consequently, the particle acceleration process can continue also after the external pump wave is removed, for a time of the order of the characteristic time of gas-dynamic expansion of the cumulative configuration (~100 nsec).

Thus, in connection with the facts observed in our experiments, such as the abrupt increase of the opacity zone, the appearance of long hard x-ray pulses from the plasma and of a long electron pulse from the probe in the case when the flux of the high-power laser radiation is increased from $5 \cdot 10^{11}$ to $5 \cdot 10^{13}$-10^{14} W/cm^2, and also in connection with the effects, known from the literature, of the anomalously high reflection of the heating radiation from the laser plasma and the generation of harmonics of the incident radiation in the plasma, we can state that these nonlinear effects can be fully due to parametric instabilities, since the threshold intensities for the excitation of a number of these instabilities are much lower than those obtaining in our experiment, and the reciprocal growth rates are much smaller than the duration of the laser pulse. The energy of the Langmuir oscillations excited by the laser radiation is transferred to the fast electrons, the number of which is approximately 1% of the total number of electrons of the entire volume of the laser plasma. For unambiguous conclusions with respect to the character of the processes, however, it is necessary to have much more detailed experimental investigations, which should include also investigations of the distribution function of the fast particles (both the electrons and ions), exact measurements of the instability wavelength, a spectral analysis of the scattered laser radiation, as well as investigations of the anomalous absorption coefficient.

CONCLUSION

At present, as is well known, the laser is capable of applying energy to a substance at an unprecedentedly rapid rate. The presently attainable flux densities are ~10^{16} W/cm^2 when nanosecond pulses are used [91] and ~10^{17} W/cm^2 in the mode-locking regime [92]. At such high intensities, the radiation can be absorbed in the substance both as a result of the ordinary inverse bremsstrahlung process and as a result of different nonlinear mechanisms such as parametric instability, stimulated Compton scattering, etc. An investigation of processes of this type is undoubtedly of great interest, since there are practically no reliable experimental data at present on these questions.

On the other hand, it is clear that a plasma focus as a target for a high-power laser has many valuable properties. First, in contrast to the case of heating a small granule of solid substance by a laser, the requirements with respect to such an important and difficultly attainable laser parameter as the radiation contrast are greatly reduced. The point is that the weak radiation background preceding the main laser pulse capable of evaporating a granule of a solid target has practically no infleunce whatever on a cluster of hot plasma in the plasma focus. In addition, a target in the form of a plasma focus has a high initial ion temperature (~1 kV) and a large magnetic self-field (~1 MG). The plasma-focus installation makes it easy to vary the composition of the target by introducing ahead of the discharge, into a deuterium-filled chamber, various additives. As shown in the present paper, the plasma density distribution in the plasma focus at the instant of maximum compression is such that the interpretation of the experimental results on the interaction with high-power laser radiation can be carried out more consistently than in the case of heating of a solid target, where the very sharp density gradients obscure greatly the overall picture. The plasma focus makes it possible to study the interaction of laser radiation with a strongly magnetized plasma.

It should be noted that the plasma focus itself is a device that holds the record both with respect to the neutron yield (up to 10^{12} neutrons) and with respect to the x rays. As seen from the present paper, this is due to the presence of high-power electron beams in the plasma focus. By placing solid targets containing deuterium in the anode of the plasma focus and by prior heating of these targets, with the aid of a laser, to a temperature of the order of 1 keV before the arrival of the electron beams at the anode, it is possible to use the energy of these beams quite effectively for combined heating of this target. Great promise is also offered by experiments with exploding wires [93], if, prior to the instant when current is broken and the electron beam develops (as seen from the foregoing this takes place in the plasma focus also in this group of experiments), the part of the wire under the place where the current breaks is compressed with the aid of laser radiation to ultrahigh densities.

Thus, the use in a single experiment of both high-power laser radiation and high-power relativistic electron beams for plasma heating is, from our point of view, very promising, since it makes it possible to combine the good absorbing ability of the laser radiation and the high efficiency of the electron beam.

In conclusion the authors thank Academician N. G. Basov for interest and support of the work, Yu. V. Afanas'ev, V. S. Imshennik, and V. V. Pustovalov for valuable consultations, and V. Ya. Nikulin and V. M. Korzhavin for help with the experiment.

LITERATURE CITED

1. N. G. Basov, O. N. Krokhin, and G. V. Sklizkov, Pis'ma Zh. Éksp. Teor. Fiz., 6:6 (1967).
2. N. V. Filippov, T. I. Filippova, and V. P. Vinogradov, Nucl. Fusion Suppl., Pt. 2:577 (1962).
3. A. C. Kolb, Rev. Mod. Phys., 32:748 (1960).
4. N. G. Basov, O. N. Krokhin, and G. V. Sklizkov, Laser Interaction and Related Plasma Phenomena, Proc. Second Workshop, Hartford, USA (1972).
5. K. A. Bruekner, KMSF-NP5, Ann Arbor, Michigan (1972).
6. J. Nuckolls, L. Wood, A. Thiessen, and G. Zimmerman, Doubleday, Life Report II, Switzerland (1972).
7. G. Birkhoff, D. McDougall, E. Pugt, and G. Taylor, J. Appl. Phys., 19:563-582 (1948).
8. M. A. Lavrent'ev, Usp. Mat. Nauk, 12(4):41 (1957).
9. J. Meiser, Transactions of Conference on Plasma Physics and Controlled Thermonuclear Fusion [Russian translation], Vol. 2, Report CN-21/80, Culham (1965).
10. Yu. A. Kolesnikov, N. V. Filippov, and T. I. Filippova, Proc. Seventh Intern. Conf. on Ionization Phenomena in Gases, Beograd (1965).
11. V. V. Vikhrev, Zh. Prikl. Mat. Tekh. Fiz., No. 2:160 (1973).
12. I. V. Kurchatov, At. Énerg., 3:65 (1956).
13. L. A. Artsimovich, A. M. Andrianov, E. I. Dobrokhotov, S. Yu. Luk'yanov, I. M. Podgornyi, V. N. Sinitsyn, and N. V. Filippov, At. Énerg., 3:76 (1956).
14. M. A. Leontovich and S. M. Osovets, At. Énerg., 3:81 (1956).
15. V. S. Imshennik, Preprint No. 17, IPM AN SSSR (1972).
16. D. E. Potter, Phys. Fluids, 14:1911 (1971).
17. D. E. Potter and M. G. Haines, Fourth Conf. on Plasma Phys. and Controlled Nuclear Fusion Res., IAEA, CN-28/D-8, Madison, Wisconsin (1971).
18. L. A. Artsimovich, Controlled Thermonuclear Reactions [in Russian], Fizmatgiz, Moscow (1961).

19. P. D. Morgan and N. J. Peacock, Paper, Second Topical Conf. on Pulsed High Beta Plasmas, Garching (1972).

20. C. Maisonnier et al., Fourth Conf. on Plasma Physics and Controlled Nuclear Fusion Res., Pt. I, IAEA CN-28, D-1, D-2, Madison, Wisconsin (1971).

21. C. Patou and A. Simonnet, Note C. E. A., N1189, Commissariat 1'Energie Atomique (1969).

22. N. V. Filippov, V. I. Agafonov, I. F. Belyaev, V. V. Vikhrev, V. A. Gribkov, L. G. Golub-chikov, V. F. D'yachenko, V. S. Imshennik, V. D. Ivanov, O. N. Krokhin, M. P. Moiseeva, G. V. Sklizkov, and T. I. Filippova, Fourth Conf. on Plasma Physics and Controlled Nuclear Fusion Res., IAEA CN-28/D-6, Madison, Wisconsin (1971).

23. V. V. Aleksandrov, N. G. Koval'skii, S. Yu. Luk'yanov, V. A. Rantsev-Kartinov, and M. M. Stepanenko, Zh. Éksp. Teor. Fiz., 64:1222 (1973).

24. J. P. Watteau et al., Fourth Conf. on Plasma Phys. and Controlled Nuclear Fusion Res., IAEA, CN-28/D-4, Madison, Wisconsin (1971).

25. N. W. Jalufka and J. H. Lee, Phys. Fluids, 15:1954 (1972).

26. A. Bernard et al., Fifth Eur. Conf. on Controlled Fusion Phenomena and Plasma Phys., Vol. 1, No. 65; Vol. 2, No. 65, Grenoble (1972).

27. H. L. L. van Paassen and R. H. Vandre, Phys. Fluids, 13:2606 (1970).

28. H. L. L. van Paassen, Rev. Sci. Instr., 42:12, 1823 (1971).

29. J. H. Lee et al., Plasma Phys., 13:347 (1971).

30. I. A. Anderson, V. R. Baker, S. N. Colgate, et al., Reports to the Third International Conf. on Gas Discharges, Venice (1957).

31. B. A. Trubnikov, in: Plasma Physics and Problems of Controlled Thermonuclear Reactions (M. A. Leontovich, ed.) [in Russian], Vol. 4, Izd. AN SSSR (1958), p. 87.

32. W. H. Bostick et al., Fifth Eur. Conf. on Controlled Fusion Phenomena and Plasma Phys., Grenoble, Vol. 1, No. 69 (1972).

33. J. G. Linhart, Nucl. Fusion, 10:211 (1970).

34. H. Conrads and P. Cloth, Fifth Eur. Conf. on Controlled Fusion Phenomena and Plasma Phys., Grenoble, Vol. 1, No. 67 (1972).

35. O. N. Krokhin, Zh. Tekh. Fiz., 34:1324 (1964).

36. V. A. Gribkov, V. Ya Nikulin, and G. V. Sklizkov, Preprint No. 153, FIAN (1970).

37. H. Puell, H. Opower, and H. J. Neusser, Phys. Lett., 31A:4 (1970).

38. Yu. V. Afanas'ev, V. A. Gribkov, et al., Preprint No. 78, FIAN (1970).

39. N. G. Basov, Yu. S. Ivanov, O. N. Krokhin, et al., Pis'ma Zh. Éksp. Teor. Fiz., 15:598 (1972).

40. R. Huddlestone and S. Leonard (editors), Plasma Diagnostic Techniques, Academic Press, New York (1965).

41. W. Lochte-Holtgreven (editor), Plasma Diagnostics, North Holland Publishing, Amsterdam (1968).

42. N. G. Basov, V. A. Gribkov, O. N. Krokhin, and G. V. Sklizkov, Zh. Éksp. Teor. Fiz. 54: 4 (1968).

43. V. V. Korobkin and A. M. Leontovich, Zh. Éksp. Teor. Fiz., 44:1847 (1963).

44. V. L. Ginzburg, Propagation of Electromagnetic Waves in Plasmas, 2nd rev. ed., Pergamon Press (1971).

45. "Atomic transition probabilities," National Standard Reference Data Series, National Bureau of Standards, Washington (1966).

46. M. Born and E. Wolf, Principles of Optics, 5th ed., Pergamon Press (1975).

47. F. G. Jahoda et al., J. Appl. Phys., 35:8 (1964).

48. F. D. Bennett et al., J. Appl. Phys., 23:450 (1952).

49. R. U. Ladenburg et al., Physical Measurements in Gas Dyamics and during Combustion [Russian translation], IL, Moscow (1957).

50. V. A. Gribkov, V. Ya. Nikulin, and G. V. Sklizkov, Kvant. Élektron., No. 6:60 (1971).
51. V. A. Gribkov, V. Ya. Nikulin, et al., Preprint No. 53, FIAN (1972).
52. Ya. B. Zel'dovich and Yu. P. Raizer, Elements of Gas Dynamics and the Classical Theory of Shock Waves, Acaemic Press, New York (1968).
53. V. F. Dyachenko and V. S. Imshennik, in: Reviews of Plasma Physics, Vol. 5, Consultants Bureau, New York (1970).
54. V. N. Tsytovich, Theory of Turbulent Plasma, Consultants Bureau, New York (1976); V. Pustovalov, V. P. Silin, and V. T. Tikhonchuk, Preprint No. 183, FIAN (1973).
55. A. B. Mikhailovskii, Theory of Plasma Instabilities, Consultants Bureau, New York, (1974).
56. B. A. Demidov and S. D. Fanchenko, At. Énerg., 20:516 (1966).
57. R. S. Post and T. C. Marshall, Preprint, Columbia University Press, New York (1974).
58. L. V. Dubovoi, Doctoral Dissertation, NII ÉFA, Leningrad (1972).
59. S. I. Braginskii, in: Reviews of Plasma Physics, Vol. 1, Consultants Bureau, New York (1965).
60. L. D. Tsendin, Zh. Éksp. Teor. Fiz., 56:929 (1969).
61. B. Coppi, Phys. Lett., 34A:395 (1971).
62. V. A. Gribkov, Candidate's Dissertation, Physics Institute, Academy of Sciences of the USSR, Moscow (1974).
63. G. Benford and D. Book, in: Advances in Plasma Phys., Vol. 4, (1971), pp. 125-172.
64. L. Lee and M. Lampe, Phys. Rev. Lett., 31:23, 1390 (1973).
65. A. A. Ivanov and L. I. Rudakov, Zh. Éksp. Teor. Fiz., 58:1332 (1970).
66. M. V. Babykin, E. K. Zavoiskii, A. S. Ivanov, and L. I. Rudakov, Fourth Conf. on Plasma Phys. and Controlled Nuclear Fusion Res., IAEA/CN-28/D10, Madison, Wisconsin (1971).
67. V. D. Shapiro, Doctoral Dissertation, Khar'kov Physicotechnical Institute (1967).
68. D. A. Freiwald et al., Phys. Lett., 36A:297 (1971)
69. E. S. Weibel, Phys. Rev. Lett., 2:83 (1959).
70. J. Benford and B. Ecker, Phys. Rev. Lett., 28:10 (1972).
71. F. A. Baum, Ya. B. Zel'dovich, and K. P. Stanyukovich, Physics of Explosions [in Russian], Fizmatgiz, Moscow (1969).
72. F. H. Harlow and W. E. Pracht, Phys. Fluids, 9:1951 (1966).
73. V. A. Boiko et al., Preprint No. 77, FIAN (1973).
74. K. Buchl et al., Fourth Conf. on Plasma Physics and Controlled Nuclear Fusion Res., IAEA, Pt. 1, Madison, Wisconsin (1971).
75. A. Saleres, D. Cognard, and F. Floux, Fifth Eur. Conf. on Controlled Fusion Phenomena and Plasma Phys., Grenoble, No. 59 (1972).
76. W. Seka et al., J. Appl. Phys., 42:315 (1971).
77. P. Floux et al., Ninth Intern. Conf. on Phenomena in Ionized Gases, Bucharest (1969), p. 335.
78. G. V. Sklizkov et al., Preprint No. 45, FIAN (1972).
79. R. Z. Sagdeev, in: Reviews of Plasma Physics, Vol. 4, Consultants Bureau, New York (1966).
80. J. P. Somon, Proc. Enrico Fermi Intern. School Phys., Course 48 (1969).
81. B. A. Trubnikov, in: Reviews of Plasma Physics, Vol. 1, Consultants Bureau, New York (1965).
82. A. V. Vinogradov and V. V. Pustovalov, Zh. Éksp. Teor. Fiz., 62:980 (1972).
83. N. E. Andreev, Zh. Éksp. Teor. Fiz., 59:2105 (1970).
84. E. A. Jackson, Phys. Rev., 153:235 (1967).
85. M. Rosenbluth, Phys. Rev. Lett., 29:565 (1972).
86. A. A. Galeev et al., Pis'ma Zh. Éksp. Teor. Fiz., 17:48 (1973).
87. V. N. Oraevskii and R. Z. Sagdeev, Zh. Tekh. Fiz., 32:1291 (1962).
88. P. Ko et al., Kvant. Élektron., No. 3:3 (1971).

89. V. V. Pustovalov and V. P. Silin, Pis'ma Zh. Éksp. Teor. Fiz., 14:439 (1971).

90. W. L. Kruer and J. W. Dawson, Phys. Fluids, 15:446 (1972).

91. N. G. Basov et al., Zh. Éksp. Teor. Fiz., 62:202 (1972).

92. N. G. Basov et al., Transactions of Conf. "Lasers and Their Applications," Dresden, Vol. 1, June (1970), p. 7.

93. S. J. Stephanakis et al., Phys. Rev. Lett., 29:9, 568 (1972).